Advances in Experimental Medicine and Biology

Volume 1127

More information about this series at http://www.springer.com/series/5584

Andres Trostchansky • Homero Rubbo
Editors

Bioactive Lipids in Health and Disease

Springer

Editors
Andres Trostchansky
Departamento de Bioquímica and Centro de Investigaciones Biomédicas (CEINBIO)
Facultad de Medicina, Universidad de la República
Montevideo, Uruguay

Homero Rubbo
Departamento de Bioquímica and Centro de Investigaciones Biomédicas (CEINBIO)
Facultad de Medicina, Universidad de la República
Montevideo, Uruguay

ISSN 0065-2598 ISSN 2214-8019 (electronic)
Advances in Experimental Medicine and Biology
ISBN 978-3-030-11487-9 ISBN 978-3-030-11488-6 (eBook)
https://doi.org/10.1007/978-3-030-11488-6

This Springer imprint is published by the registered company Springer Nature Switzerland AG.
The registered company address is: Gewerbestrasse 11, 6330 Cham, Switzerland

Preface

Lipids exert key biologically relevant activities in cells and tissues, being part of cell signaling pathways in both health and disease. It is required a better understanding of the biological mechanisms underlying the role that enzymatic- and nonenzymatic-derived lipids have in pathophysiological processes as well their association to drug development. The lack of lipid involvement treatments for inflammatory diseases and the need of early diagnosis footprints highlight the relevance of a comprehensive understanding of the biological mechanisms of lipids' underlying changes that occur. Lipids are being determined as novel biomarkers that would lead to novel treatment strategies, earlier diagnosis, and improve patients' lives. A small, lipid-derived blood-based biomarker could act as a screening tool to identify at-risk individuals. In this way, lipidomic analysis has been focused on small-to-medium lipid molecules as final products of normal and pathological processes. Coupled with the fact that blood is relatively easily accessible, plasma metabolites are an ideal source of noninvasive biomarkers of disease. Modifications exerted to bioactive lipids may affect membrane composition, structure, and function, while changes in their biosynthesis or metabolism may have a significant role in the onset and development of disease.

The aim of this book is to analyze the role of bioactive lipids from physicochemistry to therapeutics. We present a comprehensive discussion of the chemical structure of different bioactive lipids and their role in inflammatory processes as well as in cardiovascular and neurodegenerative diseases. We expect that the current text will serve for students and senior researchers of either basic or clinical science to improve knowledge on how studying lipid metabolism may aid in understanding underlying key mechanisms of health and disease.

Montevideo, Uruguay

Andres Trostchansky
Homero Rubbo

Contents

Part I

Structure, Characterization and Physicochemical Properties of Bioactive Lipids

Diffusion and Transport of Reactive Species Across Cell Membranes

1

Matias N. Möller, Ernesto Cuevasanta, Florencia Orrico, Ana C. Lopez, Leonor Thomson, and Ana Denicola

Abstract

This chapter includes an overview of the structure of cell membranes and a review of the permeability of membranes to biologically relevant oxygen and nitrogen reactive species, namely oxygen, singlet oxygen, superoxide, hydrogen peroxide, hydroxyl radical, nitric oxide, nitrogen dioxide, peroxynitrite and also hydrogen sulfide. Physical interactions of these species with cellular membranes are discussed extensively, but also their relevance to chemical reactions such as lipid peroxidation. Most of these species are involved in different cellular redox processes ranging from physiological pathways to damaging reactions against biomolecules. Cell membranes separate and compartmentalize different processes, inside or outside cells, and in different organelles within cells. The permeability of these membranes to reactive species varies according to the physicochemical properties of each molecule. Some of them, such as nitric oxide and oxygen, are small and hydrophobic and can traverse cellular membranes virtually unhindered. Nitrogen dioxide and hydrogen sulfide find a slightly higher barrier to permeation, but still their diffusion is largely unimpeded by cellular membranes. In contrast, the permeability of cellular membranes to the more polar hydrogen peroxide, is up to five orders of magnitude lower, allowing the formation of concentration gradients, directionality and effective compartmentalization of its actions which can be further regulated by specific aquaporins that facilitate its diffusion through membranes. The compartmentalizing effect on anionic species such as superoxide and peroxynitrite is even more accentuated because of the large energetic barrier that the hydrophobic interior of membranes presents to ions that may be overcome

M. N. Möller (✉) · A. Denicola
Laboratorio de Fisicoquímica Biológica, Instituto de Química Biológica, Facultad de Ciencias, Universidad de la República, Montevideo, Uruguay

Center for Free Radical and Biomedical Research, Universidad de la República, Montevideo, Uruguay
e-mail: mmoller@fcien.edu.uy

E. Cuevasanta · L. Thomson
Laboratorio de Enzimología, Instituto de Química Biológica, Facultad de Ciencias, Universidad de la República, Montevideo, Uruguay

Center for Free Radical and Biomedical Research, Universidad de la República, Montevideo, Uruguay

F. Orrico · A. C. Lopez
Laboratorio de Fisicoquímica Biológica, Instituto de Química Biológica, Facultad de Ciencias, Universidad de la República, Montevideo, Uruguay

Laboratorio de Enzimología, Instituto de Química Biológica, Facultad de Ciencias, Universidad de la República, Montevideo, Uruguay

Center for Free Radical and Biomedical Research, Universidad de la República, Montevideo, Uruguay

A. Trostchansky, H. Rubbo (eds.), *Bioactive Lipids in Health and Disease*, Advances in Experimental Medicine and Biology 1127, https://doi.org/10.1007/978-3-030-11488-6_1

by protonation or the use of anion channels. The large difference in cell membrane permeability for different reactive species indicates that compartmentalization is possible for some but not all of them.

Keywords
Membrane permeability · Reactive nitrogen species · Cell membrane · Reactive oxygen species

1.1 Cellular Membranes

Lipid membranes are selective barriers that compartmentalize different cellular functions. They separate the cells from its surroundings and also separate organelles within cells. Cell membranes are commonly represented by the fluid mosaic model of Singer and Nicolson [1] where a fluid lipid bilayer accommodates different types of proteins. The model serves as a basic framework to understand the function of different membranes, but additional factors need to be considered to acknowledge the complexity of cell membranes. For instance, the lipid and protein composition of membranes varies depending on the cell type and the subcellular location [2]. The most abundant lipids in mammalian cells are phosphatidylcholine, phosphatidylethanolamine and cholesterol, whereas some lipids are more abundant in particular locations, such as cardiolipin in mitochondria [2]. Furthermore, the lipids in cellular membranes are distributed asymmetrically, for instance, glycolipids locate exclusively on the outer leaflet of plasma membranes, whereas phosphatidylserine is mostly found in the inner leaflet [3].

Proteins account from 18% to 76% in weight of the total membrane in myelin or mitochondria respectively, whereas in erythrocytes and liver cell plasma membrane they account for 49–45% of the total weight [2]. The proteins are responsible for most of the functions of particular membranes, thus, different membranes also show different protein composition, i.e. the proteins found in plasma membrane are different from those found in mitochondria. The proteins are also asymmetrically located in the membrane, always oriented in the same direction. Some plasma membrane proteins are glycosylated, and these modifications are located on the external face of the membrane. These carbohydrates on the surface are used in cell-cell recognition, communication and adhesion [4].

Unlike the classical representation of the fluid mosaic model, cell membranes contain a high density of transmembrane proteins that perturb lipid packing, transmembrane proteins often contain bulky extramembranous protein domains and protein-protein oligomers are a common feature in membranes, resulting in restricted lateral diffusion in membranes. As a consequence of the presence of transmembrane proteins, the thickness of the membrane is not homogenous. The lipids have to accommodate slightly different lengths of hydrophobic α-helices in the transmembrane domains of proteins, usually by increasing or decreasing the proportion of *gauche* rotamers, which can decrease or increase the length of the alkyl chains. As a result, some regions of the membrane may be thinner than others [3]. Furthermore, lateral heterogeneities are found in membranes, where domains enriched in certain lipids and proteins are formed. Also, the lipids may arrange themselves in other phases beside the liquid-crystalline lamellar phase [3].

The passage of molecules and ions across these membranes is regulated by different mechanisms. Small nonelectrolytes are capable of crossing the membrane by diffusion through the lipids, with rates that depend mostly on their size and hydrophobicity. Oxygen, that is a small molecule and mildly hydrophobic, is able to diffuse across cell membranes at high rates [5]. For polar molecules this rate may be very low and membrane proteins are involved in their transport (i.e. glucose). Fatty acids, although containing a large hydrophobic acyl chain, are amphipathic and they also rely on transporters in the membrane to be internalized [6]. Because of the apolar nature of membrane's interior, an electric charge faces a very large thermodynamic barrier that effectively limits the permeability of membranes to ions and ionic species, whose transport then relies on specific transport proteins [7].

1.2 Diffusion Across Membranes

Many small molecules can cross membranes through the lipid fraction virtually unhindered. Such is the case of dissolved gas molecules involved in respiration, such as oxygen and carbon dioxide. It is also the case of the so called "gasotransmitters", like nitric oxide, carbon monoxide and hydrogen sulfide, and of other reactive species such as nitrogen dioxide (see below). These molecules are gases in their standard state and most of them are only sparingly soluble in water. The unfavorable interactions with water give these molecules a slightly hydrophobic character in solution that translate into a favorable partitioning into membranes and very high permeation rates. The permeability to more polar or ionic species by simple diffusion is rather limited, and protein channels facilitate the diffusion across the membrane. No active transport has been reported for reactive species.

Simple diffusion requires that the molecule first enters the membrane, and then diffuses across the acyl chain region of the membrane to the other side. For polar molecules, the highest resistance to permeation is given by the low solubility at the acyl chain region. Actually, a correlation between solubility in organic solvents and permeation was observed by Overton more than 100 years ago and confirmed several times since then [8–10]. This indicates that hydrophobic molecules can traverse the membrane more easily than polar molecules by overcoming the thermodynamic barrier presented by the acyl chains. Furthermore, the tight packing of lipids presents a further selectivity to small molecules, so that small size also makes the molecules to traverse the membranes more easily than expected from polarity alone [5, 9, 11, 12]. Since diffusion coefficients are similar among small molecules, the correlation between permeability and solubility suggests that the largest barrier presented by membranes is thermodynamic rather than kinetic for small molecules.

Two regions may be identified in the membrane that show different barriers to diffusion. The first one is the interface, where the water molecules meet the polar headgroups in the lipid bilayer. The headgroups of phospholipids are bulky, ionic and polar so that there is a tight packing and high capacity to form hydrogen bonds. The other region would be the acyl chain region, characterized by a lower atom density and a high hydrophobicity. The main barrier lies in different regions depending on the physicochemical properties of the molecule. For small hydrophobic molecules this barrier is usually located in the interface region, whereas for small polar molecules this barrier is located in the acyl chain region [13].

As a general rule, the diffusion coefficient of small molecules increases towards the center of the bilayer, regardless of its physicochemical properties [14]. However, solubility is very different for either hydrophobic or polar molecules, and pose an important thermodynamic barrier that slows down the passage of polar molecules. In the case of hydrophobic molecules, the headgroup region presents a small thermodynamic barrier, whereas for polar molecules the acyl chain region presents a large thermodynamic barrier [13, 14].

Membranes may be thought of as a selective filter where individual molecules cross the membrane at basically the same speed, but there are millions more holes for hydrophobic than for polar small molecules.

1.3 The Permeability Coefficient

The permeability coefficient (P_m) indicates how fast molecules cross a membrane. The permeability coefficient can be used to compare the ability of different molecules to traverse membranes and therefore identify groups of molecules that are more or less capable to diffuse across a membrane [15].

Basically, P_m (expressed in cm s^{-1}) is related to the flux of molecules through the membrane. For uncharged small molecules, it is related to the diffusion coefficient of the molecule in the membrane (D_m), as well as its solubility in the membrane (represented by the partition coefficient between membrane and water, K_P).

In its simplest form, P_m is related to the diffusion coefficient, by:

$$P_m = D_m \ K_P / \delta$$

Where δ is the thickness of the membrane, usually approximated to 4 nm. A more complex equation can be used to take into account that both diffusion and solubility change at different depths in the membrane and that interfacial resistance may be significant [16].

1.4 Interactions of Reactive Species with Membranes

Reactive oxygen and nitrogen species (ROS and RNS) are very general and vague terms that include a wide range of molecules with different physicochemical properties. Most ROS and RNS derive from oxygen and nitric oxide (Fig. 1.1) and they do have in common the ability to directly or indirectly cause chemical damage to biomolecules. Many of these reactive species are produced under controlled conditions by specialized proteins to accomplish physiological functions but they can also derive from environmental sources such as smog, smoke, xenobiotics and exposure to ultraviolet radiation.

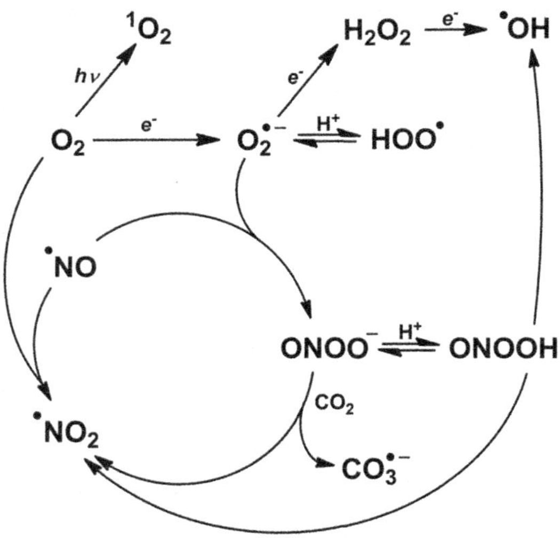

Fig. 1.1 Reactive Oxygen and Nitrogen Species. Oxygen (O_2) and nitric oxide (•NO) are the precursors of ROS and RNS. Photosensitization can give rise to singlet oxygen ($^1O_2$) whereas partial reduction of O_2 produces superoxide ($O_2^{\bullet-}$), that can be protonated to give the hydroperoxyl radical (HOO•) or further reduced to hydrogen peroxide (H_2O_2) and hydroxyl radical (HO•). The reaction of •NO with O_2 (autoxidation) generates nitrogen dioxide (•NO_2), while the reaction with $O_2^{\bullet-}$ generates peroxynitrite anion ($ONOO^-$) that can either protonate to the peroxynitrous acid (ONOOH) that can then homolyze and give •NO_2 and HO• as products (5–30% yield) or react with CO_2 and give •NO_2 and carbonate radical ($CO_3^{\bullet-}$ 35% yield)

Accomplishing either a physiological function or exerting a toxic effect, will depend largely on the ability of the reactive species to cross the cellular membranes, so a directionality and compartmentalization of action may be established. Some of these reactive species are able to cross membranes rapidly by simple diffusion, so that they can diffuse practically unhindered in a large volume of tissue. In this case, chemical reactions rather than membranes will be responsible for decreasing its diffusion distance. Other reactive molecules, however, find high energy barriers to traverse membranes thus their reactions and actions can be more effectively compartmentalized.

Furthermore, cellular membranes are not just spectators of the flow of reactive species, but are also targets of their reactivity. The most reactive targets are polyunsaturated fatty acids, which oxidation may start a lipid peroxidation chain reaction, but other lipids may also be oxidized by specific reactive species. The permeability of membranes to the different reactive species also indicates how easy it is for these reactive species to oxidize the lipids and embedded proteins in a membrane.

The following section will focus on membrane interactions and potential reactions with biologically relevant reactive species. The parameters used to evaluate these interactions will allow us to compare and classify them according to how easily they can travel across membranes.

1.4.1 Oxygen

Molecular oxygen (O_2) is one of the essential molecules that define aerobic life. Because of oxygen, much more energy can be extracted from organic matter such as glucose and sustain the metabolism of larger lifeforms. Its usefulness comes with the associated risk of forming reactive species in mitochondria by partial reduction of oxygen. It is calculated that 0.1–0.5% of all the oxygen consumed by mitochondria is converted to reactive oxygen species such as superoxide and hydrogen peroxide [17].

Oxygen is only sparingly soluble in water (1.28 mM/atm at 25 °C for pure O_2 and 0.27 mM/atm at 25 °C for air [18]) and therefore larger animals need proteins to transport it through the body such as hemoglobin. On the other side, oxygen is around 10 times more soluble in organic solvents than in water [19]. A slight hydrophobic character combined with a small size allow it to simply diffuse through lipid membranes [5]. Oxygen also plays an essential role in lipid oxidation. The initial formation of a lipid radical leads to the addition of an O_2 molecule and the formation of the longer lived lipid peroxyl radical, that then can oxidize another molecule of lipid and propagate the oxidative damage [20]. These chain reactions that occur in the membrane are usually inhibited by the membrane's main antioxidant, α-tocopherol [21].

The diffusion of oxygen through membranes has been studied by several approaches including fluorescence quenching, electron paramagnetic resonance, nuclear magnetic resonance and molecular dynamics [13, 22–24]. Each technique has different advantages and disadvantages (discussed in [16]), but taken together yield a complete picture of the diffusion of oxygen through membranes at a molecular level.

Apparent diffusion coefficients (D_{app}) are experimentally accessible, and provide useful information about the rate of permeation and reactions of oxygen in the membrane [16]. These values derive from quantifying the collisions between probes inserted in the lipid membrane and oxygen, that depend on both the diffusion coefficient inside the membrane and the local concentration of oxygen in the membrane. Most times the concentration of O_2 inside the membrane is not known, so the aqueous concentration of O_2 is taken as a reference [16]. The apparent diffusion coefficient is thus the product of the true diffusion coefficient in the membrane times the partition coefficient ($D_{app} = D_m.K_P$). The permeability coefficient can then be easily obtained from the equation in sect. 1.3. On the down side, apparent diffusion coefficients yield an incomplete mechanistic picture of the permeation process, because diffusion cannot be separated from partition. Nevertheless, some attempts have been done in this direction [5].

Different methods have been developed to determine the partition coefficients for membranes. Most of them rely on quantifying how much O_2 can be dissolved in membrane suspensions relative to the aqueous buffer alone [5, 25–27]. To summarize, it was found that O_2 is 3–4 times more soluble in fluid lipid membranes than in water, and that the solubility depends on the temperature but more importantly on the physical state of the membrane and the available free volume [5]. Phospholipid membranes in the gel state showed a very low solubility to O_2 ($K_P < 1$), with a sharp increase above the transition temperature [5, 25, 26]. Although no results are yet available for actual plasma membranes, there is evidence suggesting that some domains in cellular membranes that are more rigid and tightly packed may prevent the partition of O_2 and limit the permeability.

The independent determination of apparent diffusion coefficients and partition coefficients permitted the calculation of the diffusion coefficient of O_2 in the membrane [5]. Unexpectedly, a virtually unique $D_m = 1.6 \times 10^{-5}$ cm^2 s^{-1} was found for membranes of different compositions, in different physical states. Those studies were conducted by pyrene fluorescence quenching and a plausible explanation is that the bulky pyrene probe is disrupting the local lipid structure, creating a similar microenvironment in the different conditions and preventing the study of D_m by this approach [5].

A more detailed picture of the interaction of O_2 with membranes can be obtained from molecular dynamics simulations. In this case the solubility and diffusion can be obtained separately at different depths in the membrane, and show that both the solubility and diffusion have a bell-shaped profile with its maximum in the middle of the bilayer, with only a low barrier at the headgroups region [13, 14, 28].

The permeability coefficients determined so far indicate that the diffusion of O_2 through lipid membranes is extremely rapid (P_m ~ 40 cm s^{-1}), similar to diffusing through an equally thick layer of water (Table 1.1) [5, 15]. Therefore, no protein channels are needed to facilitate O_2 diffusion that will diffuse unhindered over a large number of cells. The only case where O_2 diffusion through membranes decreased more than 2 orders of magnitude was in compressed phospholipid monolayers [29]. We proposed that lung surfactant may behave in a similar way, creating a "closed valve effect" when compressed during respiration [5], but this remains to be proven.

1.4.2 Singlet Oxygen

The main difference between singlet oxygen ($^1O_2$) and oxygen at the basal triplet state is the electronic configuration. In the singlet state $^1\Delta_g$ the two electrons at the highest energy state are in

Table 1.1 Permeability coefficients for different membranes and reactive species

Permeant	Membrane	T (°C)	P_m (cm/s)	References
O_2	4 nm slab of water	20	53[a]	
O_2	DMPC	18	12	[118]
	DMPC	29	125	[118]
	DOPC	30	114	[118]
	POPC	35	157	[119]
	POPC:Chol (50% Chol)	35	50	[119]
	RBC (human)	20	38	[41]
	CHO cells	20	21	[120]
	CHO cells	37	42	[120]
$^{\bullet}NO$	EYPC	20	73	[42]
	EYPC:Chol (30% Chol)	20	66	[42]
	RBC (human)	20	18	[41]
$^{\bullet}NO_2$	EYPC	25	~5	[19]
H_2S	E.coli lipids	23	≥0.5	[78]
	DLPC	25	~3	[77]
H_2O	EYPC	25	3.3×10^{-3}	[121]
H_2O_2	RBC (horse)	20	6×10^{-4}	[99]
	Peroxisome (rat liver)		3×10^{-3}	[122]
	RBC (rat)		1.2×10^{-2}	[103]
	Jurkat T cells	37	2×10^{-4}	[100]
	Chara corallina		3.6×10^{-4}	[107]
	Escherichia coli	37	1.6×10^{-3}	[102]
	PC12 cells		4×10^{-4}	[123]
	HUVEC cells		1.6×10^{-3}	[123]
	IMR-90 cells		1.1×10^{-3}	[123]
	HeLa cells		4.4×10^{-4}	[124]
$HOO^{\bullet}$	EYPC	23	4.9×10^{-4}	[88]
$O_2^{\bullet-}$	SBPC	25	2.1×10^{-6}	[125]
	EYPC	23	7.6×10^{-8}	[88]
ONOOH	DMPC	23	8×10^{-4}	[83]
	EYPC	21	1.3×10^{-3}	[60]
	DMPC	21	6.3×10^{-4}	[60]
	DPPC	21	4×10^{-4}	[60]

Abbreviations: *DMPC*, dimyristoylphosphatidylcholine; *DOPC*, dioleoylphosphatidylcholine; *POPC*, palmitoyloleoylphosphatidylcholine; *Chol*, cholesterol; *RBC*, red blood cells; *DLPC*, dilauroylphosphatidylcholine; *DPPC*, dipalmitoylphosphatidylcholine.

[a]Calculated $P_w = D_w^{O2} / \delta = 2.1 \times 10^{-5}$ cm^2 s^{-1}/ 4×10^{-7} cm [5]

the same molecular orbital with paired (opposite) spins, whereas in the triplet state $^3\Sigma$ the electrons are located in separated orbitals with the same spin. This leads to a dramatic change in reactivity. Triplet oxygen reacts exclusively with paramagnetic species, such as some metals and radicals, but singlet oxygen can react with other singlet state, non-radical, and electron-rich molecules containing double bonds [30, 31]. Singlet oxygen is produced mostly by photochemical reactions involving energy transfer from a sensitizer in the excited state to triplet oxygen. There are endogenous sensitizers that can produce $^1O_2$ when exposed to UV light, such as protoporphyrin IX, FMN, FAD, NADH and NADPH, whereas exogenous sensitizers can be found in cosmetics, food additives and drugs [30].

In lipid membranes, singlet oxygen is expected to share most of O_2 physicochemical properties, except for reactivity. Singlet oxygen reacts with double bonds in the acyl chains to produce lipid hydroperoxides directly [30, 31], and with membrane proteins to produce protein hydroperoxides and other reaction products [32, 33]. These oxidative reactions can be prevented by carotenoids that are very effective $^1O_2$ quenchers [34, 35].

1.4.3 Nitric Oxide

Nitric oxide ($^{\bullet}NO$) is synthesized by the enzymes nitric oxide synthases (NOS) of which three isoforms are known: endothelial, neuronal and inducible. Nitric oxide is used as a signaling molecule in vasorelaxation and also in neurotransmission. The signaling is accomplished mostly by interacting with the hemeprotein soluble guanylate cyclase in the target cells [36]. Moreover, when produced in large quantities by the inducible NOS it has cytostatic and cytocidal activity. The toxicity depends mostly on the formation of secondary more reactive species such as nitrogen dioxide and peroxynitrite [37].

Nitric oxide has a low dipole moment (0.159 D [38]), so it has weak intermolecular interactions and it is a gas at 1 atm and 25 °C. It shares some similarities with oxygen in that it is only sparingly soluble in water (1.94 ± 0.03 mM [39]), but about ten times more soluble in organic solvents [40]. Because of packing effects that decrease the solubility in membranes, the partition coefficient in EYPC liposomes and human low density lipoprotein is 4.4 and 3.4, respectively at 25 °C [27].

The diffusion through membranes is very rapid and requires the use of either fluorescence or EPR probes inserted in the membrane that change their parameters upon collision with $^{\bullet}NO$. The use of fluorescent pyrene probes to study $^{\bullet}NO$ diffusion has recently been reviewed [16]. These measurements have shown that lipid membranes do not limit $^{\bullet}NO$ diffusion significantly and at most can decrease the diffusion to a half that in water [27, 41, 42]. This is because of the combination of a favorable partition coefficient and a high diffusion in the membrane. The permeability coefficients of lipid membranes to $^{\bullet}NO$ range from 18 to 73 cm s^{-1} [41, 42], in the order of an equally thick layer of water (Table 1.1).

The favorable partitioning of $^{\bullet}NO$ into lipids leads to an acceleration of the autoxidation reaction, to form the oxidizing and nitrosating nitrogen dioxide and dinitrogen trioxide [43, 44]. Nitric oxide is extremely efficient at inhibiting lipid peroxidation [45–47]. The antioxidant properties of $^{\bullet}NO$ in lipids are supported by the diffusion-controlled reaction with peroxyl radicals [48], aided by the favorable partitioning in lipids and the very high diffusion [27]. The products include nitrosated and nitrated lipids that show several biological effects [46, 49].

1.4.4 Nitrogen Dioxide

Nitrogen dioxide ($^{\bullet}NO_2$) is one of the oxidizing products of $^{\bullet}NO$, either from autoxidation or from peroxynitrite decomposition [43, 50]. Alternatively it can be formed from the oxidation of nitrite by myeloperoxidase [51] and it has been proposed to be an important component in the new therapeutics by cold atmospheric plasma [52]. It is also an important pollutant in urban areas and one of the main components of smog. It is paramagnetic and reactive with lipids, proteins and DNA bases [50]. It can undergo different types of reactions, including radical recombination, addition to double bonds, hydrogen abstraction and electron transfer [53]. The characteristic products include nitrated proteins and lipids, which have been dealt in detail elsewhere [54, 55].

Nitrogen dioxide has an angle of 134.4° and a low dipole moment (0.29–0.58 D [56]) and is also a gas at 25° and 1 atm. It is more soluble than ${}^{\bullet}NO$ in water (12 ± 4 mM at 1 atm, 20 °C) [57, 58], but rapidly decomposes because of a fast dimerization and reaction with water to produce nitrite and nitrate [57, 58]. The solubility has been measured in a few organic solvents, including decane, chloroform and carbon tetrachloride (reviewed in [59]) and it has been estimated by quantum mechanical calculations that ${}^{\bullet}NO_2$ is 2.7 times more soluble in octanol than in water, and 1.5 times more soluble in lipid membranes than water [19]. It was estimated a permeability coefficient of 5 cm s^{-1} (Table 1.1) [19], in the upper range of that estimated by Khairutdinov *et al.* [60]. Molecular dynamics has provided a more detailed picture of ${}^{\bullet}NO_2$ interacting with lipid membranes and shown a favorable interaction between ${}^{\bullet}NO_2$ and the lipid membrane interior, in agreement with previous estimations, with the permeation barrier located at the headgroups region [52, 61].

1.4.5 Hydrogen Sulfide

Hydrogen sulfide (H_2S) has been related to the origin and evolution of life [62]. Not only it is necessary for cysteine synthesis in plants and microorganisms but it can also be a source of electrons for respiration in chemolithotrophy and anoxygenic photosynthesis. Nonetheless, high levels of hydrogen sulfide represent a threat to mammals because it can inhibit cytochrome c oxidase and mitochondrial respiration [63]. Surprisingly, it was found that hydrogen sulfide is synthesized in mammals from cysteine or homocysteine at moderately high rates by three enzymes: cystathionine β-synthase, cystathionine γ-lyase and 3-mercaptopyruvate sulfurtransferase, whose relative contribution is tissue-dependent [64]. Hydrogen sulfide is also efficiently detoxified by an enzymatic system located in the mitochondria [65]. Physiological effects of administering low levels of H_2S include neuromodulation, vasodilation, protection in ischemia-reperfusion injury events and even induction of a suspended animation state in mice [66–69]. In spite of the diversity of observed physiological effects, the mechanisms for sulfide signaling are still elusive and a subject of intensive research [70].

The molecular structure of hydrogen sulfide is analogous to the molecule of water, but with a smaller angle between bonds (92°) and a lower dipole moment (0.97 D) because of the lower electronegativity of sulfur relative to oxygen. Moreover, H_2S is not able to form hydrogen bonds, so, under normal temperature and pressure conditions, it is a gas (boiling temperature: -60 °C). It is fairly soluble in water (101.3 mM / atm at 25 °C) hydrated like a hydrophobic solute [71] and, as weak acid, is able to donate protons to the solvent. Hydrogen sulfide is in fast equilibria with hydrosulfide anion (HS^-) and sulfide anion (S^{2-}) with pK_as of 6.98 at 25 °C for the first dissociation step and >17 for the second one [72]. Hydrosulfide anion is the main species at pH 7.4 (~70%) while H_2S accounts for the remaining and the concentration of sulfide anion is negligible.

Hydrosulfide anion is a nucleophile able to react towards biological oxidants. It reacts with two electron oxidants (hydrogen peroxide [73], peroxynitrite [74] and hypochlorite [73, 75]), as well as with one electron oxidants (${}^{\bullet}OH$, $CO_3^{\bullet-}$, ${}^{\bullet}NO_2$ and $O_2^{\bullet-}$ [70]), forming diverse sulfur oxidation products. Among other cellular targets, disulfides (RSSR) and sulfenic acids (RSOH) are candidates for signaling transduction [76]. These reactions produce persulfides (RSSH), reactive sulfane derivatives of thiols, of increasing interest [70].

As a first approach to the hydrophobicity of H_2S, partition coefficients of 2.1 ± 0.2 and 1.9 ± 0.5 at 25 °C were reported for H_2S in octanol and hexane [77]. Due to the pK_a of H_2S, dissociation to form HS^- determines a distribution coefficient of 0.64 ± 0.05 in octanol at pH 7.4 and 25 °C. Using multilamellar liposomes of DLPC as a model for biological membranes, a partition coefficient of 2.0 ± 0.6 at 25 °C for H_2S was determined [77]. This affinity of H_2S for hydrophobic milieu suggests a high permeability in membranes.

The ability of H_2S to traverse biological membranes has been assessed through different approximations. Mathai et al. studied the resistance imposed to H_2S transport by planar bilayers prepared from *E. coli* lipids [78]. They concluded that the diffusion through the membrane was not the rate-limiting step, allowing the estimation of a lower boundary for the permeability coefficient of 0.5 cm s^{-1}, probably limited by the diffusion through the unstirred water layers. Moreover, the inclusion of cholesterol or sphingomyelin in the composition of the bilayer did not restrict the diffusion through the membrane. This estimation lead to the conclusion that neither aquaporins nor other protein channels are needed to facilitate the transport of H_2S. Accordingly, permeability of H_2S was also too rapid to be observed by stopped-flow using fluorophores encapsulated in DMPC:cholesterol 1:1 unilamellar liposomes [77]. Semi-theoretical approaches comparing partition and permeability coefficients for similarly sized molecules, enabled the estimation of a permeability coefficient ≥ 3 cm s^{-1} (Table 1.1) [77]. Molecular dynamics simulations on lipid membranes of DPPC showed that the barrier for H_2S permeation was negligible according to the Gibbs energy profiles, and a permeability coefficient of 11.9 ± 0.7 cm s^{-1} was estimated [79].

Although H_2S shows higher solubility in biological bilayers than in water, most of the reactions involving this molecule are triggered by HS^-, the nucleophile species, which is essentially not soluble in hydrophobic environments. Unless a relevant biological reaction is being considered for undissociated H_2S within the bilayer, it is concluded that the favorable partition does not assist the acceleration of reactions but contribute to permeation and spreading through tissues.

1.4.6 Peroxynitrite Anion and Peroxynitrous Acid

Peroxynitrite ($ONOO^-$) is formed by the diffusion-controlled reaction between nitric oxide and superoxide [80]. It is continually generated under basal conditions, but its formation increases with higher rates of superoxide and $^{\bullet}NO$ production such as those found in inflammation [80]. It is considered one of the effector molecules in the cytotoxic effect of $^{\bullet}NO$ [37]. Peroxynitrite has a pK_a of 6.8, and the peroxynitrous acid is unstable and either isomerizes to nitrate or homolyzes to yield $^{\bullet}NO_2$ and $^{\bullet}OH$. The most rapid reactions of peroxynitrite in a biological context are with CO_2, in a reaction that yields an intermediary that homolyzes to $^{\bullet}NO_2$ and carbonate radical with a 35% yield. In addition, it rapidly reacts with peroxidatic cysteines of the antioxidant proteins peroxiredoxins, yielding nitrite and an oxidized peroxiredoxin that can then be reduced by thioredoxin [80]. Peroxynitrite is also able to nitrate proteins [54] and oxidize and nitrate lipids [81, 82].

It has been well established that peroxynitrite is able to cross lipid membranes through different pathways. Peroxynitrous acid can cross directly through the lipid fraction, and this has been shown in pure phospholipid membranes [60, 83]. Estimated permeability coefficients of phosphatidylcholine membranes to ONOOH range from 4×10^{-4} in gel state membranes to 1.3×10^{-3} cm s^{-1} in fluid state membranes (Table 1.1) [60, 83]. Recent molecular dynamics studies show that peroxynitrous acid interacts favorably with the headgroups in phosphatidylcholine membranes but finds an energetic barrier in the acyl chain region, and the free energy barrier experienced by ONOOH is similar to that of ethanol [61].

Peroxynitrite anion has to deal with a very large energetic barrier to diffusion through the hydrophobic fraction of the membrane, but it can cross membranes through protein channels. In erythrocytes it was demonstrated that peroxynitrite could use the anion exchanger 1 (band 3) to traverse the membrane [7] and that this path accounted for 50% of the transport of peroxynitrite into red blood cells.

Carbonate radical may be an important product derived from peroxynitrite. Based on its short life given by its high reactivity, its negative charge, and that CO_2 decreases lipid peroxidation by peroxynitrite, the permeability of membranes to carbonate radical is expected to be very low, even through anion channels [50].

1.4.7 Superoxide and Hydroperoxyl Radical

Superoxide ($O_2^{\bullet -}$) is the one-electron reduction product of O_2 and can be produced enzymatically by NADPH oxidases, xanthine oxidases, and also as a byproduct in mitochondrial respiration [84]. Superoxide by itself is not very reactive, but is a precursor of more reactive species, such as peroxynitrite and hydrogen peroxide (Fig. 1.1).

Superoxide has a pKa of 4.7, and the conjugated acid the hydroperoxyl radical ($HOO^\bullet$) is more oxidizing than superoxide [85]. Both superoxide and $HOO^\bullet$ have been observed to initiate lipid peroxidation [86], leading to cell lysis (erythrocyte ghost membranes) [87].

The permeability of phospholipid membranes to superoxide was studied by Gus'kova *et al.*, who found that egg yolk phosphatidylcholine liposome membranes had a permeability of 4.9×10^{-4} cm s^{-1} to $HOO^\bullet$, and 7.6×10^{-8} cm s^{-1} to superoxide at 23 °C (Table 1.1)[88].

Because of its charged nature, superoxide does not interact favorably with membranes. Molecular dynamics show that superoxide remains in the aqueous phase and is excluded even from the headgroup region [13], as expected for a charged particle. Even though the passage through the lipid fraction is highly unfavorable, superoxide can use anion channels in the membrane such as anion exchange 1 (band 3) in erythrocytes [89] and thus also traverses plasma membranes. Unlike superoxide, $HOO^\bullet$ is not charged and interacts favorably with the headgroup region, by means of hydrogen bonds with acyl chain carbonyls, and has a lower energy barrier to traverse the membrane than hydrogen peroxide [13, 90]. It can reach the unsaturated bonds in the lipids and therefore it may be the main responsible for superoxide-initiated lipid peroxidation [13, 90].

1.4.8 Hydrogen Peroxide

Hydrogen peroxide (H_2O_2) is generated from many sources, but mainly NADPH oxidases, xanthine oxidase and from the dismutation of superoxide. Cellular sources are therefore located in the plasma membrane, mitochondria, endoplasmic reticulum peroxisomes and phagolysosomes. Although mitochondria is considered the major site of production under physiological conditions, it is generated in large amounts during neutrophil activation and the respiratory burst by the activation of NADPH oxidases [84]. The role of H_2O_2 as a signaling molecule is starting to be unveiled. It was found that H_2O_2 affects cell proliferation, growth, migration, apoptosis and survival [91]. The reactivity of H_2O_2 is limited to metal centers, selenoproteins and specific thiol proteins [92]. The most reactive proteins include catalase, glutathione peroxidase and peroxiredoxins [92, 93]. Peroxiredoxins are present in cells at high concentrations, thus keep intracellular levels of H_2O_2 very low [93–95]. The molecular mechanisms of H_2O_2-induced signaling are under study but suggest that peroxiredoxins play a key role in relaying oxidation equivalents to secondary proteins [95, 96].

Hydrogen peroxide itself is not able to oxidize lipids, but can yield other products such as hydroxyl radical that can initiate lipid peroxidation [97]. It is also the fundamental substrate of peroxidases such as myeloperoxidase, which uses it to make more reactive species such as hypochlorous acid [92].

The idea of a permeability barrier imposing a limit to the entry of hydrogen peroxide into the cell was first introduced by Clayton [98] and Nicholls [99], by studying the catalase activity in intact and disrupted *Rhodopseudomonas spheroides* cells and horse erythrocytes, respectively. Later on, the presence of gradients across the membranes from mammalian cell lines, bacteria and yeast was reported, confirming that biological membranes generally prevent the free diffusion of H_2O_2 [100–102]. Notably there are very few reports of permeability of pure lipid membranes to H_2O_2. Although changes in diffusion rates depending on composition, temperature and compressibility have been observed, no permeability coefficients are reported for liposomes [103–105].

Most of the reported H_2O_2 permeability coefficients come from studies with cell membranes and lie between 1×10^{-3} and 4×10^{-4} cm/s, very similar to the values reported for water (Table 1.1). Water and H_2O_2 have similar molecular properties, almost the same dipole moment, dielectric constant, molecular diameter and ability to form hydrogen bonds [106]. In fact, it is believed that H_2O_2 passes into the cell the same way as water, both by simple diffusion and by facilitated diffusion through specific aquaporins. The first study that showed that aquaporins facilitate H_2O_2 diffusion was done by Henzler and Steudle, in the plant cell model *Chara corallina* [107].

Aquaporins are tetrameric proteins, where each monomer acts as a functional channel composed of six membrane-spanning helices connected by five loops (A to E). Loops B and E contain the highly conserved asparagine-proline-alanine (NPA) sequence and these motifs meet in the middle of the membrane, forming a narrow hydrophobic pathway. A second selectivity filter is provided by the aromatic/arginine (ar/R) region, formed by four aminoacids, which works as a size exclusion barrier and generates the adequate environment for the establishment of the hydrogen bonds necessary to transport the substrate [106].

The importance of aquaporins in facilitating the diffusion of H_2O_2 was further established by Bienert et al., who used *Saccharomyces cerevisiae* to express different aquaporins. Overexpression of human aquaporin 8 (hAQP8) and TIP1;1 and TIP1;2 from *Arabidopsis thaliana* lead to a decrease in growth and survival of the yeasts exposed to H_2O_2 [108]. In contrast, hAQP1, hAQP2, rAQP3, rAQP4, hAQP5, hAQP9 and other aquaporins from *Arabidopsis thaliana* did not affect the survival in the presence of H_2O_2, suggesting that these isoforms were not permeable to H_2O_2 [106, 108]. Similar assays identified AQP1 from rat and multiple other aquaporins from plants, including PIP2;1 from *Arabidopsis thaliana* as facilitators of H_2O_2 diffusion through the membrane [106, 109]. It was suggested that PIP2 type aquaporins are efficient H_2O_2 channels, and the main regulation of H_2O_2 passage is given by the ar/R region [110]. The idea that H_2O_2 is transported across the cell membrane by aquaporins is also supported by computational simulations, using bovine AQP1 and PIP2;1 from *Spinacia oleracia* as models [111].

Studies in mammalian cell lines were also conducted and they showed further evidence that confirmed the role of hAQP8 in H_2O_2 transport. Human aquaporin 3 was also proposed as a transporter with the characteristic that it can endure changes in its expression levels in order to modulate the accumulation of H_2O_2 and thus regulate cellular signaling cascades [112–114]. On the other hand, studies carried out in *S. cerevisiae* and erythrocytes suggest that human aquaporin 1 is unable to transport H_2O_2 [103, 108]. In these experiments it is shown that the H_2O_2 permeability of the cells does not change in the absence or presence of mercurial compounds, which are well known aquaporin 1 inhibitors. This potential selectivity for different aquaporins suggests that the expression profiles of aquaporins could dictate the susceptibility of a particular cell or tissue to external H_2O_2 signaling [112].

A decrease in the capacity of red blood cells to metabolize exogenously added H_2O_2 during storage for transfusion has recently been described that was independent of the loss of antioxidant thiols [115]. It is proposed that the decrease in H_2O_2 metabolism is related to changes in the permeability of the membrane to H_2O_2.

1.4.9 Hydroxyl Radical

Hydroxyl radical ($^{\bullet}OH$) is one of the most reactive molecules [85]. It can be formed from H_2O_2 by reaction with a reductant such as Fe^{2+} (Fenton reaction) and from the homolysis of ONOOH

[54, 116]. This radical reacts at diffusion-controlled rates with most organic compounds by electron transfer or addition [117]. Hydroxyl radical is responsible for starting lipid peroxidation in metal catalyzed reactions, and also in peroxynitrite induced lipid peroxidation [81].

The diffusion of •OH is limited by its very high reactivity and therefore extremely short lifetime. Molecular dynamics show that it can interact with the membrane headgroup region, and suggest that •OH has a lower energy barrier to permeation than H_2O_2 [13, 52]. As pointed out before, the high reactivity of •OH will prevent its diffusion across membranes, and rather favor reactions near the headgroup region.

1.5 Considerations About the Permeability of Membranes to Reactive Species

Cell membranes are not equally permeable to all reactive species (Table 1.1). The small and more hydrophobic molecules can freely diffuse across cell membranes whereas the more polar molecules find significantly higher barriers to diffusion that permit their compartmentalization (Fig. 1.2).

The high permeability of cellular membranes to O_2, •NO and H_2S has the biological advantage that no specific transport proteins are needed, and that these molecules will be able to diffuse large distances in tissues unrestrictedly [77]. The limiting factor will be given by the chemical reactions that consume them and decrease their concentration and activity range. Therefore, the diffusion of O_2 in tissues is limited mostly by its consumption by mitochondria. For •NO, one of the most relevant reactions decreasing its lifetime and diffusion distances in tissues is with oxyhemoglobin in red blood cells [126, 127]. On the other side, the same properties that make cell membranes permeable to beneficial molecules may also make cells more susceptible to damage by •NO_2. Although the lifetime of •NO_2 is short because of rapid reactions with endogenous antioxidants and even hydrolysis of the •NO_2 dimer by water [50], diffusion of •NO_2 through lipid membranes will be fast and likely contribute to cellular damage.

In contrast to the aforementioned reactive species, the permeability of lipid membranes to hydrogen peroxide, peroxynitrite and superoxide is significantly lower (Fig. 1.2), and fast reactions inside the cells will create a steep concentration gradient across the membrane [100]. Furthermore, these molecules may use protein channels to access the cells, and specific aquaporins have been found to facilitate the transport of H_2O_2, whereas anion channels such as band 3 in erythrocytes have been found to facilitate the transport of the anions peroxynitrite and superoxide (Fig. 1.2).

A low membrane permeability allows for true compartmentalization so that higher concentrations of one of these reactive species can be achieved in a membrane-enclosed organelle such as lysosome, phagosomes, peroxisomes or endoplasmic reticulum. Furthermore, a low membrane permeability will allow only a small fraction of the reactive species to leak into the cytosol, so that antioxidant defenses are not overwhelmed. There will also be a clear directionality in the production of a given reactive species to a given compartment. For instance the production of H_2O_2 by NADPH oxidases in the plasma membrane occurs towards the extracellular space, and in the absence of specific aquaporins the H_2O_2 will tend to diffuse away from the cell rather than consume its antioxidant defenses. Another example is given by the formation of peroxynitrite. Conversely to the freely diffusible •NO, the low membrane permeability of superoxide limits the formation of peroxynitrite to the same compartment [80]. The concentration of peroxynitrite will therefore be significantly higher in intracellular sites of superoxide formation such mitochondria and phagosomes [80].

There is still much to uncover about how cell membranes regulate the transport, especially of this last group of reactive species. The membrane permeability is a fundamental parameter when trying to understand the metabolism or signaling

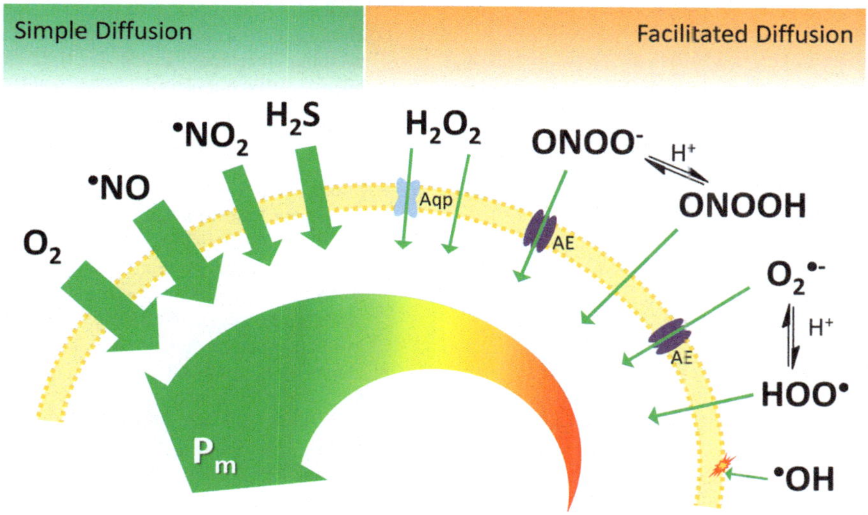

Fig. 1.2 **Permeability of lipid membranes to different reactive species.** The size of the arrows indicates how permeable cell membranes are to the different molecules. Two groups may be differentiated based on the permeability of the membrane: 1) Oxygen, $^{\bullet}NO$, $^{\bullet}NO_2$ and H_2S can traverse lipid membranes virtually unhindered by simple diffusion, 2) molecules which permeation rates have been observed to increase by the presence of specific proteins that facilitate their diffusion across cellular membranes: H_2O_2 (can use specific aquaporins, (Aqp), $ONOO^-$ and $O_2^{\bullet-}$ (can use anion exchange protein channels, AE). The permeability (P_m) to these species in their electrically neutral form through pure phospholipid membranes is five orders of magnitude lower than that of O_2. Cellular adaptations to decrease simple diffusion by H_2O_2 had been observed, indicating that cell membranes could tightly regulate the transport of this group of reactive molecules. Hydroxyl radical is in a different category because it is so reactive that reacts with basically any component near the surface of the cellular membrane and therefore cannot cross it

properties of these molecules. For instance, the steady state concentration of H_2O_2 in the vascular system (still a matter of debate), will be determined by both the rate of formation and the rate of decomposition, that will largely depend on the permeability of red blood cells to it, that is still not known in human cells.

Acknowledgements We are grateful to Dr. Gerardo Ferrer-Sueta and Dr. Beatriz Alvarez for excellent suggestions and critical discussions.

This work was partially supported by grants C38–432 from Comisión Sectorial de Investigación Científica from Universidad de la República (CSIC, UdelaR) to AD, FCE_1_2017_1_136043 Fondo Clemente Estable from Agencia Nacional de Investigación e Innovación (ANII) to LT and MNM and Fondo Vaz Ferreira from Ministerio de Educación y Cultura to EC. MNM, LT, EC and AD also acknowledge Sistema Nacional de Investigadores and Programa de Desarrollo de las Ciencias Básicas. EC is a research fellow of Comisión Académica de Posgrado from Universidad de la República (CAP, UdelaR), FO is a graduate fellow of CAP, and ACL from ANII.

References

1. Singer SJ, Nicolson GL (1972) The fluid mosaic model of the structure of cell membranes. Science 175:720–731
2. Gennis RB (1989) Biomembranes. Molecular structure and function. Springer, New York/Harrisonbourg
3. Goñi FM (2014) The basic structure and dynamics of cell membranes: an update of the Singer–Nicolson model. Biochimica et Biophysica Acta (BBA)-Biomembranes 1838:1467–1476
4. Varki A (2017) Biological roles of glycans. Glycobiology 27:3–49
5. Möller MN, Li Q, Chinnaraj M, Cheung HC, Lancaster JR Jr, Denicola A (2016) Solubility and diffusion of oxygen in phospholipid membranes. Biochim Biophys Acta Biomembr 1858:2923–2930
6. Glatz JF, Luiken JJ (2017) From fat to FAT (CD36/SR-B2): understanding the regulation of cellular fatty acid uptake. Biochimie 136:21–26
7. Denicola A, Souza JM, Radi R (1998) Diffusion of peroxynitrite across erythrocyte membranes. Proc Natl Acad Sci 95:3566–3571
8. Missner A, Pohl P (2009) 110 years of the Meyer–Overton rule: predicting membrane permeability of

gases and other small compounds. ChemPhysChem 10:1405–1414
9. Walter A, Gutknecht J (1986) Permeability of small nonelectrolytes through lipid bilayer membranes. J Membr Biol 90:207–217
10. Lieb WR, Stein WD (1986) Simple diffusion across the membrane bilayer. In: Stein WD (ed) Transport and diffusion across cell membranes. Academic, Orlando, pp 69–112
11. Xiang TX, Anderson BD (1998) Influence of chain ordering on the selectivity of dipalmitoylphosphatidylcholine bilayer membranes for permeant size and shape. Biophys J 75:2658–2671
12. Lieb WR, Stein WD (1986) Non-Stokesian nature of transverse diffusion within human red cell membranes. J Membr Biol 92:111–119
13. Cordeiro RM (2014) Reactive oxygen species at phospholipid bilayers: distribution, mobility and permeation. Biochimica et Biophysica Acta (BBA)-Biomembranes 1838:438–444
14. Marrink SJ, Berendsen HJC (1996) Permeation process of small molecules across lipid membranes studied by molecular dynamics simulations. J Phys Chem 100:16729–16738
15. Möller MN, Lancaster JR Jr, Denicola A (2008) The interaction of reactive oxygen and nitrogen species with membranes. In: Matalon S (ed) Free radical effects on membranes, vol 61. Academic, Amsterdam, pp 23–43
16. Möller MN, Denicola A (2018) Diffusion of nitric oxide and oxygen in lipoproteins and membranes studied by pyrene fluorescence quenching. Free Radic Biol Med 128:137–143
17. Brand MD (2016) Mitochondrial generation of superoxide and hydrogen peroxide as the source of mitochondrial redox signaling. Free Radic Biol Med 100:14–31
18. Wilhelm E, Battino R, Wilcock RJ (1977) Low-pressure solubility of gases in liquid water. Chem Rev 77:219–262
19. Signorelli S, Moller MN, Coitino EL, Denicola A (2011) Nitrogen dioxide solubility and permeation in lipid membranes. Arch Biochem Biophys 512:190–196
20. Yin H, Xu L, Porter NA (2011) Free radical lipid peroxidation: mechanisms and analysis. Chem Rev 111:5944–5972
21. Niki E, Yoshida Y, Saito Y, Noguchi N (2005) Lipid peroxidation: mechanisms, inhibition, and biological effects. Biochem Biophys Res Commun 338:668–676
22. Fischkoff S, Vanderkooi JM (1975) Oxygen diffusion in biological and artificial membranes determined by the fluorochrome pyrene. J Gen Physiol 65:663–676
23. Subczynski WK, Hyde JS, Kusumi A (1991) Effect of alkyl chain unsaturation and cholesterol intercalation on oxygen transport in membranes: a pulse ESR spin labeling study. Biochemistry 30:8578–8590
24. Al-Abdul-Wahid MS, Evanics F, Prosser RS (2011) Dioxygen transmembrane distributions and partitioning thermodynamics in lipid bilayers and micelles. Biochemistry 50:3975–3983
25. Smotkin ES, Moy FT, Plachy WZ (1991) Dioxygen solubility in aqueous phosphatidylcholine dispersions. Biochim Biophys Acta Biomembr 1061:33–38
26. Subczynski WK, Hyde JS (1983) Concentration of oxygen in lipid bilayers using a spin-label method. Biophys J 41:283–286
27. Möller M, Botti H, Batthyany C, Rubbo H, Radi R, Denicola A (2005) Direct measurement of nitric oxide and oxygen partitioning into liposomes and low density lipoprotein. J Biol Chem 280:8850–8854
28. Dotson RJ, Smith CR, Bueche K, Angles G, Pias SC (2017) Influence of cholesterol on the oxygen permeability of membranes: insight from atomistic simulations. Biophys J 112:2336–2347
29. Borden MA, Longo ML (2004) Oxygen permeability of fully condensed lipid monolayers. J Phys Chem B 108:6009–6016
30. Girotti AW (2001) Photosensitized oxidation of membrane lipids: reaction pathways, cytotoxic effects, and cytoprotective mechanisms. J Photochem Photobiol B Biol 63:103–113
31. Min D, Boff J (2002) Chemistry and reaction of singlet oxygen in foods. Compr Rev Food Sci Food Saf 1:58–72
32. Davies MJ (2003) Singlet oxygen-mediated damage to proteins and its consequences. Biochem Biophys Res Commun 305:761–770
33. Möller MaN, Hatch DM, Kim H-YH, Porter NA (2012) Superoxide reaction with tyrosyl radicals generates para-hydroperoxy and para-hydroxy derivatives of tyrosine. J Am Chem Soc 134:16773–16780
34. Di Mascio P, Kaiser S, Sies H (1989) Lycopene as the most efficient biological carotenoid singlet oxygen quencher. Arch Biochem Biophys 274:532–538
35. Stratton SP, Liebler DC (1997) Determination of singlet oxygen-specific versus radical-mediated lipid peroxidation in photosensitized oxidation of lipid bilayers: effect of β-carotene and α-tocopherol. Biochemistry 36:12911–12920
36. Murad F (2006) Nitric oxide and cyclic GMP in cell signaling and drug development. N Engl J Med 355:2003–2011
37. Beckman JS, Koppenol WH (1996) Nitric oxide, superoxide, and peroxynitrite: the good, the bad, and ugly. Am J Phys Cell Phys 271:C1424–C1437
38. Lide DR (2004) In: Lide DR (ed) Permittivity (dielectric constant) of gases. CRC Press, Boca Raton, p 174
39. Zacharia IG, Deen WM (2005) Diffusivity and solubility of nitric oxide in water and saline. Ann Biomed Eng 33:214–222
40. Shaw AW, Vosper AJ (1976) Solubility of nitric oxide in aqueous and nonaqueous solvents. Journal of the Chemical Society, Faraday Transactions 1: Physical Chemistry in Condensed Phases 8:1239–1244

41. Denicola A, Souza JM, Radi R, Lissi E (1996) Nitric oxide diffusion in membranes determined by fluorescence quenching. Arch Biochem Biophys 328:208–212
42. Subczynski WK, Lomnicka M, Hyde JS (1996) Permeability of nitric oxide through lipid bilayer membranes. Free Radic Res 24:343–349
43. Möller MN, Li Q, Lancaster JR Jr, Denicola A (2007) Acceleration of nitric oxide autoxidation and nitrosation by membranes. IUBMB Life 59:243–248
44. Möller MN, Li Q, Vitturi DA, Robinson JM, Lancaster JRJ, Denicola A (2007) Membrane "lens" effect: focusing the formation of reactive nitrogen oxides from the NO/O_2 reaction. Chem Res Toxicol 20:709–714
45. Rubbo H, Radi R, Anselmi D, Kirk M, Barnes S, Butler J, Eiserich JP, Freeman BA (2000) Nitric oxide reaction with lipid peroxyl radicals spares α-tocopherol during lipid peroxidation. J Biol Chem 275:10812–10818
46. Rubbo H, Radi R, Trujillo M, Telleri R, Kalyanaraman B, Barnes S, Kirk M, Freeman BA (1994) Nitric oxide regulation of superoxide and peroxynitrite-dependent lipid peroxidation. Formation of novel nitrogen-containing oxidized lipid derivatives. J Biol Chem 269:26066–26075
47. Hogg N, Kalyanaraman B, Joseph J, Struck A, Parthasarathy S (1993) Inhibition of low-density lipoprotein oxidation by nitric oxide. Potential role in atherogenesis. FEBS Lett 334:170–174
48. Padmaja S, Huie RE (1993) The reaction of nitric oxide with organic peroxyl radicals. Biochem Biophys Res Commun 195:539–544
49. Freeman BA, Pekarova M, Rubbo H, Trostchansky A (2017) Chapter 16 – Electrophilic nitro-fatty acids: nitric oxide and nitrite-derived metabolic and inflammatory signaling mediators. In: Ignarro LJ, Freeman BA (eds) Nitric oxide, 3rd edn. Academic, Amsterdam, pp 213–229
50. Augusto O, Bonini MG, Amanso AM, Linares E, Santos CCX, De Menezes SIL (2002) Nitrogen dioxide and carbonate radical anion: two emerging radicals in biology. Free Radic Biol Med 32:841–859
51. Byun J, Mueller DM, Fabjan JS, Heinecke JW (1999) Nitrogen dioxide radical generated by the mieloperoxidase-hydrogen peroxide-nitrite system promotes lipid peroxidation of low density lipoprotein. FEBS Lett 455:243–246
52. Razzokov J, Yusupov M, Cordeiro RM, Bogaerts A (2018) Atomic scale understanding of the permeation of plasma species across native and oxidized membranes. J Phys D Appl Phys 51:365203
53. Huie RE (1994) The reaction kinetics of NO2(.). Toxicology 89:193–216
54. Ferrer-Sueta G, Campolo N, Trujillo M, Bartesaghi S, Carballal S, Romero N, Alvarez B, Radi R (2018) Biochemistry of peroxynitrite and protein tyrosine nitration. Chem Rev 118:1338–1408
55. O'Donnell VB, Eiserich JP, Bloodsworth A, Chumley PH, Kirk M, Barnes S, Darley-Usmar VM, Freeman BA (1999) [47] Nitration of unsaturated fatty acids by nitric oxide-derived reactive species In: Methods in enzymology, vol 301. Academic, San Diego, pp 454–470
56. Gray P, Yoffe A (1955) The reactivity and structure of nitrogen dioxide. Chem Rev 55:1069–1154
57. Lee Y, Schwartz S (1981) Reaction kinetics of nitrogen dioxide with liquid water at low partial pressure. J Phys Chem 85:840–848
58. Cheung J, Li Y, Boniface J, Shi Q, Davidovits P, Worsnop D, Jayne J, Kolb C (2000) Heterogeneous interactions of NO2 with aqueous surfaces. Chem A Eur J 104:2655–2662
59. Squadrito GL, Postlethwait EM (2009) On the hydrophobicity of nitrogen dioxide: could there be a "lens" effect for NO2 reaction kinetics? Nitric Oxide 21:104–109
60. Khairutdinov RF, Coddington JW, Hurst JK (2000) Permeation of phospholipid membranes by peroxynitrite. Biochemistry 39:14238–14249
61. Cordeiro RM (2018) Reactive oxygen and nitrogen species at phospholipid bilayers: peroxynitrous acid and its homolysis products. J Phys Chem B
62. Olson KR, Straub KD (2016) The role of hydrogen sulfide in evolution and the evolution of hydrogen sulfide in metabolism and signaling. Physiology (Bethesda) 31:60–72
63. Nicholls P, Marshall DC, Cooper CE, Wilson MT (2013) Sulfide inhibition of and metabolism by cytochrome c oxidase. Biochem Soc Trans 41:1312–1316
64. Vitvitsky V, Kabil O, Banerjee R (2012) High turnover rates for hydrogen sulfide allow for rapid regulation of its tissue concentrations. Antioxid Redox Signal 17:22–31
65. Hildebrandt TM, Grieshaber MK (2008) Three enzymatic activities catalyze the oxidation of sulfide to thiosulfate in mammalian and invertebrate mitochondria. FEBS J 275:3352–3361
66. Abe K, Kimura H (1996) The possible role of hydrogen sulfide as an endogenous neuromodulator. J Neurosci 16:1066–1071
67. Hosoki R, Matsuki N, Kimura H (1997) The possible role of hydrogen sulfide as an endogenous smooth muscle relaxant in synergy with nitric oxide. Biochem Biophys Res Commun 237:527–531
68. Elrod JW, Calvert JW, Morrison J, Doeller JE, Kraus DW, Tao L, Jiao X, Scalia R, Kiss L, Szabo C, Kimura H, Chow C-W, Lefer DJ (2007) Hydrogen sulfide attenuates myocardial ischemia-reperfusion injury by preservation of mitochondrial function. Proc Natl Acad Sci U S A 104:15560–15565
69. Blackstone E, Morrison M, Roth MB (2005) H2S induces a suspended animation-like state in mice. Science (New York) 308:518
70. Cuevasanta E, Möller MN, Alvarez B (2017) Biological chemistry of hydrogen sulfide and persulfides. Arch Biochem Biophys 617:9–25
71. Riahi S, Rowley CN (2014) Solvation of hydrogen sulfide in liquid water and at the water-vapor inter-

face using a polarizable force field. J Phys Chem B 118:1373–1380
72. Filipovic MR, Zivanovic J, Alvarez B, Banerjee R (2018) Chemical biology of H2S signaling through persulfidation. Chem Rev 118:1253–1337
73. Carballal S, Trujillo M, Cuevasanta E, Bartesaghi S, Möller MN, Folkes LK, García-Bereguiaín MA, Gutiérrez-Merino C, Wardman P, Denicola A, Radi R, Alvarez B (2011) Reactivity of hydrogen sulfide with peroxynitrite and other oxidants of biological interest. Free Radic Biol Med 50:196–205
74. Cuevasanta E, Zeida A, Carballal S, Wedmann R, Morzan UN, Trujillo M, Radi R, Estrin DA, Filipovic MR, Alvarez B (2015) Insights into the mechanism of the reaction between hydrogen sulfide and peroxynitrite. Free Radic Biol Med 80:93–100
75. Nagy P, Winterbourn CC (2010) Rapid reaction of hydrogen sulfide with the neutrophil oxidant hypochlorous acid to generate polysulfides. Chem Res Toxicol 23:1541–1543
76. Cuevasanta E, Lange M, Bonanata J, Coitiño EL, Ferrer-Sueta G, Filipovic MR, Alvarez B (2015) Reaction of hydrogen sulfide with disulfide and sulfenic acid to form the strongly nucleophilic persulfide. J Biol Chem 290:26866–26880
77. Cuevasanta E, Denicola A, Alvarez B, Möller MN (2012) Solubility and permeation of hydrogen sulfide in lipid membranes. PLoS One 7:e34562
78. Mathai JC, Missner A, Kügler P, Saparov SM, Zeidel ML, Lee JK, Pohl P (2009) No facilitator required for membrane transport of hydrogen sulfide. Proc Natl Acad Sci 106:16633–16638
79. Riahi S, Rowley CN (2014) Why can hydrogen sulfide permeate cell membranes? J Am Chem Soc 136:15111–15113
80. Ferrer-Sueta G, Radi R (2009) Chemical biology of peroxynitrite: kinetics, diffusion, and radicals. ACS Chem Biol 4:161–177
81. Radi R, Beckman JS, Bush KM, Freeman BA (1991) Peroxynitrite-induced membrane lipid peroxidation: the cytotoxic potential of superoxide and nitric oxide. Arch Biochem Biophys 288:481–487
82. Rubbo H, Trostchansky A, O'Donnell VB (2009) Peroxynitrite-mediated lipid oxidation and nitration: mechanisms and consequences. Arch Biochem Biophys 484:167–172
83. Marla SS, Lee J, Groves JT (1997) Peroxynitrite rapidly permeates phospholipid membranes. Proc Natl Acad Sci 94:14243–14248
84. Holmström KM, Finkel T (2014) Cellular mechanisms and physiological consequences of redox-dependent signalling. Nat Rev Mol Cell Biol 15:411
85. Buettner GR (1993) The pecking order of free radicals and antioxidants: lipid peroxidation, α-tocopherol, and ascorbate. Arch Biochem Biophys 300:535–543
86. Kellogg EW, Fridovich I (1975) Superoxide, hydrogen peroxide, and singlet oxygen in lipid peroxidation by a xanthine oxidase system. J Biol Chem 250:8812–8817
87. Lynch RE, Fridovich I (1978) Effects of superoxide on the erythrocyte membrane. J Biol Chem 253:1838–1845
88. Gus'kova RA, Ivanov II, Kol'tover VK, Akhobadze VV, Rubin AB (1984) Permeability of bilayer lipid membranes for superoxide (O2−) radicals. Biochimica et Biophysica Acta (BBA)-Biomembranes 778:579–585
89. Lynch RE, Fridovich I (1978) Permeation of the erythrocyte stroma by superoxide radical. J Biol Chem 253:4697–4699
90. Van der Paal J, Verheyen C, Neyts EC, Bogaerts A (2017) Hampering effect of cholesterol on the permeation of reactive oxygen species through phospholipids bilayer: possible explanation for plasma cancer selectivity. Sci Rep 7:39526
91. Breton-Romero R, Lamas S (2014) Hydrogen peroxide signaling in vascular endothelial cells. Redox Biol 2:529–534
92. Winterbourn CC (2013) The biological chemistry of hydrogen peroxide. In: Methods in enzymology, vol 528. Elsevier, Amsterdam, pp 3–25
93. Orrico F, Möller MN, Cassina A, Denicola A, Thomson L (2018) Kinetic and stoichiometric constraints determine the pathway of H2O2 consumption by red blood cells. Free Radic Biol Med 121:231–239
94. Manta B, Hugo M, Ortiz C, Ferrer-Sueta G, Trujillo M, Denicola A (2009) The peroxidase and peroxynitrite reductase activity of human erythrocyte peroxiredoxin 2. Arch Biochem Biophys 484:146–154
95. Randall LM, Ferrer-Sueta G, Denicola A (2013) Chapter three – Peroxiredoxins as preferential targets in H2O2-induced signaling. In: Cadenas E, Packer L (eds) Methods in enzymology, vol 527. Academic, San Diego, pp 41–63
96. Stöcker S, Van Laer K, Mijuskovic A, Dick TP (2018) The conundrum of hydrogen peroxide signaling and the emerging role of peroxiredoxins as redox relay hubs. Antioxid Redox Signal 28:558–573
97. Halliwell B, Chirico S (1993) Lipid peroxidation: its mechanism, measurement, and significance. Am J Clin Nutr 57:715S–725S
98. Clayton RK (1959) Permeability barriers and the assay of catalase in intact cells. Biochim Biophys Acta 36:35–39
99. Nicholls P (1965) Activity of catalase in the red cell. Biochimica et Biophysica Acta (BBA)-Enzymology and Biological Oxidation 99:286–297
100. Antunes F, Cadenas E (2000) Estimation of H2O2 gradients across biomembranes. FEBS Lett 475:121–126
101. Branco MR, Marinho HS, Cyrne L, Antunes F (2004) Decrease of H2O2 plasma membrane permeability during adaptation to H2O2 in Saccharomyces cerevisiae. J Biol Chem 279:6501–6506
102. Seaver LC, Imlay JA (2001) Hydrogen peroxide fluxes and compartmentalization inside growing Escherichia coli. J Bacteriol 183:7182–7189

103. Mathai JC, Sitaramam V (1994) Stretch sensitivity of transmembrane mobility of hydrogen peroxide through voids in the bilayer. Role of cardiolipin. J Biol Chem 269:17784–17793
104. Sitaramam V, Mathai JC, Rao N, Block LH (1989) Hydrogen peroxide permeation across liposomal membranes: a novel method to assess structural flaws in liposomes. Mol Cell Biochem 91:91–97
105. Abuin E, Lissi E, Ahumada M (2012) Diffusion of hydrogen peroxide across DPPC large unilamellar liposomes. Chem Phys Lipids 165:656–661
106. Bienert GP, Chaumont F (2014) Aquaporin-facilitated transmembrane diffusion of hydrogen peroxide. Biochim Biophys Acta Gen Sub 1840:1596–1604
107. Henzler T, Steudle E (2000) Transport and metabolic degradation of hydrogen peroxide in Chara corallina: model calculations and measurements with the pressure probe suggest transport of H2O2 across water channels. J Exp Bot 51:2053–2066
108. Bienert GP, Møller AL, Kristiansen KA, Schulz A, Møller IM, Schjoerring JK, Jahn TP (2007) Specific aquaporins facilitate the diffusion of hydrogen peroxide across membranes. J Biol Chem 282:1183–1192
109. Almasalmeh A, Krenc D, Wu B, Beitz E (2014) Structural determinants of the hydrogen peroxide permeability of aquaporins. FEBS J 281:647–656
110. Dynowski M, Schaaf G, Loque D, Moran O, Ludewig U (2008) Plant plasma membrane water channels conduct the signalling molecule H2O2. Biochem J 414:53–61
111. Cordeiro RM (2015) Molecular dynamics simulations of the transport of reactive oxygen species by mammalian and plant aquaporins. Biochim Biophys Acta Gen Sub 1850:1786–1794
112. Miller EW, Dickinson BC, Chang CJ (2010) Aquaporin-3 mediates hydrogen peroxide uptake to regulate downstream intracellular signaling. Proc Natl Acad Sci 107:15681–15686
113. Hara-Chikuma M, Satooka H, Watanabe S, Honda T, Miyachi Y, Watanabe T, Verkman AS (2015) Aquaporin-3-mediated hydrogen peroxide transport is required for NF-κB signalling in keratinocytes and development of psoriasis. Nat Commun 6:7454
114. Dalla Sega FV, Zambonin L, Fiorentini D, Rizzo B, Caliceti C, Landi L, Hrelia S, Prata C (2014) Specific aquaporins facilitate Nox-produced hydrogen peroxide transport through plasma membrane in leukaemia cells. Biochimica et Biophysica Acta (BBA)-Molecular Cell Research 1843:806–814
115. Amen F, Machin A, Touriño C, Rodríguez I, Denicola A, Thomson L (2017) N-acetylcysteine improves the quality of red blood cells stored for transfusion. Arch Biochem Biophys 621:31–37
116. Koppenol WH (2017) Hydrogen peroxide, a molecule with a Janus face: its history, chemistry, and biology. In: Hydrogen peroxide metabolism in health and disease, Boca Raton : CRC Press pp 3–16
117. Buxton GV, Greenstock CL, Helman WP, Ross AB (1988) Critical review of rate constants for reactions of hydrated electrons, hydrogen atoms and hydroxyl radicals (· OH/· O−) in aqueous solution, J Phys Chem Ref Data 17:513–886
118. Subczynski WK, Hyde JS, Kusumi A (1989) Oxygen permeability of phosphatidylcholine-cholesterol membranes. Proc Natl Acad Sci 86:4474–4478
119. Widomska J, Raguz M, Subczynski WK (2007) Oxygen permeability of the lipid bilayer membrane made of calf lens lipids. Biochim Biophys Acta Biomembr 1768:2635–2645
120. Subczynski WK, Hopwood LE, Hyde JS (1992) Is the mammalian cell plasma membrane a barrier to oxygen transport? J Gen Physiol 100:69–87
121. Fettiplace R, Haydon DA (1980) Water permeability of lipid membranes. Physiol Rev 60:510–550
122. De Duve C (1965) The separation and characterization of subcellular particles. Harvey Lect 59:49–87
123. Makino N, Sasaki K, Hashida K, Sakakura Y (2004) A metabolic model describing the H2O2 elimination by mammalian cells including H2O2 permeation through cytoplasmic and peroxisomal membranes: comparison with experimental data. Biochim Biophys Acta Gen Subj 1673:149–159
124. Lim JB, Langford TF, Huang BK, Deen WM, Sikes HD (2016) A reaction-diffusion model of cytosolic hydrogen peroxide. Free Radic Biol Med 90:85–90
125. Takahashi M-A, Asada K (1983) Superoxide anion permeability of phospholipid membranes and chloroplast thylakoids. Arch Biochem Biophys 226:558–566
126. Liu X, Miller MJ, Joshi MS, Sadowska-Krowicka H, Clark DA, Lancaster JR (1998) Diffusion-limited reaction of free nitric oxide with erythrocytes. J Biol Chem 273:18709–18713
127. Liu X, Samouilov A, Lancaster JR, Zweier JL (2002) Nitric oxide uptake by erythrocytes is primarily limited by extracellular diffusion not membrane resistance. J Biol Chem 277:26194–26199

Characterization of Hydroxy and Hydroperoxy Polyunsaturated Fatty Acids by Mass Spectrometry

2

Priscilla Bento Matos Cruz Derogis, Adriano B. Chaves-Fillho, and Sayuri Miyamoto

Abstract

Lipids containing unsaturated doubles bonds are oxidized by enzymatic and non-enzymatic mechanisms yielding hydroperoxides and hydroxides as primary products.

This process has been implicated in physiological and pathological mechanisms. Thus, precise characterization and quantification of lipid oxidation products in biological samples can provide important mechanistic insights. In this context, the use of liquid chromatography coupled to tandem mass spectrometry (LC-MS/MS) to characterize the primary products of polyunsaturated fatty acids, like hydroxides and hydroperoxides, is a sensitive, specific and versatile tool. Here we will focus on the characterization and specific analysis of hydroxy and hydroperoxy regioisomers of linoleic acid, arachidonic acid, eicosapentaenoic acid and docosahexaenoic acid by LC-MS/MS.

Keywords

Polyunsaturated fatty acids · Fatty acid hydroperoxides · Fatty acid hydroxides · Mass spectrometry

2.1 Introduction

In the last decades, free radical processes delineated an interdisciplinary field linking chemistry to biology and medicine [26]. Free radicals derived primarily from oxygen have been implicated in the pathophysiology of a wide variety of human diseases including atherosclerosis [77], neurodegenerative diseases [7, 25], cancer [33], and even the normal aging process [5, 8].

Lipids, in particular unsaturated fatty acids, such as linoleic acid (C18:2, ω-6) (LA), arachidonic acid (C20:4, ω-6), eicosapentaenoic acid (20:5, ω-3) (EPA) and docosahexaenoic acid (C22:6, ω-3) (Fig. 2.1), are susceptible to oxidation by free radicals producing a wide number of hydroperoxides as primary products [26, 87, 91]. Lipid hydroperoxides are also produced by photosensitized oxidation involving singlet oxygen [23] as well as by enzymatic oxidation mediated by enzymes, such as lipoxygenases (LOX) and cyclooxygenases (COX) [69].

Lipid hydroperoxides are readily reduced to corresponding hydroxides by antioxidant enzymes like glutathione peroxidases (GPx) [10, 23], glutathione S-transferases (GSTA1-1 and

P. B. M. C. Derogis
Departamento de Bioquímica, Instituto de Química – Universidade de São Paulo, São Paulo, SP, Brazil

Hospital Israelita Albert Einstein, São Paulo, SP, Brazil

A. B. Chaves-Fillho · S. Miyamoto (✉)
Departamento de Bioquímica, Instituto de Química – Universidade de São Paulo, São Paulo, SP, Brazil
e-mail: miyamoto@iq.usp.br

A. Trostchansky, H. Rubbo (eds.), *Bioactive Lipids in Health and Disease*, Advances in Experimental Medicine and Biology 1127, https://doi.org/10.1007/978-3-030-11488-6_2

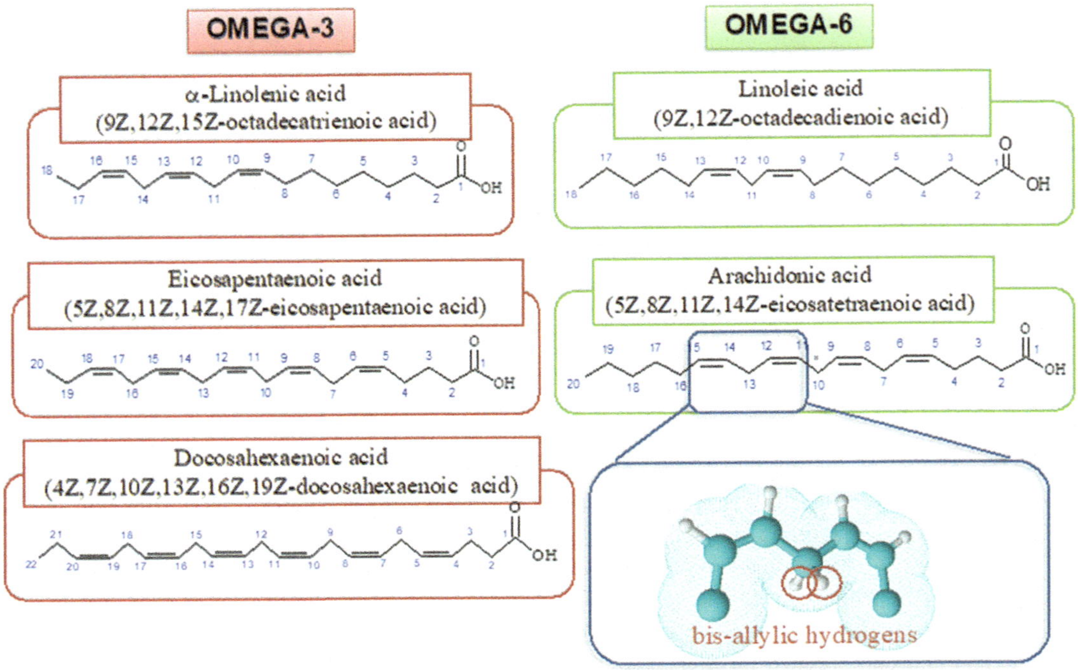

Fig. 2.1 Structures of omega-3 and omega-6 PUFAs. The double bond results in a weakening of C-H bond energy. In focus, the bis-allylic hydrogens

GSTA2-2) [90], and peroxiredoxins (Prx) [64]. Recently, studies have uncovered the essential role of GPx4 as a master regulator of a novel form of cell death termed as ferroptosis [34]. If not efficiently reduced, lipid hydroperoxides can undergo reactions leading to membrane disruption [4] and generation of secondary products, including short-chain electrophilic species (e.g. MDA, malondialdehyde and HNE, 4-hydroxy-2-nonenal) and a series of bioactive cyclization products, such as isoprostanes [66, 91]. With increasing evidences indicating the involvement of lipid peroxidation in various disorders and diseases, the identification of hydroperoxy and hydroxy-fatty acids can often provide valuable mechanistic information [60].

Lipid oxidation products can be analyzed by several methods, including immunoassays, colorimetric, spectroscopic and chromatographic methods [60, 75]. Recent tools to study these products include the lipidomic analysis approaches in which different molecular species of lipids and its oxidized products are comprehensively analyzed by chromatographic techniques coupled to mass spectrometry (MS), such as gas chromatography and/or liquid chromatography coupled to MS (GC-MS or LC-MS, respectively) [38, 74, 82, 84]. In this chapter we will describe recent applications of LC-MS techniques used to characterize the primary oxidation products derived from polyunsaturated fatty acids (PUFA), specifically the hydroxy- and hydroperoxy-fatty acids. Before going into the details of oxidized product analysis, we will briefly describe the main mechanisms leading to fatty acid hydroperoxide formation.

2.2 Fatty Acid Hydroperoxide Formation

Fatty acid hydroperoxides are prominent primary intermediates of oxidative reactions induced by the attack of reactive oxygen and/or nitrogen species on unsaturated fatty acids. PUFAs are defined as fatty acids that have more than two carbon-

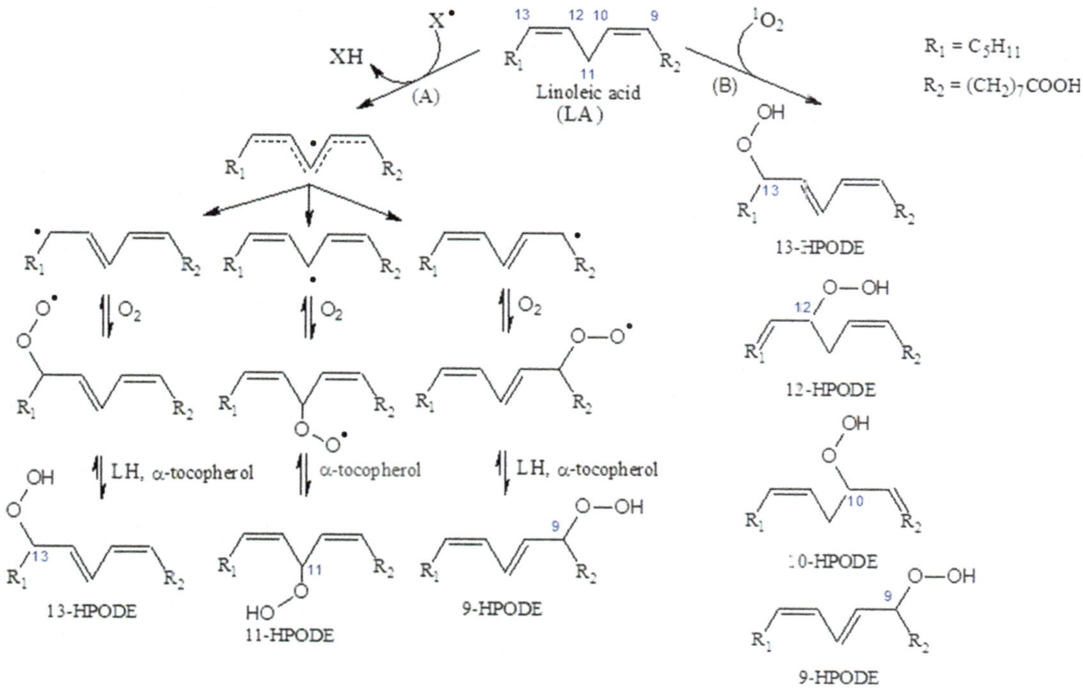

Fig. 2.2 Two main non-enzymatic oxidation process of PUFA. The oxidation of LA was used as an example to illustrate the oxidation of unsaturated fatty acids by free radicals (A) and singlet oxygen (B). Free-radical induced oxidation starts with bis-allylic hydrogen atom abstraction by a free radical (X•) resulting in a pentadienyl radical, which combines with O_2 generating a peroxyl radical. The peroxyl radical abstracts a hydrogen atom from another hydrogen atom donor, for instance a fatty acid or α-tocopherol, generating the LA hydroperoxy-fatty acids (HPODE). Major products of LA free radical oxidation are the 9- and 13-HPODE (*E,Z* and *E,E*). The 11-HPODE is observed only in the presence of high concentrations of antioxidants such as the α-tocopherol. In the photooxidation, singlet oxygen ($^1O_2$) reacts with the carbon-carbon double bond of LA to form the 9, 10, 12 and 13-HPODEs

carbon double bonds. They are classified according to the original fatty acids from which they are synthesized into two distinct families: ω-6 PUFA, which is derived from LA (C18:2, ω-6); and ω-3 PUFA, which is synthesized from α-linolenic acid (C18:3, ω-3) (ALA). Contrary to other fatty acids, LA and ALA are essential fatty acids that must be obtained in adequate amounts from diet [73]. LA is metabolized to AA, whereas ALA is metabolized to EPA and DHA, increasing the chain length and degree of unsaturation by adding extra double bonds to the carboxyl end of the fatty acid molecule [42].

Unsaturated fatty acids are very susceptible to lipid peroxidation due to the presence of double bonds, which causes a weakening of the C-H bond energy in the neighbor carbon (allylic hydrogen, BDE 88 kcal/mol), especially if there is a double bond on either side of C-H bond, giving *bis*-allylic hydrogens (BDE 78–80 kcal/mol) [87] (Fig. 2.1). Thus, PUFA peroxidation induced by free radicals begins with the abstraction of bis-allylic hydrogen atom by a free radical (X•) (Fig. 2.2A). In the case of PUFAs containing more than two double bonds, the hydrogen abstraction occurs at the *bis*-allylic hydrogens resulting in the formation of a carbon centered radical. This radical undergoes rearrangements yielding a penta-dienyl radical that combines rapidly with O_2 to generate a peroxyl radical (LOO•), and this propagates the radical chain reaction by abstracting a hydrogen atom from another lipid molecule or other H-donors.

The peroxyl radical is involved in several competing reactions, such as β-scission and cyclization by intramolecular rearrangement. In this manner, LA free radical oxidation proceeds by a hydrogen abstraction from the doubly allylic methylene on carbon-11 to produce a delocalized penta-dienyl radical. Oxygen (O_2) rapidly adds to the penta-dienyl end positions to produce equal amounts of the conjugated 9- (9-HPODE) and 13-hydroperoxide (13-HPODE) isomers with *trans, cis-* (Z,E) and *trans-trans* (E,E) configuration [19, 20, 91]. Additionally, oxidation of LA conducted in the presence of efficient H atom-donors (e.g. α-tocopherol) yield, besides the 9-, 13-hydroperoxides (9-HPODE, 13-HPODE), the *bis*-allylic 11-hydroperoxide (11-HPODE) (Fig. 2.2A) [70, 91]. The number of positional isomeric hydroperoxides (2n-2) that can be formed upon PUFA peroxidation depends on the number of double bonds (n) [16]. Hence, the oxidation of unsaturated fatty acids containing more than three double bonds (such as AA, EPA and DHA) usually occurs by more complicated mechanisms [92]. Such mechanisms include, among others, an intramolecular peroxyl addition reactions forming numerous products, including prostaglandin-type cyclic peroxides and hydroperoxides [66, 91, 92].

Photosensitized oxidation of PUFAs proceeds by a different non-radical mechanism (Fig. 2.2B). In this process, there is a direct reaction of singlet molecular oxygen ($^1O_2$) with the carbon-carbon double bond by "*ene*" addition, and hydroperoxides are formed at each side of the double bond. Thus, for example, $^1O_2$ promoted oxidation of LA produces 4 isomers: two conjugated 9- and 13-hydroperoxides (as free radical mediated oxidation) and 2 unconjugated 10- and 12-hydroperoxides (10-HPODE and 12-HPODE) [19]. It is important to note that some hydroperoxides are specifically formed upon $^1O_2$ mediated oxidation and therefore they can be used as a fingerprint for $^1O_2$ in biological system [78]. This is the case of 10- and 12-hydroperoxides of LA (Table 2.1). The relevance of studying these hydroperoxides is based on the fact that $^1O_2$ can be formed in biologically relevant processes, such as phagocytosis in inflammatory conditions and also reactions involving biological hydroperoxides, enzymes (e.g. myeloperoxidase, lipoxygenase), peroxynitrite, HOCl, and metal ions [51, 52]. Singlet oxygen is also an important oxidant produced in photosensitized oxidation [4] and optogenetic approaches [86].

Hydroperoxide metabolites are also formed through the action of cyclooxygenases (COX) and lipoxygenases (LOX) [62, 69, 76]. For example, several studies have demonstrated that mammalian LOX are regio- and stereo-specific, producing specifically the *R* or *S* enantiomer at the targeted positions of oxidation on their PUFA substrates [69]. Much of the knowledge on product specificities comes from studies using arachidonic acid or linoleic acid, which are typical substrates for animal and plant LOX, respectively [69]. In contrast, non-enzymatic PUFA oxidation produces the same hydroperoxides as those catalyzed by LOX but with equal quantities of both stereoisomers ($R = S$, racemic mixture) at all positions of oxidation [18, 60]. The hydroperoxides formed through non-enzymatic and enzymatic oxidation of PUFAs are listed in Table 2.1.

Oxygenated fatty acid metabolites display signaling roles in physiological and pathological conditions. For example, enzymatic oxidation of AA and DHA generates a series of bioactive mono- and poly-oxygenated products that are collectively known as "eicosanoids" and "docosanoids", respectively [89]. Leukotrienes, prostaglandins and thromboxanes are subclasses of eicosanoids that play an important role in inflammatory processes by promoting neutrophil recruitment, inducing allergic reactions and platelet aggregation [29, 43, 65]. Docosanoids have received increased scientific attention due to their role as anti-inflammatory agents. Among the latter, resolvins, maresins and neuroprotectins are implicated in the inhibition of neuron death by apoptosis, in tissue regeneration and downregulation of pro-inflammatory mediator levels [41, 71, 72]. Oxidized lipids are also known to promote protein lipoxidation by reacting with nucleophilic residues, such as histidine, lysine and cysteine residues [1]. Chemical modifications of proteins may result in alterations of folding and biological function, leading to

Table 2.1 Hydroperoxy-polyunsaturated fatty acids that can be formed by three different oxidation mechanisms

PUFAs	Autoxidation	Photooxidation	Enzymatic oxidation*
Linoleic acid	9-HPODE[a,b,c]	9-HPODE[c,d]	9-HPODE[e]
	11-HPODE[c]	10-HPODE[c,d]	11-HPODE[f]
	13-HPODE[a,b,c]	12-HPODE[c,d]	13-HPODE[e,f]
		13-HPODE[c,d]	
Arachidonic acid	5-HPETE[g]	5-HPETE[d]	
	8-HPETE[g]	6-HPETE[d]	
	9-HPETE[g]	9-HPETE[d]	5-HPETE[h]
	11-HPETE[g]	11-HPETE[d]	12-HPETE[h]
	12-HPETE[g]	12-HPETE[d]	15-HPETE[h]
	15-HPETE[g]	14-HPETE[d]	
		15-HPETE[d]	
Eicosapentaenoic acid		5-HPEPE[j]	
	5-HPEPE[i,j]	6-HPEPE[j]	
	8-HPEPE[i,j]	8-HPEPE	
	9-HPEPE[i,j]	9-HPEPE[j]	5-HPEPE[h,k,l]
	11-HPEPE[i,j]	11-HPEPE[j]	8-HPEPE[h,k,l]
	12-HPEPE[i,j]	12-HPEPE	12-HPEPE[h,k,l]
	14-HPEPE[i,j]	14-HPEPE[j]	15-HPEPE[h,k,l]
	15-HPEPE[i,j]	15-HPEPE[j]	
	18-HPEPE[i,j]	17-HPEPE[j]	
		18-HPEPE[j]	
Docosahexaenoic acid		4-HPDHA[m]	
	4-HPDHA[m]	5-HPDHA[m]	
	7-HPDHA[m]	7-HPDHA[m]	
	8-HPDHA[m]	8-HPDHA[m]	
	10-HPDHA[m]	10-HPDHA[m]	
	11-HPDHA[m]	11-HPDHA[m]	17-HPDHA[n]
	13-HPDHA[m]	13-HPDHA[m]	14-HPDHA[n]
	14-HPDHA[m]	14-HPDHA[m]	
	16-HPDHA[m]	16-HPDHA[m]	
	17-HPDHA[m]	17-HPDHA[m]	
	20-HPDHA[m]	19-HPDHA[m]	
		20-HPDHA[m]	

*The isomers shown are the most widely studied and characterized in the literature

HPODE hydroperoxy-octadecadienoic acid, *HPETE* hydroperoxyeicosatetraenoic acid, *HPEPE* hydroperoxyeicosapentaenoic acid, *HPDHA* hydroperoxydocosahexaenoic acid. [a]Ref. [19]; [b]Ref. [20]; [c]Ref. [68]; [d]Ref. [80]; [e]Ref. [27]; [f]Ref. [62]; [g]Ref. [56]; [h]Ref. [46]; [i]Ref. [88]; [j]Ref. [92]; [k]Refs. [12]; [l]Refs. [47]; [m]Ref. [45]; Ref. [30]

impaired proteostasis and loss of cellular homeostasis [40].

In this context, the identification and structural characterization of lipid derived hydro(pero)xides formed by enzymatic and non-enzymatic transformations continue to reveal novel enzymes, products and biological actions [37, 68]. In this chapter we will focus on describing the chemical characterization of PUFA hydro(pero)xides by mass spectrometry.

2.3 Hydroperoxide- and Hydroxy-Polyunsaturated Fatty Acids Analysis

Measurement of fatty acid hydro(pero)xides and related species have been mostly carried out using GC-MS and LC-MS based techniques [11, 48, 85]. Historically, GC and GC-MS have been widely used for the analysis of lipid oxidation products [53, 81, 85]. However, GC-based methods require derivatization steps that are not suitable for hydroperoxides analysis since they can be decomposed during sample preparation. Therefore, strategies used to analyze hydroperoxides usually involve their conversion to more stable hydroxides using a reducing agent, such as sodium borohydride or triphenylphosphine, and then their derivatization to methyl esters. The hydroperoxide concentration can be calculated as the difference between pre- and post-reduction hydroxy-fatty acid concentrations [36].

Compared to GC, liquid chromatography (LC)-based methods offer the advantage of separating lipid hydroperoxides and hydroxides without derivatization [11]. Lipid hydro(pero)xide regioisomers can be separated on either normal-phase or reversed-phase chromatographic methods. Additionally, enantiomers can be separated by chiral-columns [22]. The simplest way to detect hydroperoxy- and hydroxy-PUFA is accomplished through the use of LC-coupled to UV detectors by measuring conjugated-diene absorbance at 234 nm (ε = 23,000–27,000 M^{-1}. cm^{-1}). However, conjugated-diene detection by UV is not very specific or sensitive enough to detect lipid hydro(pero)xides in biological samples [24]. More sensitive methods to detect hydro(pero)xides are based on LC-coupling to chemiluminescence, electrochemical, fluorescence or mass spectrometer detectors [58]. Although chemiluminescence and electrochemical detectors have much higher sensitivity than UV, these techniques do not allow the detection of lipid hydroxides and suffer interferences from sample components [59].

The use of mass spectrometry (MS) technology provided important advances in the analysis of lipid oxidation products. This technique, coupled to the wide range of ionization methods and high-resolution mass analyzers, has expanded the capacity to detect and analyze these compounds in biological samples. Great advancement on the analysis of lipid oxidation products was also obtained by the combination of different modes of mass analysis to separation science, including conventional or high-speed LC (HPLC or UHPLC) coupled to tandem mass spectrometry (LC-MS/MS). These systems have promoted a great increase in sensitivity and efficiency in the identification and quantification of such products, being the method of choice for oxidized lipid species analysis [3, 28, 32, 35, 49, 55, 74, 83].

Low-energy ionization techniques, such as electrospray ionization, usually generate deprotonated ($[M-H]^-$) or protonated ($[M + H]^+$) ions, which can be selected and further cleaved by collision-induced dissociation (CID) in MS/MS analysis. The spectra of the fragmented ions are then used to identify and elucidate the structures of oxidized fatty acid species [44, 54, 55]. For quantitative and targeted lipidomic analysis, the mostly used MS methods are single ion monitoring (SIM) and multiple reaction monitoring (MRM) on triple-quadrupole or quadrupole-ion trap mass spectrometers. In any of these methods the individual molecules are identified by their retention time and *m/z* values of the precursor and fragment ions. Of note, in SIM and MRM analysis, the target molecules to be analyzed must be defined in advance, and the data of their precursor ion masses and their fragments are necessary to set the analytical conditions [79]. Moreover, accurate quantification requires the use of an internal standard that are added into the samples before lipid extraction. The best internal standard is the stable isotopically labeled version of the compound (^2H or ^{13}C-labeled) or an isotopic analogue.

In the next sections we will describe the main fragmentations pathways and fragment specificities for LA, ARA, EPA and DHA hydroperoxides and hydroxides observed using different set LC-MS^n systems, columns, solvents and ionization sources.

2.3.1 Characterization of Linoleic Acid Hydro(pero)xides

9- and 13- Hydroperoxy-octadecadienoic acid (HPODE) are products of non-enzymatic and enzymatic oxidation of LA (Table 2.1). Non-enzymatically, these hydroperoxides are formed in reactions involving free radicals and singlet molecular oxygen. Enzymatically, they are formed in the presence of enzymes, such as LOX. These enzymes catalyze the dioxygenation of the *cis*, *cis*-1,4-pentadiene moiety of fatty acids [69]. The ferric form of LOX is catalytically active and catalyzes the stereospecific abstraction of one hydrogen from the bis-allylic methylene group of the (1Z,4Z)-pentadiene structure of the substrate as the initial step. Iron-LOX form 9*S*- and 13*S*-hydroperoxides of LA [27].

MacMillan and Murphy [46] described a detailed characterization of PUFA hydroperoxides, including 9- and 13-HPODE, by negative ion ESI-MS. Full scan MS spectra of all hydroperoxides typically show two abundant ions, corresponding to the deprotonated ion ($[M\text{-}H]^-$) and the dehydration product ($[M\text{-}H\text{-}H_2O]^-$). In the case of linoleate hydroperoxides, these ions appear at *m/z* 311 and *m/z* 293, respectively [46, 50]. The products expected to be formed after the loss of water are keto or epoxy-acids (Fig. 2.3A). Although abundant, the dehydrated ion does not give structural information about the position of the hydroperoxide moiety. Thus, structurally important fragment ions are obtained by collision induced dissociation (CID) of the precursor ion. For instance, MS/MS spectra obtained for 9- and 13-HPODE show characteristic fragments that are formed as a result of carbon-carbon cleavages. The MS/MS spectra of 9-HPODE show two characteristic fragments: one at *m/z* 185 (dehydration and cleavage of double bond allylic to the hydroperoxide group after a two proton shift) and the other at *m/z* 125 (cleavage of C_8-C_9 bond after two

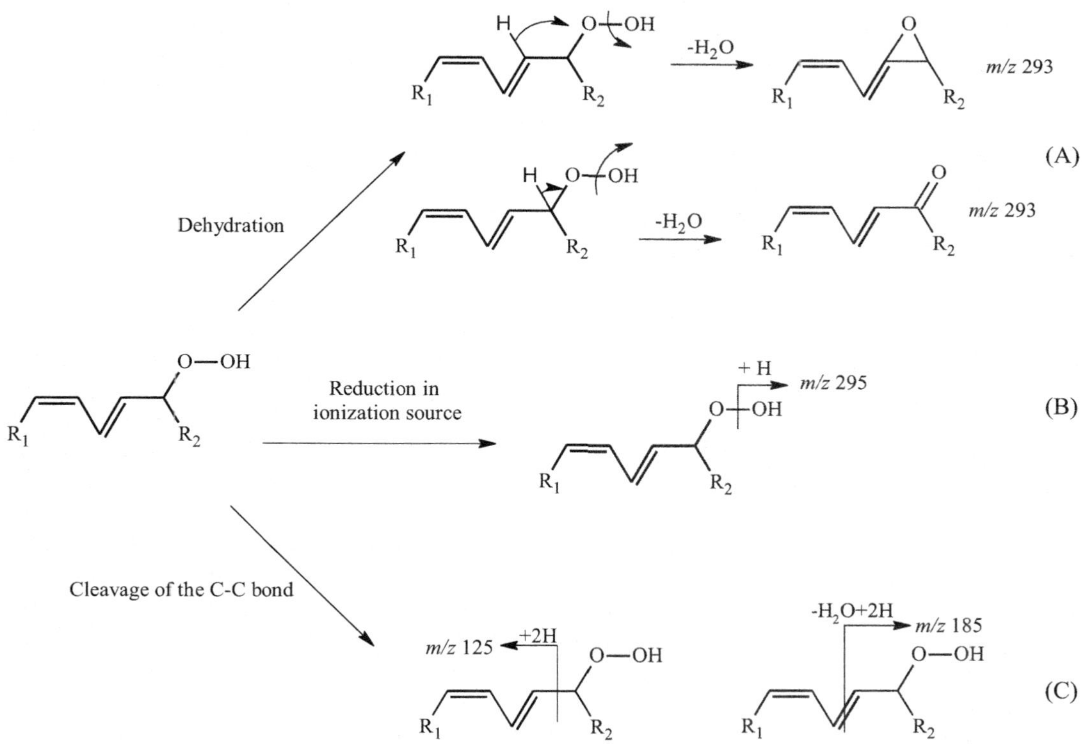

Fig. 2.3 Common fragmentations observed for PUFA hydroperoxides in the mass spectrometer. As an example, the fragmentations observed for 9-HPODE ($R_1 = C_5H_{11}$ and $R_2 = (CH_2)_7COOH$) are shown. (A) Loss of water ($-H_2O$) is a general fragmentation observed in the full scan MS and MS/MS spectra of all hydroperoxides. Dehydration can occur by two mechanisms, one leading to epoxides and the other to ketones. (B) Hydroperoxides can also undergo reduction in the ionization source. (C) Collision induced dissociation of the carboxylated anions yields characteristic fragments derived from cleavage of carbon bonds vicinal to the hydroxyl/hydroperoxyl group and/or the cleavage of the double bond allylic to the hydroperoxide

proton shift) (Fig. 2.3C) [46]. Similarly, 13-HPODE also shows two characteristic fragments, one at *m/z* 195 (cleavage of the C_{12}-C_{13} bond after a single proton shift) and the other at *m/z* 113 (loss of water and a cleavage of the double bond allylic to the hydroperoxide after a single proton shift) [46].

HPODE analyses are usually conducted in the negative ion mode. As an alternative method, Schneider et al. [70] described an LC-ESI-MS in the positive ion mode. LC-ESI-MS analysis of pure 9- and 13-*S*-HPODE in methanol:water:ammonium acetate revealed prominent adduct ions $[M + NH_4]^+$ at *m/z* 330 for 9/13-*S*-HPODE.

Oliw et al. [27], described a detailed analysis of Mn-LOX oxidation products by LC coupled to ion trap mass analyzer (MS^n). Mn-LOX acts by a different mechanism from iron-LOX that results in the formation of 11*S*-HPODE and 13*R*-HPODE [27]. They used reversed phase-HPLC columns eluted with methanol:water:acetic acid. The eluent first passed in an ultraviolet detector and then get into an ion trap mass spectrometer, where it was subjected to atmospheric pressure chemical ionization (APCI) or ESI. In both cases, negative ions were monitored. APCI-MS resulted in fragmentation of the hydroperoxides (M) with three strong signals, $[M\text{-}1]^-$, $[M\text{-}19]^-$ and $[M\text{-}17]^-$, which corresponded to the carboxylate anion ($[M\text{-}H]^-$), a fragment due to the loss of water ($[M\text{-}H\text{-}H_2O]^-$), and the formation of a reduced form of the hydroperoxide during ionization ($[M\text{-}H\text{-}O]^-$). The reduction of the hydroperoxide to the alcohol might be due to the high temperature (450 °C) of the APCI source (Fig. 2.3B). In contrast, ESI-MS showed only two of these signals, the $[M\text{-}1]^-$ and $[M\text{-}19]^-$. APCI-MS analysis of both 11*S*-and 13*R*-HPODE showed signals at *m/z* 311 ($[M\text{-}H]^-$), 295 ($[M\text{-}H\text{-}O]^-$) and 293 $[M\text{-}H\text{-}H_2O]^-$, while ESI-MS spectra showed abundant ions at *m/z* 311 and 293. In order to get structural information about the position of the hydroperoxide moiety, the carboxylate anion was subjected to CID fragmentation and the full scan spectrum of the fragments was acquired (MS/MS or MS^2). By analogy, a MS/MS/MS (MS^3) analysis was performed by selecting a daughter ion and submitting it to CID fragmentation. The MS^2 (311 → full-scan) for both HPODEs showed mainly the loss of water, while the MS^3 (311 → 293 → full-scan) analysis yielded several fragments that were used to identify the hydroperoxide group position. 11-HPODE has shown characteristic ions at *m/z* 177 and 151 and 13-HPODE at *m/z* 179 and 113 [22].

Oliw et al. [27] also performed analysis of the hydroxy derivatives. APCI and ESI-MS spectra for both 11- and 13-HODE showed characteristic signals at *m/z* 295 $[M\text{-}H]^-$ and 277 $[M\text{-}H\text{-}H_2O]^-$. MS^3 analysis of 13-HODE showed characteristic ions at *m/z* 195 and 179 and of 11-HODE at *m/z* 179 and 169. The formation of these ions can be explained by cleavage at either side of carbon linked to the hydroxide group.

Other HPODEs of biological relevance are the 10- and 12-HPODEs, which are products formed exclusively by singlet oxygen [19]. In contrast to the 9-, 11- and 13-isomers, these hydroperoxides do not have UV absorption at 234 nm, due to the absence of conjugated double bounds in their structure. Therefore, their detection is usually conducted at 200–205 nm, a region where both conjugated and non-conjugated diene absorb. Garscha et al. [22] analyzed these hydroperoxides from LA by LC coupled to ion trap ESI-MS^n. According to them, the characteristic and intense ions formed during MS^3 analysis (311 → 293 → full scan) were *m/z* 139 and 153 for 10-HPODE, and 165 and 191 for 12-HPODE. Nuñez et al. [61] studied hydroperoxides formed from linoleate methyl ester by using electron impact (EI) ionization. The 10-HPODE methyl ester EI spectrum presents mass ions at *m/z* 308 $[M\text{-}H_2O]^-$, 293 $[M\text{-}H\text{-}CH_3OH]^-$ and 261, which are consistent with an HPODE methylated isomer, 197 (loss of water and cleavage of C_{10}-C_{11}) and 165 (loss of CH_3OH group from ion 197). The EI spectrum of 12-HPODE methyl ester showed ions at *m/z* 310 $[M\text{-}H]^-$, 308, 293, 261, and 111 (loss of water and cleavage of C_{11}–C_{12}).

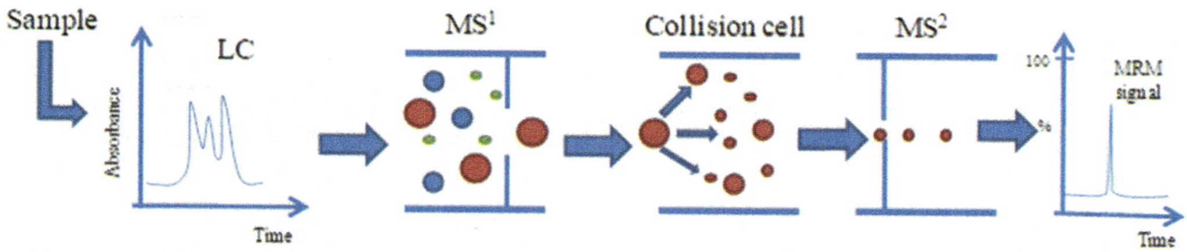

PUFAs	Isomers (hydroxides)	MRM transitions	PUFAs	Isomers (hydroxides)	MRM transitions
Linoleic acid	9-HODE 10-HODE 11-HODE 13-HODE	295 → 171 295 → 183 295 → 151 295 → 195	Arachidonic acid	5-HETE 6-HETE 8-HETE 9-HETE 11-HETE 12-HETE 14-HETE 15-HETE	319 → 115 319 → 191 319 → 155 319 → 151 319 → 167 319 → 208 319 → 207 319 → 219
Docosahexaenoic acid	4-HDHA 5-HDHA 7-HDHA 8-HDHA 10-HDHA 11-HDHA 13-HDHA 14-HDHA 16-HDHA 17-HDHA 19-HDHA 20-HDHA	343 → 101 343 → 85 343 → 141 343 → 189 343 → 153 343 → 165 343 → 193 343 → 205 343 → 233 343 → 245 343 → 229 343 → 285	Eicosapentaenoic acid	5-HEPE 8-HEPE 9-HEPE 11-HEPE 12-HEPE 15-HEPE 18-HEPE	317 → 115 317 → 155 317 → 149 317 → 167 317 → 179 317 → 219 317 → 259

Fig. 2.4 LC-MS/MS analysis of hydroxy-fatty acids. The liquid chromatography (LC) separates compounds chromatographically before they are introduced into the ion source of mass spectrometer. In this case the mass spectrometer is formed by two mass analyzers, the MS^1 and MS^2. The first and second mass analyzers scan across the spectrum. Structural characterization of each hydroxy-fatty acid is accomplished by analysing the fragments ions at MS^2. This is done by selecting the precursor ion of MS^1 and submitting it to CID. The MRM analysis is accomplished by specifying the mass of the parent and the fragment ions of the compound

Based on the method and results described by Oliw et al. [27], Reverberi et al. [63] developed a MRM method to quantify the hydroxy- and hydroperoxy-linoleic acid in order to investigate how lipid peroxidation affects the ochratoxin A biosynthesis in *Aspergillus ochraceus*. The transitions monitored for LA hydroperoxides were: *m/z* 311 to 185 for 9-HPODE, *m/z* 311 to 183 for 10-HPODE, *m/z* 311 to197 for 11-HPODE and *m/z* 311 to 195 for 13-HPODE. For the corresponding hydroxides compounds, they were: *m/z* 295 to 171 for 9-HODE, *m/z* 295 to 183 for 10-HODE, *m/z* 295 to 151 for 11-HODE, and *m/z* 295 to 195 for 13-HODE. Masoodi et al. [47] described the same MRM transitions for 9-HODE and 13-HODE (Fig. 2.4).

2.3.2 Characterization of Arachidonic Acid Hydro(pero)xides

AA is a precursor of local hormones (prostaglandins, thromboxanes and leukotrienes) involved in the AA cascade [14, 21, 57]. Inflammatory cells typically contain a high proportion of AA, which is a major substrate for eicosanoid synthesis. Free AA can be a substrate for several isoforms of LOX, being oxidized to 5-, 12- and 15-hydroperoxyeicosatetraenoic acid (HPETE) by 5-, 12- and 15-LOX, respectively [9]. This PUFA can also undergo non-enzymatic oxidation promoted by free radicals and singlet molecular oxygen. As described for LA, a different mixture of isomeric hydroperoxides is generated by these two processes. Free radical mediated oxidation yields 5-, 8-, 9-, 11-, 12-, 15-, 20-HPETE isomers, while singlet oxygen yields all these hydroperoxides and the 6- and 14-HPETE isomers [80]. AA can also be oxidized enzymatically by cytochrome P450 monooxygenase (CYP450), yielding, for example, 12-, 15-HETE [13] and 16-, 17-, 18-, 19- or 20-HETE [6, 17].

MacMillan and Murphy [46] separated the HPETE formed by LOX oxidation on a C18-column eluted isocratically with a mixture of acetic acid, acetonitrile and methanol. Tandem mass spectrometry of HPETE generated a unique and characteristic spectrum with some features common to all PUFA hydroperoxides, such as abundant $[M\text{-}H]^-$ ions and its dehydrated ions. Cleavage of the double bond allylic to the hydroperoxide formed structurally characteristic ions at *m/z* 129, 153 and 113 from 5-, 12- and 15-HPETE, respectively. Moreover, charge-driven allylic fragmentation led to the formation of *m/z* 203, 179 and 219 from 5-, 12-, and 15-HPETE, respectively. All HPETE isomers were monitored by a MRM method transition from *m/z* 335 (the carboxylate anion, $[M\text{-}H]^-$) to *m/z* 317 (the dehydrated product, $[M\text{-}H\text{-}H_2O]^-$). In addition, 15-, 12- and 5-HPETE were specifically detected by monitoring the MRM transitions from *m/z* 335 to 113, 179 and 129, respectively (Fig. 2.4).

There is a great interest in hydroxyeicosatetraenoic acids (HETEs) mainly due to their well-documented role in human inflammation and related immune responses. Kerwin and Torkvik [39] reported detailed identification of 5-, 8-,9-, 12- and 15-HETEs by negative-ion ESI-MS/MS. The characteristic ions observed for each HETE are shown in parenthesis: 8-HETE (*m/z* 127, 155 and 163); 9-HETE (*m/z* 139, 179, 151, 167); 12-HETE (*m/z* 135, 179, 208); and 15-HETE (*m/z* 121, 175, 219).

Nakamura et al. [56] analyzed the epoxyeicosatrienoic acids and HETE isolated from human red blood cells' membranes followed by base hydrolysis. EET are isobaric with HETE regioisomers. Analysis of these compounds by LC-ESI-MS/MS provided abundant signals for the carboxylate anions at *m/z* 319 for each EET (5,6-, 8,9-, 11,12- and 14,15-EET) and HETE (5-, 8-, 9-, 11-, 12- and 15-HETE). Specific analysis of HETE was achieved using the MRM mode of the mass spectrometer. The following transitions were analyzed: *m/z* 319 to 115 for 5-HETE; *m/z* 319 to 155 for 8-HETE; *m/z* 319 to 151 for 9-HETE; *m/z* 319 to 167 for 11-HETE; *m/z* 319 to 208 for 12-HETE; and *m/z* 319 to 219 for 15-HETE (Fig. 2.4). [^{18}O]-12-HETE was used as an internal standard to quantify each HETE isomer. The MRM analysis for [^{18}O]-12-HETE was done by monitoring the transition from *m/z* 323 to 183, using a collision energy of 11eV.

Masoodi et al. [47] also developed a LC-ESI-MS/MS methodology for the simultaneous analysis of hydroxy-fatty acids and other bioactive lipid mediators. Separation was conducted on a C-18 column and a methanol-based isocratic system. They described MRM transitions with some differences from Nakamura et al. [56]: *m/z* 319 to 123 for 9-HETE, *m/z* 319 to 179 for 12-HETE and *m/z* 319 to 175 for 15-HETE (Fig. 2.4). Again the 12-HETE-*d*8 was used as internal standard and the MRM transition monitored for it was *m/z* 328 to 185.

In our search, we were unable to find LC-MS analysis for isomers 6- and 14-HPETE, which are specifically formed in the oxidation of AA by singlet molecular oxygen.

2.3.3 Characterization of Eicosapentaenoic Acid Hydro(pero)xides

EPA is one of the most abundant omega-3 fatty acids found in marine oils. However, in mammals EPA is found in much lower abundance compared to DHA. EPA autoxidation generates eight regioisomers of 5-, 8-, 9-, 11-, 12-, 14-, 15- and 18- hydroperoxyeicosapentaenoic acid (HPEPE). In contrast, EPA oxidation mediated by singlet molecular oxygen yields ten monohydroperoxides: the same eight formed in autoxidation and the 6- and 17-HPEPE (non-conjugated isomers) [81]. The latter two are considered specific photooxidation products of EPA. However, we were unable to find any description of LC-MS analysis for 6- and 17-H(P)EPE. Enzymatic oxidation products of EPA have also been characterized. CYP450 forms the isomers 19- and 20-HEPE [6]. EPA oxidation mediated by 15-, 12- and 5-LOX has been shown to yield 15-, 12- and 5-HPEPE, respectively [12].

Morrow's group [92] studied several oxidation products formed from EPA autoxidation. The authors described the analysis by HPLC of eight hydroperoxides, whose structures were determined by Ag^+-coordination ionspray (LC-Ag^+-CIS-MS) based on specific Hock-cleavage fragments induced by silver ion coordination. This technique has been proven to be powerful for the characterization of PUFA hydroperoxides and their cyclization products, namely monocyclic and bicyclic endoperoxides (precursor to 3-series isoprostanes). However, LC-Ag^+CIS-MS showed poor sensitivity to detect EPA oxidation products *in vivo* and APCI-MS has been described to have better sensitivity for this purpose [92].

EPA hydroxides (hydroxyeicosapentaenoic acids, HEPE) analysis by LC-MS has been described [2, 47, 83, 93]. Specific transitions described for 5-, 8-, 9-, 12- and 15-HEPE are: *m/z* 317 to 115 for 5-HEPE, *m/z* 317 to 155 for 8-HEPE, *m/z* 317 to 149 for 9-HEPE, *m/z* 317 to 167 for 11-HEPE, *m/z* 317 to 179 for 12-HEPE, m/z 317 to 219 for 15-HEPE and *m/z* 317 to 259 for 18-HEPE (Fig. 2.4).

2.3.4 Characterization of Docosahexaenoic Acid Hydro(pero)xides

DHA are highly enriched in the cerebral cortex, retina, testis and sperm [67]. DHA oxidation produces hydroperoxydocosahexaenoic acids (HPDHA) as primary oxidation products. Free radical mediated oxidation yields ten monohydroperoxides: 4-, 7-, 8-, 10-, 11-, 13-, 14-, and 16-, 17- and 20-HPDHA, all having a conjugated diene. In contrast, the photooxidation of DHA yields 12 isomers: the same ten formed by free radical oxidation and the 5- and 19-HPDHA isomers, which do not present conjugated-dienes [15].

Enzymatically, DHA oxidation by 15-, 12- and 5-LOX have been characterized [30]. 15-LOX converts DHA to 17S-H(p)DHA; 12-LOX activity wields two isomers 14S- and 11S-HPDHA; and the 4S-H(p)DHA 7S-H(p) DHA are metabolites of leukocyte 5-LOX. Among these products the17-HpDHA is the most described hydroperoxide formed by enzymatic oxidation of DHA and is an important precursor of the pro-resolving compounds resolvin D1 and protectin D1 [31].

LC-MS/MS analysis of HPDHA isomers shows common fragments ions derived from the loss of water $[M\text{-}H\text{-}H_2O]^-$ and CO_2 $[M\text{-}H\text{-}H_2O\text{-}CO_2]^-$ at *m/z* 341 and 297, respectively. Characteristic fragments are obtained by the cleavage of the carbon adjacent to the hydroperoxide group and also by the cleavage of the α- and β-carbon (one carbon away from the hydroperoxide carbon) (Fig. 2.3C). A detailed analysis of the MS/MS spectra obtained for all 12 hydroperoxide isomers formed by DHA photooxidation [15], allowed to identify the most specific fragments as follow: 20-HPDHA at *m/z* 285 and 71, 19-HPDHA at *m/z* 273 and 229, 17-HPDHA at *m/z* 111, 16-HPDHA at *m/z* 233, 14-HPDHA at *m/z* 205 and 151, 13-HPDHA at *m/z* 193 and 121, 11-HPDHA at *m/z* 243, 10-HPDHA at *m/z* 161 and 188, 8-HPDHA at m/z 108 and 171, 7-HPDHA at *m/z* 201 and 228, 5-HPDHA at *m/z* 281 and 147 and 4-HPDHA at *m/z* 115.

DHA hydroxy derivatives can be specifically analyzed based on the characteristic fragmentation products [15]. Thus, the 17-, 14-, 11-, 10-, 13-, 16-hydroxydocosahexaenoic acid (HDHA) can be identified based on the detection of fragments shown: 20-HDHA, *m/z* 285 and 241; 19-HDHA, *m/z* 229 and 273, 17-HDHA, *m/z* 245 and 201; 16-HDHA, *m/z* 233 and 261; 14-HDHA, *m/z* 205 and 161; 13-HDHA, *m/z* 193 and 221; 11-HDHA, *m/z* 165 and 149; 10-HDHA, *m/z* 153 and 181; 8-HDHA, *m/z* 189 and 113; 7-HDHA, *m/z* 141 and 109; 5-HDHA, *m/z* 85 and 93, 4-HDHA, *m/z* 101 and 115.

2.4 Conclusion

In this chapter, an overview of LC-MS analysis for fatty acid hydroperoxides and hydroxides was presented. Recent advances in MS technologies and improvements in chromatography have greatly enhanced the identification of different fatty acid oxidation products. The utmost important advantage for LC-MS technique is that it allows analysis without prior derivatization. The most used ionization mode for fatty acid hydroperoxide is ESI. The main ions yielded by ESI are $[M-H]^-$ corresponding to the carboxylate anion of hydroperoxide and $[M-H-H_2O]^-$, a fragment ion formed by the loss of water. None of these ions yields structural information on the hydro(pero)xide position. Therefore, characteristic fragments are obtained by CID, which mainly induces the cleavage of carbons vicinal to the hydroxyl/hydroperoxyl group and the cleavage of the double bond allylic to the hydroperoxide.

Specific analysis of isobaric hydroperoxide and hydroxide fatty acids can be accomplished by monitoring characteristic fragmentations of the original compound by doing the LC-MS/MS analysis in the MRM mode. This technique increases both specificity and sensitivity of the analysis. Nonetheless, it is noteworthy to mention that accurate quantification requires the use of appropriate internal standards for each class of products. More recently, the use of high-resolution mass analyzers (HR-MS: q-TOF or q-Orbitrap) is expanding the field of oxidized lipid analysis (oxylipidomics). HR-MS confers some advantages including higher specificity and more detailed structural characterization due to higher mass accuracy. Using this type of instrument, oxidized lipids can be monitored by high resolution MRM or parallel reaction monitoring (PRM).

In conclusion, the use of LC-MS based techniques for hydro(pero)xy-fatty acid analysis offers the possibility to expand our understanding on lipid metabolism, especially the role of oxidized lipids to physiological and pathological conditions.

Acknowledgments The authors acknowledge the Brazilian research funding institutions Fundação de Amparo à Pesquisa do Estado de São Paulo [FAPESP, CEPID-Redoxoma 13/07937-8 and 10/50891-0]; Conselho Nacional de Desenvolvimento Científico e Tecnológico [CNPq, Universal 424094/2016-9]; NAP-Redoxoma; Pro-Reitoria de Pesquisa USP; and Coordenação de Aperfeiçoamento de Pessoal de Nível Superior [CAPES- Finance Code 001].

References

1. Aldini G, Domingues MR, Spickett CM, Domingues P, Altomare A, Sánchez-Gómez FJ, Oeste CL, Pérez-Sala D (2015) Protein lipoxidation: detection strategies and challenges. Redox Biol 5:253–266
2. Arita M (2012) Mediator lipidomics in acute inflammation and resolution. J Biochem 152:313–319
3. Astarita G, Kendall AC, Dennis EA, Nicolaou A (2015) Targeted lipidomic strategies for oxygenated metabolites of polyunsaturated fatty acids. Biochim Biophys Acta (BBA) – Mol Cell Biol Lipids 1851:456–468
4. Bacellar IOL, Oliveira MC, Dantas LS, Costa EB, Junqueira HC, Martins WK, Durantini AM, Cosa G, Di Mascio P, Wainwright M, Miotto R, Cordeiro RM, Miyamoto S, Baptista MS (2018) Photosensitized membrane permeabilization requires contact-dependent reactions between photosensitizer and lipids. J Am Chem Soc 140:9606–9615
5. Balaban RS, Nemoto S, Finkel T (2005) Mitochondria, oxidants, and aging. Cell 120:483–495
6. Barbosa-Sicard E, Markovic M, Honeck H, Christ B, Muller DN, Schunck WH (2005) Eicosapentaenoic acid metabolism by cytochrome P450 enzymes of the CYP2C subfamily. Biochem Biophys Res Commun 329:1275–1281
7. Barnham KJ, Masters CL, Bush AI (2004) Neurodegenerative diseases and oxidative stress. Nat Rev Drug Discov 3:205–214
8. Beckman KB, Ames BN (1998) The free radical theory of aging matures. Physiol Rev 78:547–581
9. Brash AR (1999) Lipoxygenases: occurrence, functions, catalysis, and acquisition of substrate. J Biol Chem 274:23679–23682
10. Brigelius-Flohé R (1999) Tissue-specific functions of individual glutathione peroxidases. Free Radic Biol Med 27:951–965
11. Browne RW, Armstrong D (2000) HPLC analysis of lipid-derived polyunsaturated fatty acid peroxidation products in oxidatively modified human plasma. Clin Chem 46:829–836
12. Calder PC (2006) n-3 Polyunsaturated fatty acids, inflammation, and inflammatory diseases. Am J Clin Nutr 83:1505S–1519S
13. Capdevila JH, Falck JR, Harris RC (2000) Cytochrome P450 and arachidonic acid bioactivation: molecular

and functional properties of the arachidonate monooxygenase. J Lipid Res 41:163–181
14. Davies P, Bailey PJ, Goldenberg MM, Ford-Hutchinson AW (1984) The role of arachidonic acid oxygenation products in pain and inflammation. Annu Rev Immunol 2:335–357
15. Derogis PBMC, Freitas FP, Marques ASF, Cunha D, Appolinario PP, de Paula F, Lourenco TC, Murgu M, Di Mascio P, Medeiros MHG, Miyamoto S (2013) The development of a specific and sensitive LC-MS-based method for the detection and quantification of hydroperoxy- and hydroxydocosahexaenoic acids as a tool for lipidomic analysis. PLoS One 8:e77561
16. Esterbauer H (1993) Cytotoxicity and genotoxicity of lipid-oxidation products. Am J Clin Nutr 57:779S–785S
17. Fer M, Dreano Y, Lucas D, Corcos L, Salaun JP, Berthou F, Amet Y (2008) Metabolism of eicosapentaenoic and docosahexaenoic acids by recombinant human cytochromes P450. Arch Biochem Biophys 471:116–125
18. Folcik VA, Nivar-Aristy RA, Krajewski LP, Cathcart MK (1995) Lipoxygenase contributes to the oxidation of lipids in human atherosclerotic plaques. J Clin Investig 96:504–510
19. Frankel E (1984) Lipid oxidation: mechanisms, products and biological significance. J Am Oil Chem Soc 61:1908–1917
20. Frankel EN (1991) Recent advances in lipid oxidation. J Sci Food Agric 54:495–511
21. Funk CD (2001) Prostaglandins and leukotrienes: advances in eicosanoid biology. Science 294:1871–1875
22. Garscha U, Nilsson T, Oliw EH (2008) Enantiomeric separation and analysis of unsaturated hydroperoxy fatty acids by chiral column chromatography-mass spectrometry. J Chromatogr B Analyt Technol Biomed Life Sci 872:90–98
23. Girotti AW (1998) Lipid hydroperoxide generation, turnover, and effector action in biological systems. J Lipid Res 39:1529–1542
24. Guichardant M, Lagarde M (2009) Analysis of biomarkers from lipid peroxidation: a comparative study. Eur J Lipid Sci Technol 111:75–82
25. Halliwell B (2006) Oxidative stress and neurodegeneration: where are we now. J Neurochem 97:1634–1658
26. Halliwell B, Gutteridge JMC (2015) Free radicals in biology and medicine, 5th edn. Oxford University Press, New York
27. Hamberg M, Su C, Oliw E (1998) Manganese lipoxygenase – discovery of a bis-allylic hydroperoxide as product and intermediate in a lipoxygenase reaction. J Biol Chem 273:13080–13088
28. Han X, Gross RW (2005) Shotgun lipidomics: electrospray ionization mass spectrometric analysis and quantitation of cellular lipidomes directly from crude extracts of biological samples. Mass Spectrom Rev 24:367–412
29. Harizi H, Corcuff JB, Gualde N (2008) Arachidonic-acid-derived eicosanoids: roles in biology and immunopathology. Trends Mol Med 14:461–469
30. Hong S, Gronert K, Devchand PR, Moussignac RL, Serhan CN (2003) Novel docosatrienes and 17S-resolvins generated from docosahexaenoic acid in murine brain, human blood, and glial cells – Autacoids in anti-inflammation. J Biol Chem 278:14677–14687
31. Hong S, Lu Y, Yang R, Gotlinger KH, Petasis NA, Serhan CN (2007) Resolvin D1, protectin D1, and related docosahexaenoic acid-derived products: analysis via electrospray/low energy tandem mass spectrometry based on spectra and fragmentation mechanisms. J Am Soc Mass Spectrom 18:128–144
32. Hu C, Wang M, Han X (2017) Shotgun lipidomics in substantiating lipid peroxidation in redox biology: methods and applications. Redox Biol 12:946–955
33. Hussain SP, Hofseth LJ, Harris CC (2003) Radical causes of cancer. Nat Rev Cancer 3:276–285
34. Ingold I, Berndt C, Schmitt S, Doll S, Poschmann G, Buday K, Roveri A, Peng X, Porto Freitas F, Seibt T, Mehr L, Aichler M, Walch A, Lamp D, Jastroch M, Miyamoto S, Wurst W, Ursini F, Arnér ESJ, Fradejas-Villar N, Schweizer U, Zischka H, Friedmann Angeli JP, Conrad M (2018) Selenium Utilization by GPX4 Is Required to Prevent Hydroperoxide-Induced Ferroptosis. Cell 172:409–422. e421
35. Ivanova PT, Milne SB, Myers DS, Brown HA (2009) Lipidomics: a mass spectrometry based systems level analysis of cellular lipids. Curr Opin Chem Biol 13:526–531
36. Jira W, Spiteller G, Schramm A (1996) Increase in hydroxy fatty acids in human low density lipoproteins with age. Chem Phys Lipids 84:165–173
37. Kagan VE, Mao G, Qu F, Angeli JPF, Doll S, Croix CS, Dar HH, Liu B, Tyurin VA, Ritov VB, Kapralov AA, Amoscato AA, Jiang J, Anthonymuthu T, Mohammadyani D, Yang Q, Proneth B, Klein-Seetharaman J, Watkins S, Bahar I, Greenberger J, Mallampalli RK, Stockwell BR, Tyurina YY, Conrad M, Bayır H (2017) Oxidized arachidonic and adrenic PEs navigate cells to ferroptosis. Nat Chem Biol 13:81–90
38. Kawai Y, Takeda S, Terao J (2006) Lipidomic analysis for lipid peroxidation-derived aldehydes using gas chromatography-mass spectrometry. Chem Res Toxicol 20:99–107
39. Kerwin JL, Torvik JJ (1996) Identification of monohydroxy fatty acids by electrospray mass spectrometry and tandem mass spectrometry. Anal Biochem 237:56–64
40. Korovila I, Hugo M, Castro JP, Weber D, Höhn A, Grune T, Jung T (2017) Proteostasis, oxidative stress and aging. Redox Biol 13:550–567
41. Kuda O (2017) Bioactive metabolites of docosahexaenoic acid. Biochimie 136:12–20

42. Leonard AE, Pereira SL, Sprecher H, Huang Y (2004) Elongation of long-chain fatty acids. Prog Lipid Res 43:36–54
43. Liu M, Yokomizo, T (2015) The role of leukotrienes in allergic diseases. Allergol Int 64:17–26
44. Lu Y, Hong S, Tjonahen E, Serhan CN (2005) Mediator-lipidomics: databases and search algorithms for PUFA-derived mediators. J Lipid Res 46:790–802
45. Lyberg AM, Adlercreutz P (2006) Monitoring monohydroperoxides in docosahexaenoic acid using high-performance liquid chromatography. Lipids, 41:67–76
46. MacMillan DK, Murphy RC (1995) Analysis of lipid hydroperoxides and long-chain conjugated keto acids by negative ion electrospray mass spectrometry. J Am Soc Mass Spectrom 6:1190–1201
47. Masoodi M, Mir AA, Petasis NA, Serhan CN, Nicolaou A (2008) Simultaneous lipidomic analysis of three families of bioactive lipid mediators leukotrienes, resolvins, protectins and related hydroxy-fatty acids by liquid chromatography/electrospray ionisation tandem mass spectrometry. Rapid Commun Mass Spectrom 22:75–83
48. Masoodi M, Nicolaou A (2006) Lipidomic analysis of twenty-seven prostanoids and isoprostanes by liquid chromatography/electrospray tandem mass spectrometry. Rapid Commun Mass Spectrom 20:3023–3029
49. Massey KA, Nicolaou A (2013) Lipidomics of oxidized polyunsaturated fatty acids. Free Radic Biol Med 59:45–55
50. Miyamoto S, Martinez GR, Martins APB, Medeiros MHG, Di Mascio P (2003) Direct evidence of singlet molecular oxygen production in the reaction of linoleic acid hydroperoxide with peroxynitrite. J Am Chem Soc 125:4510–4517
51. Miyamoto S, Martinez GR, Medeiros MHG, Di Mascio P (2014) Singlet molecular oxygen generated by biological hydroperoxides. J Photochem Photobiol B Biol 139:24–33
52. Miyamoto S, Ronsein GE, Prado FM, Uemi M, Correa TC, Toma IN, Bertolucci A, Oliveira MCB, Motta FD, Medeiros MHG, Di Mascio P (2007) Biological hydroperoxides and singlet molecular oxygen generation. IUBMB Life 59:322–331
53. Moore K, Roberts LJ (1998) Measurement of lipid peroxidation. Free Radic Res 28:659–671
54. Murphy RC, Barkley RM, Berry KZ, Hankin J, Harrison K, Johnson C, Krank J, McAnoy A, Uhlson C, Zarini S (2005) Electrospray ionization and tandem mass spectrometry of eicosanoids. Anal Biochem 346:1–42
55. Murphy RC, Fiedler J, Hevko J (2001) Analysis of nonvolatile lipids by mass spectrometry. Chem Rev 101:479–526
56. Nakamura T, Bratton DL, Murphy RC (1997) Analysis of epoxyeicosatrienoic and monohydroxyeicosatetraenoic acids esterified to phospholipids in human red blood cells by electrospray tandem mass spectrometry. J Mass Spectrom 32:888–896
57. Needleman P, Truk J, Jakschik BA, Morrison AR, Lefkowith JB (1986) Arachidonic acid metabolism. Annu Rev Biochem 55:69–102
58. Niki E (2009) Lipid peroxidation: physiological levels and dual biological effects. Free Radic Biol Med 47:469–484
59. Niki E (2014) Biomarkers of lipid peroxidation in clinical material. Biochim Biophys Acta Gen Subj 1840:809–817
60. Niki E, Yoshida Y, Saito Y, Noguchi N (2005) Lipid peroxidation: mechanisms, inhibition, and biological effects. Biochem Biophys Res Commun 338:668–676
61. Nunez A, Foglia TA, Piazza GJ (2001) Characterization of lipoxygenase oxidation products by high-performance liquid chromatography with electron impact-mass spectrometric detection. Lipids 36:851–856
62. Oliw EH, Hornsten L, Sprecher H, Hamberg M (1993) Oxygenation of 5,8,11-eicosatrienoic acid by prostaglandin endoperoxide synthase and by cytochrome P450 monooxygenase: structure and mechanism of formation of major metabolites. Arch Biochem Biophys 305:288–297
63. Reverberi M, Punelli F, Scarpari M, Camera E, Zjalic S, Ricelli A, Fanelli C, Fabbri AA (2010) Lipoperoxidation affects ochratoxin A biosynthesis in Aspergillus ochraceus and its interaction with wheat seeds. Appl Microbiol Biotechnol 85:1935–1946
64. Rhee SG, Chae HZ, Kim K (2005) Peroxiredoxins: a historical overview and speculative preview of novel mechanisms and emerging concepts in cell signaling. Free Radic Biol Med 38:1543–1552
65. Ricciotti E, FitzGerald GA (2011) Prostaglandins and inflammation. Arterioscler Thromb Vasc Biol 31:986–1000
66. Roberts LJ, Morrow JD (2000) Measurement of F2-isoprostanes as an index of oxidative stress in vivo. Free Radic Biol Med 28:505–513
67. Salem N, Litman B, Kim H, Gawrisch K (2001) Mechanisms of action of docosahexaenoic acid in the nervous system. Lipids 36:945–959
68. Schneider C (2009) An update on products and mechanisms of lipid peroxidation. Mol Nutr Food Res 53:315–321
69. Schneider C, Pratt DA, Porter NA, Brash AR (2007) Control of oxygenation in lipoxygenase and cyclooxygenase catalysis. Chem Biol 14:473–488
70. Schneider C, Schreier P, Herderich M (1997) Analysis of lipoxygenase-derived fatty acid hydroperoxides by electrospray ionization tandem mass spectrometry. Lipids 32:331–336
71. Serhan CN (2014) Pro-resolving lipid mediators are leads for resolution physiology. Nature 510:92–101
72. Serhan CN, Chiang N, Van DykeTE (2008) Resolving inflammation: dual anti-inflammatory and pro-resolution lipid mediators. Nat Rev Immunol 8:349-361

73. Simopoulos AP (2002) The importance of the ratio of omega-6/omega-3 essential fatty acids. Biomed Pharmacother 56:365–379
74. Spickett CM, Pitt AR (2015) Oxidative lipidomics coming of age: advances in analysis of oxidized phospholipids in physiology and pathology. Antioxid Redox Signal 22:1646–1666
75. Spickett CM, Wiswedel I, Siems W, Zarkovic K, Zarkovic N (2010) Advances in methods for the determination of biologically relevant lipid peroxidation products. Free Radic Res 44:1172–1202
76. Spiteller G (1996) Enzymic lipid peroxidation-[alpha] consequence of cell injury? Free Radic Biol Med 21:1003–1009
77. Stocker R, Keaney JF Jr (2004) Role of oxidative modifications in atherosclerosis. Physiol Rev 84:1381–1478
78. Stratton SP, Liebler DC (1997) Determination of singlet oxygen-specific versus radical-mediated lipid peroxidation in photosensitized oxidation of lipid bilayers: effect of β-carotene and α-tocopherol. Biochemistry 36:12911–12920
79. Taguchi R, Houjou T, Nakanishi H, Yamazaki T, Ishida M, Imagawa M, Shimizu T (2005) Focused lipidomics by tandem mass spectrometry. J Chromatogr B Analyt Technol Biomed Life Sci 823:26–36
80. Terao J, Matsushita S (1977) Products formed by photosensitized oxidation of unsaturated fatty-acid esters. J Am Oil Chem Soc 54:234–238
81. Thomas DW, van Kuijk FJGM, Dratz EA, Stephens RJ (1991) Quantitative determination of hydroxy fatty acids as an indicator of in vivo lipid peroxidation: gas chromatography-mass spectrometry methods. Anal Biochem 198:104–111
82. Tyurin VA, Tyurina YY, Jung M, Tungekar MA, Wasserloos KJ, BayIr H, Greenberger JS, Kochanek PM, Shvedova AA, Pitt B, Kagan VE (2009) Mass-spectrometric analysis of hydroperoxy- and hydroxy-derivatives of cardiolipin and phosphatidylserine in cells and tissues induced by pro-apoptotic and pro-inflammatory stimuli. J Chromatogr B Analyt Technol Biomed Life Sci 877:2863–2872
83. Wang Y, Armando AM, Quehenberger O, Yan C, Dennis EA (2014) Comprehensive ultra-performance liquid chromatographic separation and mass spectrometric analysis of eicosanoid metabolites in human samples. J Chromatogr A 1359:60–69
84. Wenk MR (2005) The emerging field of lipidomics. Nat Rev Drug Discov 4:594–610
85. Wilson R, Smith R, Wilson P, Shepherd MJ, Riemersma RA (1997) Quantitative gas chromatography-mass spectrometry isomer-specific measurement of hydroxy fatty acids in biological samples and food as a marker of lipid peroxidation. Anal Biochem 248:76–85
86. Wojtovich AP, Foster TH (2014) Optogenetic control of ROS production. Redox Biol 2:368–376
87. Xu L, Davis TA, Porter NA (2009) Rate constants for peroxidation of polyunsaturated fatty acids and sterols in solution and in liposomes. J Am Chem Soc 131:13037–13044
88. Yamauchi R, Yamada T, Kato K, Ueno Y (1983) Monohydroperoxides formed by autoxidation and photosensitized oxidation of methyl eicosapentaenoate. Agric Biol Chem 47:2897–2902
89. Yang R, Chiang N, Oh SF, Serhan CN (2011) Metabolomics-lipidomics of eicosanoids and docosanoids generated by phagocytes. Curr Protoc Immunol Chapter 14, Unit 14.26
90. Yang Y, Cheng J, Singhal SS, Saini M, Pandya U, Awasthi S, Awasthi YC (2001) Role of glutathione S-transferases in protection against lipid peroxidation. Overexpression of hGSTA2-2 in K562 cells protects against hydrogen peroxide-induced apoptosis and inhibits JNK and Caspase 3 activation. J Biol Chem 276:19220–19230
91. Yin H, Xu L, Porter NA (2011) Free radical lipid peroxidation: mechanisms and analysis. Chem Rev 111:5944–5972
92. Yin HY, Brooks JD, Gao L, Porter NA, Morrow JD (2007) Identification of novel autoxidation products of the omega-3 fatty acid eicosapentaenoic acid in vitro and in vivo. J Biol Chem 282:29890–29901
93. Zhang X, Yang N, Ai D, Zhu Y (2015) Systematic metabolomic analysis of eicosanoids after omega-3 polyunsaturated fatty acid supplementation by a highly specific liquid chromatography–tandem mass spectrometry-based method. J Proteome Res 14:1843–1853

Part II

Signaling Cascades Mediated by Bioactive Lipids

Understanding Peroxisome Proliferator-Activated Receptors: From the Structure to the Regulatory Actions on Metabolism

3

María Lamas Bervejillo and Ana María Ferreira

Abstract

Peroxisome proliferator-activated receptors (PPARs) are multi-domains proteins, belonging to the superfamily of nuclear receptors, which mainly act as ligand-activated transcription factors. A variety of lipophilic molecules, including long-chain polyunsaturated fatty acids and eicosanoids, are capable of binding to PPAR, although the nature of the physiological ligands is still under debate. PPARs regulate the expression of a set of genes involved in glucose and lipid metabolism as well as in the control of inflammatory responses. Herein we review the main molecular and cellular events associated with the activation of PPARs and their effects on metabolism.

Keywords

Peroxisome proliferator-activated receptors · PPARα · PPARβ/δ · PPARγ · Transactivation · Transrepression · Lipid metabolism

M. Lamas Bervejillo · A. M. Ferreira (✉)
Cátedra de Inmunología, Facultad de Ciencias/Facultad de Química, Universidad de la República, Montevideo, Uruguay
e-mail: aferrei@fq.edu.uy

Abbreviations

15d-PGJ$_2$	15-deoxy-$\Delta^{12,14}$-prostaglandinJ$_2$
AF1 and AF2	Activation function 1 and 2
AP-1	Activating protein-1
BCL-6	B-cell lymphoma 6 protein
C/EBPα and C/EBPβ	CCAAT/enhancer-binding protein α and β
CBP	CREB-binding protein
DBD	DNA-binding domain
DR1	Direct repeat-1
FA	Fatty acid
FATP-1	Fatty acid transport protein-1
FGF21	Fibroblastic growth factor 21
FoxO1	Forkhead box O1
Glut4	Glucose transporter type 4
H12	Helix 12
HDL	High density lipoprotein
HETE	Hydroxyeicosatetraenoic acid
HMGCS	Hydroxymethylglutaryl-CoA synthase
HODE	Hydroxyoctadecadienoic acid
LBD	Ligand-binding domain
L-FABP	Liver fatty acid binding protein
LPL	Lipoprotein lipase
LXR	Liver X receptor
N-CoR	Nuclear receptor co-repressor
NF-κB	Nuclear factor-kappa B
NR	Nuclear receptors
PDK4	Pyruvate dehydrogenase kinase 4
Pgtz	Pioglitazone

A. Trostchansky, H. Rubbo (eds.), *Bioactive Lipids in Health and Disease*, Advances in Experimental Medicine and Biology 1127, https://doi.org/10.1007/978-3-030-11488-6_3

PPARs	Peroxisome proliferator-activated receptors
PPRE	Peroxisome proliferator response elements
PUFA	Long-chain polyunsaturated fatty acids
Rgtz	Rosiglitazone
RXR	Retinoid X receptor
SMRT	Silencing mediator of retinoid and thyroid signalling
SRC	Steroid receptor co-activator
STAT3	Signal transducer and activator of transcription 3
TAGs	Triacylglycerides
TZD	Thiazolidinedione
VLDL	Very low density lipoprotein

3.1 Introduction

Lipid and carbohydrate metabolism are subject to tight regulation to maintain energy homeostasis in higher organisms. This involves a complex signalling network capable of sensing a variety of stimuli (food availability, temperature changes and other stressors) to trigger adaptive responses. Lipids and their derivatives act as inputs to this network, contributing to the genetic control of their own systemic transport, cellular uptake, biosynthesis, storage and use. This implies the existence of a variety of nuclear lipid-sensing receptors (NR) that act as transcription factors, regulating the expression of genes involved in several pathways of lipid metabolism. Among other NR, the peroxisome proliferator-activated receptor (PPAR) subfamily has a central role.

PPARs are multi-domains proteins originally named because of their ability to induce peroxisome proliferation in response to xenobiotics [1]. However, they were afterwards identified as sensors of lipophilic molecules, which can induce structural changes in these receptors to stabilize active or inactive receptor states. In their active conformation, PPARs trigger regulatory actions on lipid metabolism and inflammation (reviewed by [2]). These actions involved distinct and complex mechanisms, where PPARs promote gene expression (*transactivation*) or stop it (*transrepression*).

PPAR ligands include a structurally diverse set of lipids, including endogenous fatty acids (FA), products derived from the lipid metabolism and dietary constituents. Synthetic PPAR ligands also exist; the development of drugs targeting PPARs has been encouraged because of the relevance of these receptors as metabolic regulators. Moreover, a variety of synthetic ligands have been used as tools for elucidating PPAR role in human physiology during health and disease, and some of them are currently available for human therapy [3]. Herein we review the current knowledge on the interaction between PPARs and their ligands as well as the molecular events associated with PPAR regulatory actions, with emphasis on metabolism.

3.2 Structural Organization of PPARs

In vertebrates, PPARs exist as three subtypes, PPARα (NR1C1), PPARβ/δ (NR1C2) and PPARγ (NR1C3), which are encoded by distinct single-copy genes. In human, these genes were found to map on chromosomes 22 (*PPARA*), 6 (*PPARD*) and 3 (*PPARG*), respectively, and their protein products exhibit high homology at the amino acid level, sharing a very similar overall structure [4–6]. Indeed, PPARs are modular proteins composed by five functional domains or regions, called A to F from N- to C-terminus (Fig. 3.1) [7]. The extreme N-terminal A/B domain harbours a flexible and highly variable region, termed the activation function 1 (AF1), which provides constitutive and ligand-independent transcriptional activity. The C domain is the most conserved region among NRs and corresponds to the DNA-binding domain (DBD). Indeed, it contains two highly conserved zinc finger-like motifs folded in a globular structure that are a characteristic feature of the NR family. The DBD specifically recognizes regulatory six nucleotide sequences, located in the promotor of PPAR target genes, called peroxisome proliferator response elements (PPREs). A flexible hinge

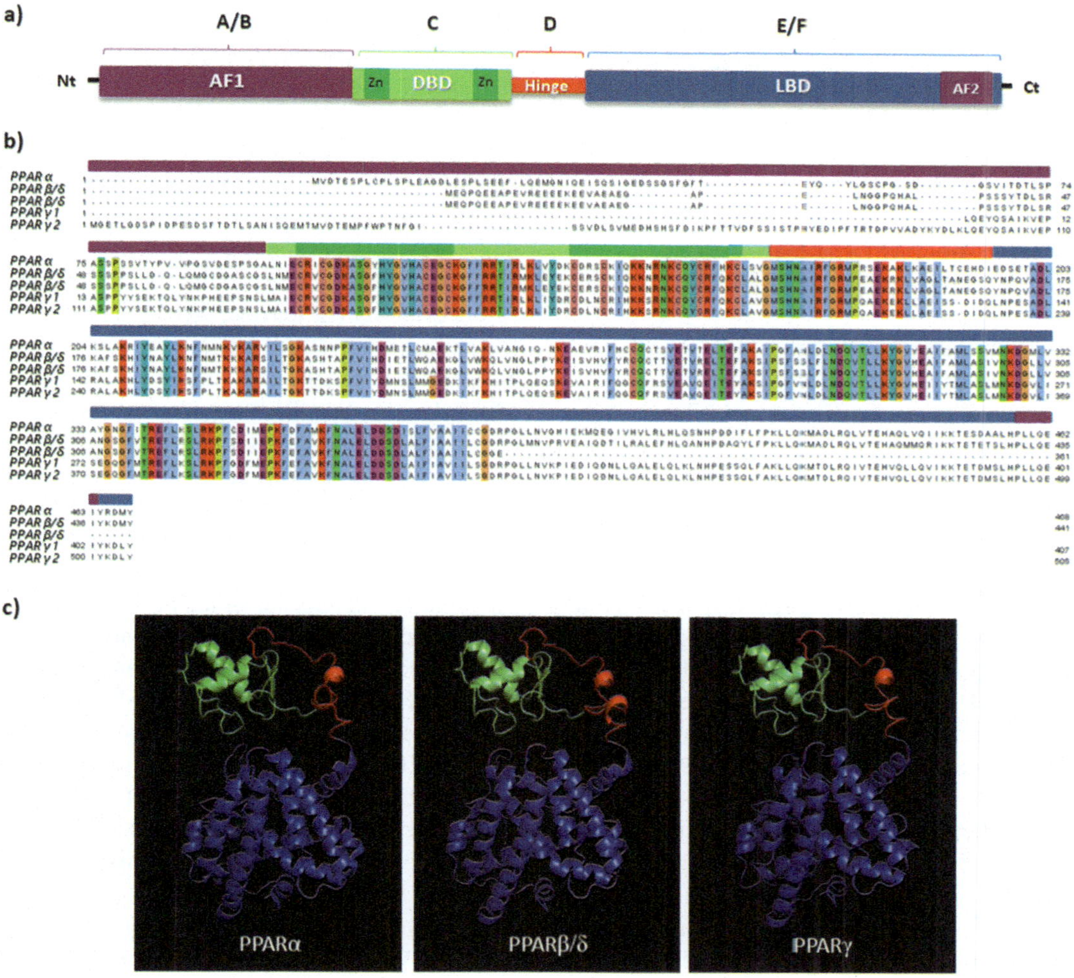

Fig. 3.1 General structure and sequence alignment of PPARs. (**a**) A representative scheme of PPARγ structure showing from N- to C-terminus the following domains: A/B domain (in purple), DBD (in green), the hinge region (in red), and the LBD (in blue). The two zinc fingers of the DBD (Zn) are shown in dark green boxes and the activation function motifs, AF1 and AF2 in violet boxes. (**b**) Sequence alignment of human PPAR subtypes, PPARα, PPARβ/δ (isoforms 1 and 2) and PPARγ (isoforms 1 and 2). Coloured residues indicate 100% of conservation between PPARs. To visualize the amino acid sequence corresponding to each PPAR domain, a representation of protein domains is included at the top of the alignment (using the same colour code as in a). The extent of each domain was taken from the description reported by [98]. (**c**) Three-dimensional structure of PPARs. Regarding PPARα and PPARβ/δ, crystallography data is available only for their LBD domains: PDB 1k71 [99] and 3ET2 [100], respectively. The DBD and hinge region of these receptors were thus modelled against PPARγ, which structure was taken from PDB 3e00 [28]. In all cases, the DBD, hinge and LBD domains are represented using the same colour code as in (a)

region (D region) connects the DBD to E/F domains. The latter include the ligand-binding domain (LBD), which contains the ligand-binding pocket, the ligand-dependent transactivation domain (AF2) and a region involved in receptor dimerization. The LBD has a lower level of conservation across PPAR subtypes than the DBD (Fig. 3.1b), with significant sequence variation in the residues that line the ligand-binding pocket. This accounts for differences in the ligand selectivity among PPAR subtypes, likely contributing to their distinct physiological role. A representation of the three-dimensional structure of PPARs along with the secondary structure elements is shown in Fig. 3.1c.

3.3 Mechanisms of Action

3.3.1 Mechanisms Dependent on Binding to PPRE

To be transcriptionally active, PPARs must form heterodimeric complexes with members of the retinoid X receptor (RXR) family. This NR family includes three subtypes called RXRα, RXRβ and RXRγ [8]. PPAR-RXR complexes bind to PPRE that exhibit the consensus sequence 5′-A(A/T)CT(A/G)GGNCAAAG(G/T)TCA-3′ (Fig. 3.2). This consensus sequence includes two elements: a DR1 motif comprising two imperfect direct repeats of the NR consensus recognition sequence (AGGTCA) separated by a single nucleotide (A) and a highly conserved A(A/T)CT sequence corresponding to the four nucleotides immediately 5′ of the DR1 motif [9, 10]. PPAR and RXR bind to the 5′ and 3′ half-sites of the PPRE, respectively, and the 5′ flanking region of PPRE strongly influences receptor binding, particularly in the case of PPARα [11–13]. PPAR-RXR binding to PPRE is required but not sufficient for transcriptional activity. This strongly depends on the presence and nature of the ligand, which influences PPAR interaction with co-regulatory proteins. These co-regulators form multicomponent protein complexes and act as helper proteins, modulating transcription factor communication with the polymerase II transcription machinery [14]. Co-activators facilitate while co-repressors block the access of this machinery to the promotor for initiating transcription. Physiologically, in the absence of a ligand, PPAR-RXR heterodimers may bind to PPRE. However, the consequences of this binding depend on the gene promoter context and the cell type. For instance, in adipocytes, unliganded PPARγ-RXR may form high-affinity complexes with nuclear co-repressors proteins, exerting active repression on some, but not all, PPARγ target genes (Fig. 3.2a) [15]. In contrast, when an activating ligand (agonist) binds to PPAR, it promotes the association of PPAR-RXR with co-activator proteins, upon co-repressors dissociation, triggering transcriptional activity (Fig. 3.2b). Ligands called antagonists would lead to the association of PPAR-RXR heterodimer with co-repressors, preventing the transcription of PPAR target genes (Fig. 3.2c).

Most of the co-activators carry multiple LXXLL motifs (L, leucine; X, any amino acid) that interact with PPAR-LBD [16]; they include members of the steroid receptor co-activator (SRC) family, the CREB-binding protein (CBP) and its paralogue p300 [17, 18]. CBP/p300 are capable of acetylating nucleosomal histones in target genes, remodelling chromatin and increasing polymerase II complex accessibility. In contrast, co-repressors use LXXXIXXXL/I motifs for binding PPARs, overlapping the surface bound by co-activators [19]; they include the NR co-repressor (N-CoR) and the silencing mediator of retinoid and thyroid signalling (SMRT), which have histone deacetylation activity [20]. The molecular basis that explains why a ligand acts as an agonist or as an antagonist is not completely elucidated yet. However, structural studies, including crystallography, NMR spectroscopy combined with biochemical co-regulator interaction analysis and molecular simulations, have shed light on the interactions between PPAR and different ligands, showing how they affect PPAR structure, leading to distinct co-regulator recruitment profiles [21, 22].

3.3.2 Mechanisms Independent on Binding to PPRE

Studies of anti-inflammatory responses mediated by PPARs have highlighted additional mechanisms of action at non-genomic level. Indeed, PPAR activation can trigger indirect repressive effects by interfering with key transcription factors. These effects are collectively known as *transrepression* and can occur by distinct mechanisms (Fig. 3.3). *Tethering* occurs when activation of PPAR results in the sequestration of transcription factors, inhibiting the

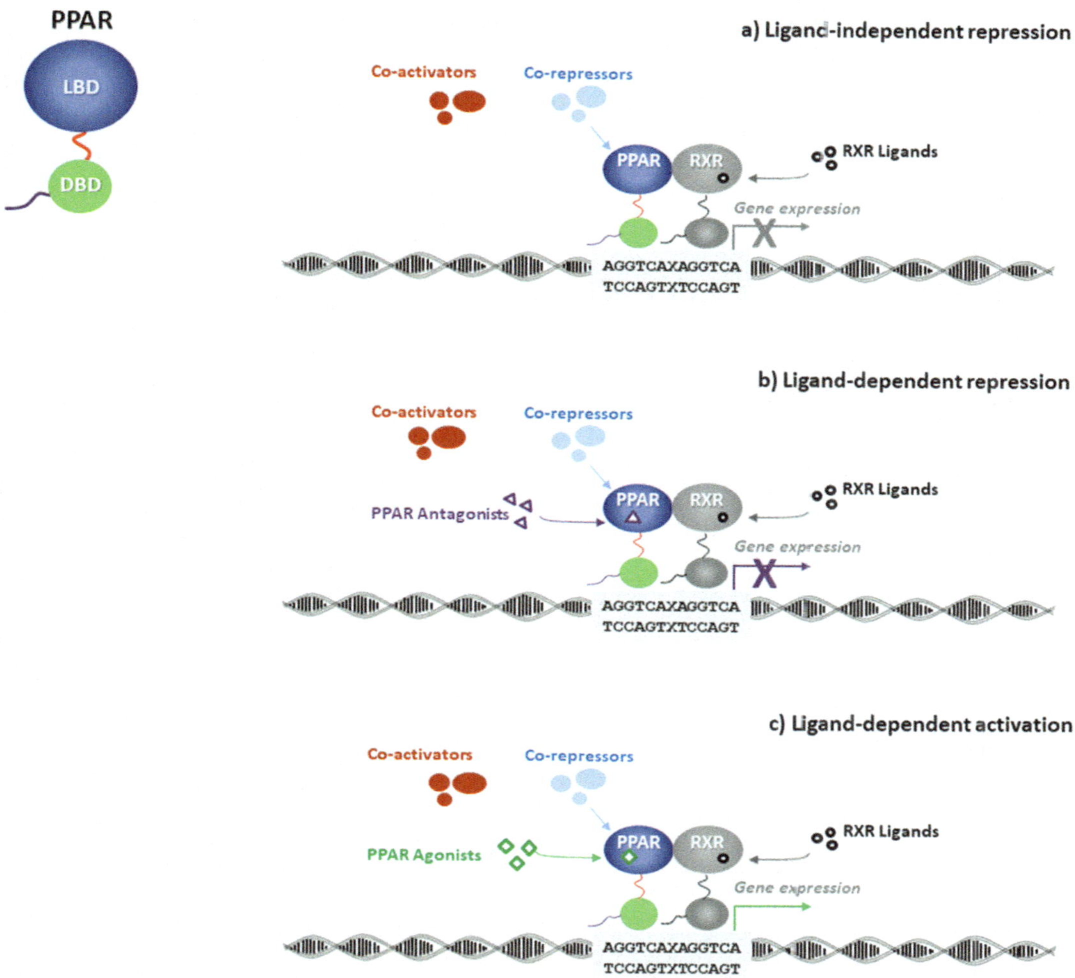

Fig. 3.2 PPAR actions dependent on binding to PPRE. The three primary mechanisms modulating PPARγ activity are illustrated. PPAR forms heterodimeric complexes with RXR and binds to PPRE. (**a**) Ligand-independent repression. In the absence of a ligand, PPAR-RXR heterodimers bind co-repressors exerting active repression on some PPAR target genes. (**b**) Ligand-dependent activation or transactivation. Binding of an activating ligand (agonist) to the PPAR-LBD induces the dissociation of co-repressors as well as the binding of co-activators, leading to the transcription of PPAR reporter genes. (**c**) Ligand-dependent repression. Binding of an antagonist induces association of co-repressors, leading to active repression on PPAR target genes

expression of their target genes. This mechanism is likely involved in the PPARα-mediated inhibition of IL-6 secretion by smooth muscle cells, since liganded PPARα was found to bind the subunits p65 and c-Jun of the nuclear factor-kappa B (NF-κB) and activating protein-1 (AP-1), respectively, which are both required for this response [23]. Alternatively, PPARs may sequestrate cofactors needed for transcriptional activity (*squelching*). Downregulation of fibrinogen and acute phase proteins in the liver was found to occur via a squelching mechanism, in which PPARα binds the co-activator GRIP1/TIF2, which is essential for C/EBPβ-driven transcription [24]. A third type of transrepressive mechanism results from the dissociation of co-repressors from activated PPAR-RXR complexes. These released

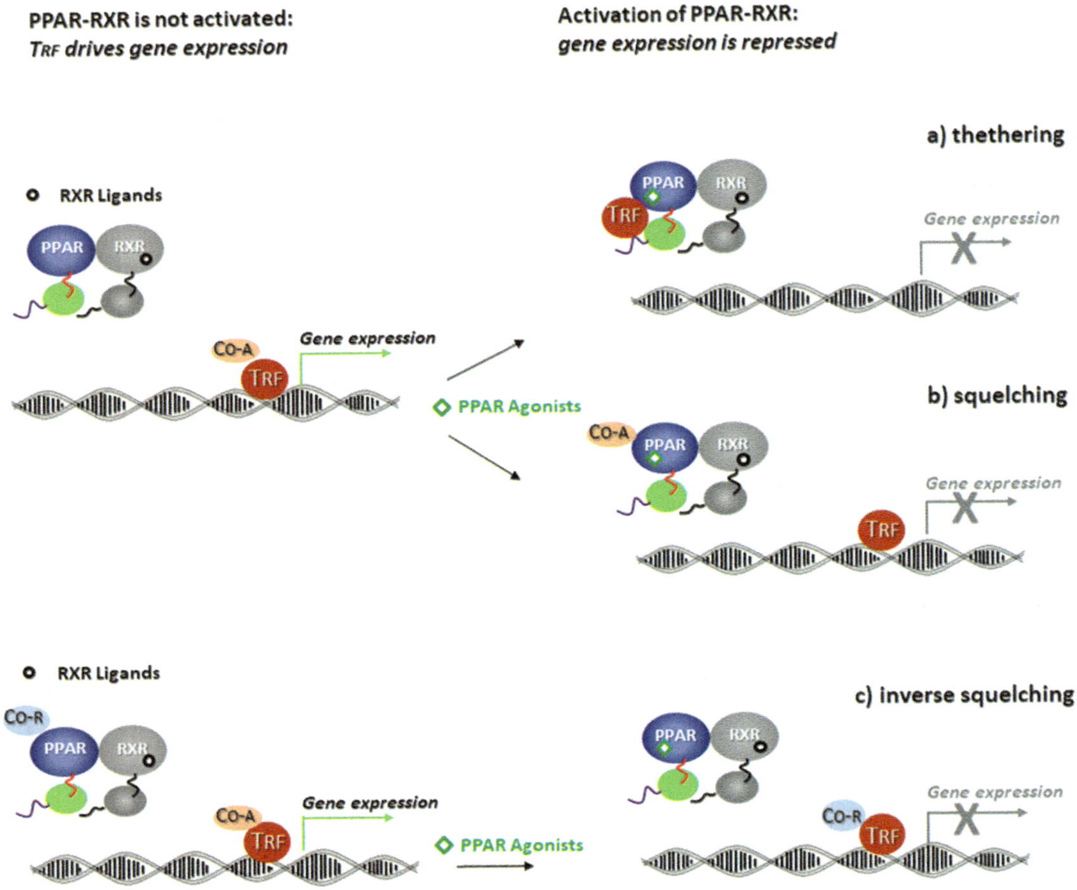

Fig. 3.3 PPAR actions independent on binding to PPRE. PPAR-RXR heterodimers can exert indirect repressive actions by interactions with factors involved in transcriptional activity. The binding of an agonist to PPAR can lead to (**a**) tethering. PPAR-RXR sequestrates a transcriptional factor (TRF) inhibiting its binding to DNA and the transcription of TRF target genes. (**b**) Squelching. PPAR-RXR binds a TRF co-activator (CO-A), interfering with TRF activity and gene transcription. (**c**) Inverse squelching. Activation of PPAR-RXR leads to dissociation of a co-repressors (CO-R), inducing its distribution among others TRF and, in turn, repressing the transcription of TRF target genes

co-repressors can, in turn, be redistributed among other transcriptional factors, inducing repressive signalling on other genes (*inverse squelching*). Examples for inverse squelching are the release of co-repressors BCL-6 and SMRT caused by PPARβ/δ and PPARγ activation, respectively, which inhibits the transcription of inflammatory genes [25, 26].

Activation of PPARs is also regulated by post-translational modifications, including phosphorylation, SUMOylation, ubiquitination, acetylation and O-GlcNAcylation. Numerous modification sites have been found in all PPAR subtypes, with consequences on protein stability, transactivation function and/or co-regulators interactions (reviewed by [27]).

3.4 The Ligand-Binding Domain

Crystallography studies have shown that PPAR-LBD folds in a single domain comprising a small three- to four-stranded β-sheet and a bundle of α-helices, named H1 to H12 plus H2′; the latter is an additional helix that is not present in other NRs ([28], reviewed by [29, 30]). These α-helices are organized in three layers that enclose a hydrophobic binding pocket, placed at the bottom half of the LBD structure. This ligand-binding pocket is a Y-shaped cavity of around 1300 to 1400 Å, which is larger than the corresponding to other NRs. Interestingly, 80% of the residues that outline the ligand-binding cavity are highly conserved across the three PPAR isotypes. The entrance of this cavity is characterized by the presence of some polar amino acids and a very flexible loop (between helices H2′ and H3), which could be adapted for allowing the entrance of larger ligands without causing a significant perturbation of the overall LBD structure. These features likely expand the set of ligands that PPARs can bind in comparison with other NRs. Regarding the interior of the ligand-binding cavity, it is branched in two arms. One, called arm I, extends towards the amphipathic α-helix H12, which seems to be highly mobile in the apo-structure and bears the AF2 domain involved in the binding of co-regulators. The second arm is located between helix H3 and the β-sheet, exhibiting hydrophobic features (arm II).

Recent studies using PPARγ-LBD have demonstrated that in the apo-form, the AF2 domain is not found as a unique structure, but rather in ensembles of structures in which the helix H12 exchanges between several conformations [22]. The presence as well as the chemical features of the ligand bound to this pocket lead to LBD structural changes that stabilize a particular conformation, determining the position adopted by H12 and, consequently, the generation of surfaces with the ability to bind co-activators or co-repressors. In general, when an agonist binds, H12 seals as a lid the ligand-binding cavity, contributing to stabilize ligand-receptor interactions. In addition to this shift in H12, structural changes (such as bending of helix H3) occur. This brings H12 into a distinct receptor environment, creating the surfaces that allow the binding of co-activators, thereby triggering AF2 transcriptional activity [31, 32]. The stabilization of H12 in this active conformation is associated with the formation of a hydrogen bonds network between the agonist and amino acid residues of arm I, which show chemical properties conserved among PPARs. These residues include Ser280, Tyr314, His440 and Tyr464 in PPARα; Thr289, His323, His449 and Tyr473 in PPARβ/δ; and Ser289, His323, His449 and Tyr473 in PPARγ [29]. The hydrogen bonds network formed between PPARγ-LBD and rosiglitazone (Rgtz, a synthetic PPARγ agonist) is shown in Fig. 3.4a. The polar head group of Rgtz forms hydrogen bonds with the side chains of PPARũFE; residues H323 (2.9°A), H449 (2.7°A) and Y473 (2.6°A) [30]. In contrast, when an antagonist binds, H12 undergoes a distinct shift towards a nonactive conformation that opens up the co-repressor binding cleft, inhibiting transcription [21]. Antagonists may also stabilize this nonactive conformation via hydrogen bonds, as shown for PPARγ-LBD and GW9662 (Figs. 3.4b). Structural superposition of PPARγ-LBD in complex with an agonist (Rgtz) and an antagonist (GW9662) showed the distinct orientation adopted by helix H12 (Fig. 3.4c).

X-ray co-crystal structural studies of PPARγ suggest that antagonists could use different mechanisms to induce this alternative conformational change in PPAR-LBD. On one hand, similarly to antagonists of other NRs, such as oestrogen receptor α and oestrogen receptor β, PPARα antagonists protrude out of the space between H3 and H11 to sterically clash with H12, forcing this helix out of the active position (a mechanism called *push and tumble*) [33–35]. On the other hand, some PPARγ antagonists – located between H3 and the β-sheet – pull H12 into the antagonist conformation. In this case, H12 is trapped in this alternative conformation (*tumble and trap* mechanism) [21].

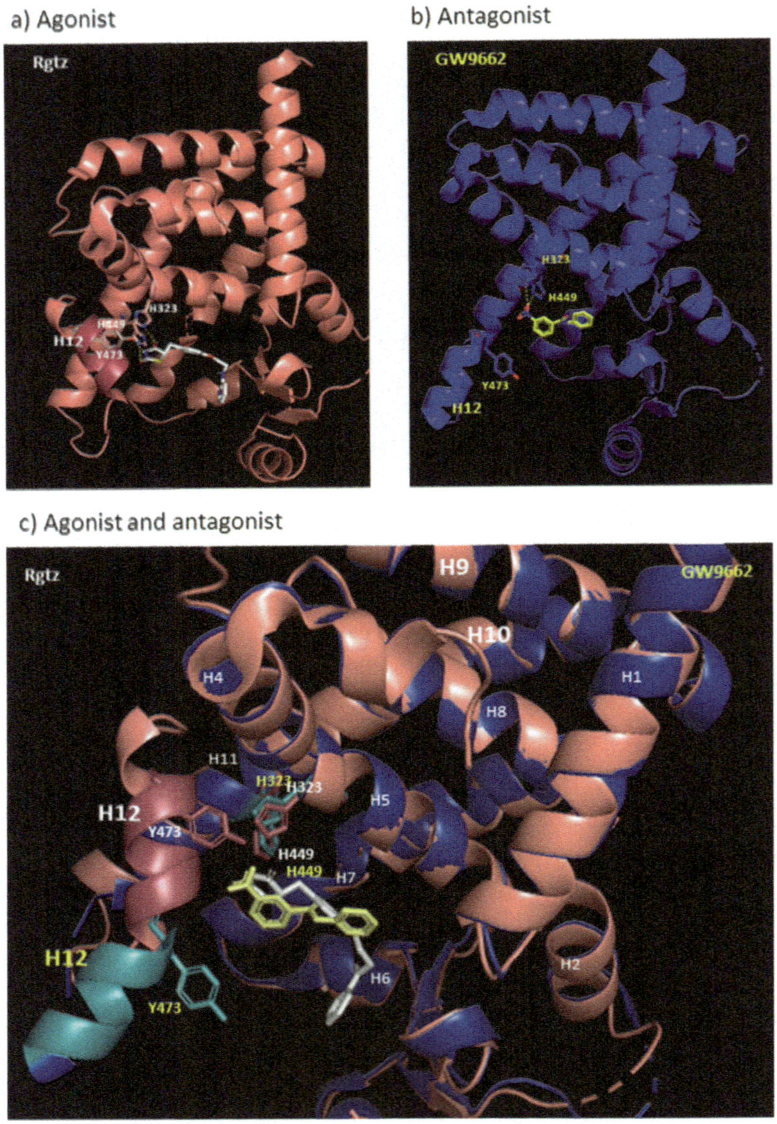

Fig. 3.4 Helix H12 orientation depends on PPAR-LBD interaction with ligands. In the absence of a ligand, PPARγ-LBD is found in ensembles of structures in which the helix H12 exchanges between several conformations. Ligand binding to the LBD induces structural changes that stabilize a particular conformation, determining the position of H12. The figure illustrates the orientation adopted by H12 after PPARγ-LBD binding to (**a**) the full agonist Rgtz (PPARγ-LBD in pink, Rgtz in white) and (**b**) the antagonist GW9662 (PPARγ-LBD in blue, GW9662 in yellow). Hydrogen bonds are indicated in yellow. Panel (**c**) corresponds to the structural superposition of PPARγ-LBD complexes with Rgtz (coloured in pink) and with GW9662 (coloured in blue/aquamarine). PPARγ-LBD structures were taken from PDB 4xld (complex with Rgtz) and PBD 3b0r (complex with GW9662). In all cases, helix H3 was hidden for a better visualization of H12. Of note, Rgtz and GW9662 stabilize helix H12 in opposite positions, generating different surfaces for interacting with co-regulators

3.5 PPAR Ligands

PPARs are capable of binding multiple endogenous molecules with moderate affinity, being more promiscuous than other NRs. This could be explained by the fact that ligands occupy a relatively small volume of the ligand-binding cavity, and only the binding of their polar head, but not of their hydrophobic moieties, is restricted within this cavity. The main endogenous molecules with the ability to bind to PPARs comprise long-chain FAs (including saturated, monounsaturated and polyunsaturated forms), certain nitrated derivatives of FAs and some eicosanoids derived from the metabolism of arachidonic acid [36–39]. In addition to endogenous molecules, rationally synthesized ligands have been developed searching for therapeutic tools.

The *efficacy* of an agonist refers to its ability to activate PPAR, promoting transcriptional activity. While the agonist binding is governed by its affinity for PPAR, the agonist efficacy depends on its ability to stabilize an active receptor conformation. This has been mostly analysed for PPARγ-LBD, for which *full* and *partial* agonists have been described [29, 30, 40]. *Full* agonists cause a shift from the inactive to a completely active conformation that leads to the maximum recruitment of co-activators and transcriptional activity. Instead, *partial* agonists induce LBD conformational changes that lead to a limited recruitment of co-activators and attenuated transcriptional activity. In consequence, they showed intermediate levels of efficacy; despite the occupancy of all receptors, the response is submaximal. Structural analysis of PPARγ-LBD in complex with full and partial agonists has revealed clear differences in the stabilization of H12. Partial agonists are mostly unable of generating a hydrogen bond network with amino acid residues of the arm I, as illustrated for the partial agonist INT131 (Fig. 3.5a). Structural superposition analysis reveals a displacement of helix H12 in the PPARγ-LBD complexed with INT131, with respect to its position in the active conformation induced by Rgtz (Fig. 3.5b). This likely impedes the formation of the hydrogen bonds network and the stabilization of an active PPARγ-LBD conformer.

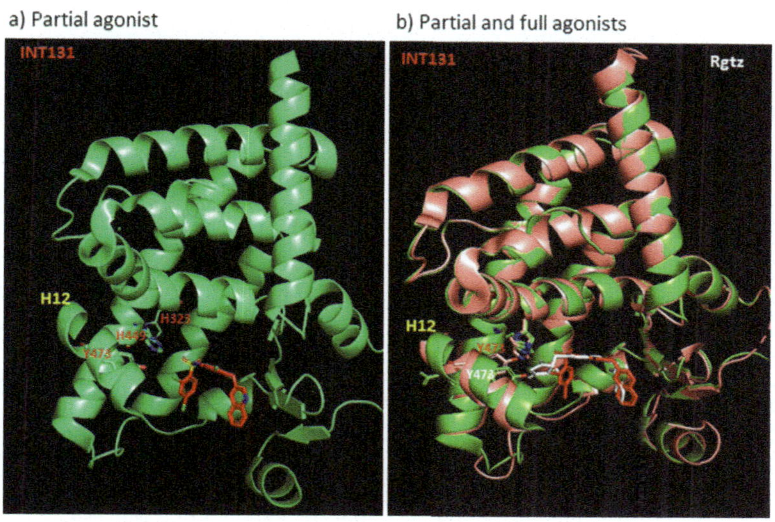

Fig. 3.5 Comparison of the structure of PPARγ-LBD in complex with partial and full agonists. (**a**) Structure of PPARγ-LBD in complex with the partial agonist INT131 (PDB 3fur, [93]). (**b**) Structural superposition of PPARγ-LBD in complex with INT131 (coloured in green, PDB 3fur) and PPARγ-LBD in complex with Rgtz (coloured in pink, PDB 4xld) showing a difference in the position of Tyr 473 between these complexes. Rgtz is shown in white and INT131 in red. Helix H12 and Tyr473 are indicated. Helix H3 was hidden for improving the visualization of H12

Table 3.1 Endogenous and synthetic PPARγ ligands

	Binding affinity $K_d/K_i/IC_{50}$	Transactivation activity EC_{50}	Structural data (PDB)
Endogenous Ligands			
Linoleic acid	$K_i > 1$ μM [39]	871 nM [73]	
	K_i 4.9–21 μM [85]		
Arachidonic acid	IC_{50} 2–20 μM [86]		
γ-Linolenic acid	$K_i = 2.28$ μM [87]		
Oleic acid		1.9 μM [73]	
Nitrolinoleic acid	$K_i = 133$ nM [39]	36 nM [73]	3CWD
Nitrooleic acid		13 nM [73]	
15d-PGJ2	K_i 2.5–13 μM [85, 89]	2 μM [88]	2ZK1
9(S)-HODE	K_i 10–20 μM [89]	~30 μM [89]	2VSR
13(S)-HODE	K_i 10–20 μM [89]	~45 μM [89]	2VST
Docosahexaenoic acid	$K_i = 2.93$ μM [87]		2VV0
cis-parinaric acid	$K_d = 600$ nM [87]		
Lysophosphatidic acid		~1 μM [90]	
Synthetic Ligands			
Agonists			
Thiazolidinediones	$K_{d\ Rgtz} = 40$ nM [91]	30–100 nM [91]	Rgtz: 5JI0, 4XLD, 4EMA. Pgtz: 5Y2O, 2XKW
	$K_{i\ Rgtz} = 53$ nM [39]		
GW1929	$K_i = 1.4$ nM [92]	10 nM [92]	
GW2331		300 nM [86]	1Y0S
INT131		4 nM [93]	3FUR
RWJ-348260		189 nM [94]	
YM440	$K_i = 4$ μM [95]	31–110 μM [95]	
Antagonists			
GW9662	$IC_{50} = 3.3$ nM [96]		3B0R
N-sulfonyl-2-indole carboxamides	$IC_{50} = 3$ nM [97]		2HFP

K_d corresponds to the dissociation constant
$K_i = IC_{50}/(1 + [L]/K_d)$, where IC_{50} is the concentration of test compound required to inhibit 50% of the specific binding of the radiolabelled ligand, and [L] and K_d are the concentration and dissociation constant of the radiolabelled ligand, respectively
EC_{50}, is the concentration of test compound required to induce 50% of the maximum transactivation activity

PPAR ligands exhibit a wide range of binding affinities and transactivation activities, from micromolar to nanomolar. Table 3.1 shows values for PPARγ ligands. In transactivation assays, cells are firstly transfected with a PPAR full-length expression plasmid and a plasmid containing a reporter gene under the control of tandem PPREs and then treated with increasing ligand concentrations to measure transcriptional activity. Alternatively, assays have employed expression plasmids for the PPAR-LBD fused to the Gal4 DNA-binding domain and for the reporter gene under regulation by Gal4 DNA-binding elements. These assays yield an EC_{50} index, which corresponds to the ligand concentration able to induce 50% of maximal activation. The relative affinities of endogenous ligands turn questionable their capacity to activate PPARs under *physiological* conditions, because they should reach high levels in target tissues (i.e. liver, muscle, adipose). Therefore, the transactivation capacities of endogenous molecules do not guarantee their condition as *physiological* ligands. In fact, we still without certainly knowing the nature of *physiological* PPAR ligands.

Thiazolidinediones (TZD), including Rgtz, are PPARγ full agonists with antidiabetic potential. However, their use has been limited because of adverse side effects (mainly hepatotoxicity and body weight gain). Interestingly, all endogenous ligands have shown lower PPARγ transactivation activity than TZD (Table 3.1). Thus, these synthetic ligands might induce a *nonphysiological* level of PPARγ activation, leading to dysregulation of signalling pathways. Development of synthetic partial agonists has been encouraged as they might reproduce physiological regulatory circuits, avoiding undesirable side effects.

3.6 PPARs Regulatory Actions on Metabolism

It has been largely accepted that PPARs regulate a wide variety of cell functions by exerting a tight control on genes involved in cell differentiation, tissue development, bioenergetics and inflammation (reviewed by [41–43]). The diversity of mechanisms under PPARs regulation is likely consequence of their wide distribution in tissues and ability to accommodate a collection of ligands within their huge ligand-binding pocket. Furthermore, PPARs ability to interact with multiple co-activators/co-repressors in distinct tissues may contribute to their wide spectrum of regulatory actions. Despite the tissue distribution and binding properties of PPAR subtypes partially overlap, each subtype has particular actions. The main PPAR regulatory actions on metabolism are reviewed below and summarized in Table 3.2.

3.6.1 PPARα

PPARα is highly expressed in tissues with high rates of FA oxidation such as the liver, heart, skeletal muscle, brown and white adipose tissue and kidney. Additionally, cells associated with inflammatory reactions (including vascular endothelium and smooth muscle cells) were found to express this receptor. PPARα ligands are mostly FA and their derivatives formed during lipolysis, lipogenesis or catabolism. They include long-chain polyunsaturated FA (PUFA), such as arachidonic, linoleic, docosahexaenoic and eicosapentaenoic acids, endocannabinoid-like molecules (oleoylethanolamide and palmitoylethanolamide), arachidonic acid derivatives (20-carboxy-arachidonic acid and leukotriene B4) and phosphatidylcholines [42, 44–47]. Dietary lipids can also regulate PPARα activity as evidenced by the increased expression of PPARα target genes in the liver upon feeding with a high-fat diet (reviewed by [42]).

PPARα actions are best characterized in the liver, where it regulates different pathways of lipid metabolism, including de novo lipogenesis, to adapt the organism to fasting or feeding conditions [42, 48]. During fasting, PPARα activation in the liver upregulates the expression of genes linked to FA uptake (FATP-1 and CD36), intracellular transport (L-FABP), mitochondrial translocation (carnitine palmitoyl transferases) and peroxisomal and mitochondrial β-oxidation (acyl-CoA oxidase-1, and medium, long and very long acyl-CoA dehydrogenases). This enhances hepatic uptake of FA (derived from peripheral tissue lipolysis) and the production of acetyl-CoA and ATP from FA β-oxidation. Upon prolonged fasting, PPARα activation facilitates the formation of ketone bodies through upregulation of key components of the ketogenic pathway, the mitochondrial HMGCS (the rate-limiting enzyme) and FGF21 [49]. Since PPARα activation promotes ATP production via FA β-oxidation, it *indirectly* modulates glucose and energy homeostasis. In mice, PPARα activation can also inhibit glycolysis and promotes gluconeogenesis [50]. In contrast, during the fed state, PPARα coordinates de novo lipogenic pathways in the liver, favouring acetyl-CoA conversion into triacylglycerides (TAGs), which are stored to be used in periods of starvation. This effect is mediated by the upregulation of proteins involved in acetyl-CoA shuttled to the cytosol (citrate carrier) and by increasing the maturation of the lipogenic factor SREBP1c from its precursor. In addition, in human hepatocytes, PPARα controls lipoprotein metabolism, increasing HDL production and decreasing plasma VLDL [51]. The rise in HDL-C production is associated with an increase in the synthesis of the major HDL-C apolipoproteins (APO-AI and APO-AII), and the induction of the phospholipid transfer protein (reviewed by [42, 48]). On the other hand, PPARα activation

Table 3.2 Main PPAR regulatory actions on metabolism

	Liver	Skeletal muscle	Cardiac muscle	White adipose tissue
PPARα	**Fasting state**			
	↑ FA uptake ↑ FA mitochondrial transport ↑ FA oxidation ↑ Ketogenesis ↓ TAGs synthesis	↑ FA uptake ↑ FA oxidation ↑ TAG lipolysis ↑ Glucose utilization	↑ FA uptake ↑ FA oxidation ↑ TAG lipolysis	↑ FA oxidation ↑ TAGs synthesis during fed state
	Fed state			
	↑ Acetyl-CoA shuttled to the cytosol ↑ *De novo* lipogenesis ↑ HDL production ↓ VLDL production			
PPARβ/δ		↑ PPARβ/δ in starvation ↑ FA transport ↑ FA oxidation ↓ Glucose utilization *Switch metabolic from glucose to FA utilization*	↑ FA oxidation ↓ Lipid accumulation ↑ Glucose transporter ↑ Glycolytic pathway ↑ Mitochondrial biogenesis	↑ FA uptake ↑ FA oxidation
PPARγ	↑ Insulin sensitivity	↑ Insulin sensitivity		↑ Adipogenesis ↑ Adipocyte survival ↑ FA uptake ↑ FA storage ↑ Adipokine secretion ↑ Insulin sensitivity

leads to a reduction in plasma VLDL by increasing lipoprotein lipase activity (LPL), enzyme that hydrolyses lipoprotein TAGs; this effect is mediated by upregulation of *LPL* and downregulation of *APO-CIII*, a specific LPL inhibitor.

Fibrates, high-affinity PPARα agonists, have been therapeutically used for decades and considered relatively safe drugs although they cause mild side effects (most common being gastrointestinal disturbances, a rise in transaminase, creatinine and homocysteine and myositis). They are well-established effective drugs for controlling dyslipidaemia, particularly when TAG-rich lipoproteins (mainly VLDL) are highly increased and HDL-C decreased, but their potential to prevent cardiovascular diseases is still uncertain and requires further studies [52–55].

3.6.2 PPARβ/δ

PPARβ/δ is the least understood of the three PPAR subtypes. It is ubiquitously expressed in tissues, with a relatively high level of expression in those tissues associated with FA metabolism, such as skeletal and cardiac muscle, hepatocytes and adipocytes. The nature of true physiological PPAR ligands is still uncertain, but several endogenous molecules are potentially capable of acting as ligands. They include PUFAs and eicosanoid metabolites (i.e. prostacyclin, 8(S) hydroxyeicosatetraenoic acid [8(S)-HETE] and 15-hydroxyeicosatetraenoic acid (15-HETE)) [38, 56, 57]. Various synthetic, high-affinity and specific agonists have been developed and used to elucidate PPARβ/δ functions (GW501516, L-165041, and carbacyclin, among others) [58–60].

Among others, PPARβ/δ major roles include the control of lipid metabolism and energy expenditure in cardiac as well as skeletal muscle. The heart is an energy-demanding organ with an active oxidative metabolism. The role played by PPARβ/δ in cardiomyocytes is evidenced in PPARβ/δ-null mice, which shows a significant decrease in myocardial FA and glucose oxidation that lead to myocardial dysfunction [61, 62]. These alterations correlate with a lower constitutive expression of genes involved in key steps of FA oxidation (mitochondrial FA uptake, malonyl-CoA metabolism, mitochondrial and peroxisomal β-oxidation) and glucose oxidation in PPARβ/δ--null cardiomyocytes. Of note, PPARα or PPARβ/δ activate different metabolic programmes in the heart [63]. Both receptors mediate responses that enhance FA β-oxidation, but activation of PPARβ/δ did not lead to lipid accumulation. In fact, genes involved in FA uptake (CD36, FATP-1), TAGs synthesis and lipogenic pathways are not target of PPARβ/δ. On the other hand, activation of PPARβ/δ, but not PPARα, induces upregulation of glucose transporter and glycolytic pathway genes, which improve myocardial glucose uptake and utilization rates. It is worth to mention that PPARβ/δ would also play a role in regulating cardiac expression of antioxidants enzymes (Cu/Zn-superoxide dismutase, Mn-superoxide dismutase and catalase), contributing to protect the heart from oxidative damage due to enhanced FA oxidation [64]. In parallel to these metabolic responses, in the adult heart, PPARβ/δ activation promotes mitochondrial biogenesis [64]. Overall, PPARβ/δ acts as a key regulatory factor controlling myocardial energy metabolism.

PPARβ/δ signalling activates different metabolic programmes in cardiomyocytes and skeletal muscle cells. Indeed, synthetic PPARβ/δ agonists drive a switch in the fuel metabolism of skeletal muscle cells, enhancing FA oxidation and decreasing carbohydrate catabolism. The increase in FA oxidation is evidenced by upregulation of muscle carnitine palmitoyl transferase Ib, uncoupling protein 3 and cytochrome c, and correlates with higher levels of serum 3-hydroxyhexadecanoic acid (an earlier intermediate of the FA oxidation pathway) [65]. On the other hand, the switch in the metabolic substrate usage, from glucose to FA utilization, correlates with upregulation of forkhead box O1 (*FOXO1*) expression. Indeed, FoxO1 promotes upregulation of *CD36* and *LPL* [66], which would contribute to cellular FA acquisition through lipoprotein internalization and lipolysis, respectively. Moreover, FoxO1 upregulates *PDK4* [66], an enzyme involved in controlling glucose oxidation by inactivating the pyruvate dehydrogenase complex. Finally, PPARβ/δ activation in the skeletal muscle favours fibre-type switching towards type I oxidative fibres, which have a higher glucose-handling capacity compared with type II fibres (reviewed by [42]). Therefore, PPARβ/δ actions in skeletal muscle promote FA utilization and spare glucose for tissues that are more dependent on it.

PPARβ/δ activity has shown efficacy for the treatment of metabolic disorders in preclinical studies and in a few clinical trials (i.e. dyslipidaemia and type 2 diabetes mellitus). However, safety issues have limited the development of drugs targeting this receptor [67]. Meta-analysis of datasets from various human cancer subtypes has revealed a negative correlation between PPARβ/δ expression and survival, the latter measured as either overall or relapse-free survival [68]. However, survival association studies cannot rule out that PPARβ/δ expression would be a consequence rather than the causative factor of developing cancer. Furthermore, PPARβ/δ has been described either as a tumour-promoting or tumour-inhibiting factor [68, 69]. Therefore, the involvement of PPARβ/δ in cancer-associated processes is currently uncertain. Further studies are needed to determine the therapeutically efficacy and safety of drugs targeting PPARβ/δ.

3.6.3 PPARγ

PPARγ is known to exist in two functional isoforms, named PPARγ1 and PPARγ2; they are generated by alternative splicing, and differential promoter usage [2]. PPARγ2 has an additional stretch of 30 amino acids at its N-terminus, resulting in a higher transcriptional activity than PPARγ1 [70]. PPARγ isoforms exhibit distinct expression pattern [71, 72]. PPARγ1 is highly expressed in adipose tissue, but it is also found in significant, but lower, levels in various tissues including the liver, muscles, pancreas, small intestine and haematopoietic cells. PPARγ2 expression is restricted to white and brown adipose tissue under physiological conditions. The expression of both isoforms is highly regulated by the diet, being strongly reduced in adipose tissue during fasting. Diverse PUFA and some electrophilic compounds, including metabolites of arachidonic acid (i.e. 15d-PGJ_2), behave as partial PPARγ ligands, but it is uncertain whether they represent genuine endogenous ligands since their physiological concentrations seem to be much lower than those required for receptor activation. Nitroalkene derivatives of mono- and polyunsaturated FA are *putative* physiological PPARγ ligands; they exhibit transactivation activity in the submicromolar range (Table 3.1) and might reach this concentration locally during inflammation [73, 74].

PPARγ plays a central role in lipid metabolism and insulin sensitivity through regulating genes involved in lipid uptake, synthesis and storage in peripheral tissues (mainly skeletal muscle, liver, and adipose tissue). Furthermore, both PPARγ isoforms, γ1 and γ2, are required for adipocyte differentiation, although PPARγ2 expression would be sufficient to induce adipogenesis. Certainly, ectopic PPARγ2 expression in fibroblasts promotes adipogenesis [75]. Moreover, a significant decrease in the size and number of adipocytes was observed in PPARγ$2^{-/-}$ mice, contributing to lipodystrophy [76]. This alteration correlates with a reduced expression of lipogenic genes, including phosphoenolpyruvate carboxykinase, adipsin, LPL and the transcription factor C/EBPα (CCAAT/enhancer-binding protein α); the latter collaborates with PPARγ in activating the full programme of adipogenesis and insulin sensitivity. Consistent with observations in mouse models, humans with dominant-negative mutations in a single allele of *PPARG* have partial lipodystrophy [77, 78].

PPARγ stimulation of adipogenesis contributes to control levels of circulating FA, protecting non-adipose tissues against excessive lipid overload and avoiding lipotoxic effects in peripheral organs, which correlate with insulin resistance [79]. In fact, mice with increased PPARγ activity are protected from obesity-induced insulin resistance, whereas mice lacking PPARγ specifically in fat, muscle, or the liver develop hyperlipidaemia and hyperglycaemia (reviewed by [79]). However, selective activation of PPARγ in mature adipocytes, but not in preadipocytes, was sufficient for improving insulin sensitization, suggesting the occurrence of PPARγ protective effects independent on adipogenesis [80]. Various mechanisms are likely involved in these effects. Firstly, *PPARγ*$2^{-/-}$ mice showed lower levels of circulating leptin and adiponectin levels, two hormones that ameliorate insulin sensitivity in peripheral tissues [76, 81, 82]. Thus, PPARγ activation in adipocytes may guarantee a balanced and adequate secretion of adipokines, maintaining insulin sensitivity of the whole body. On the other hand, PPARγ likely exert a direct control of glucose homeostasis and insulin sensitivity. In peripheral tissues such as the skeletal muscle, PPARγ activation upregulates the expression of proteins involved in glucose transport (Glut4 and c-Cbl-associated protein) and insulin response (insulin receptor substrate 1 and 2, phosphatidylinositol 3-kinase) [76, 79]. In pancreatic β-cells, PPARγ activation induces upregulation of key genes involved in glucose-stimulated insulin secretion; nevertheless, the contribution of PPARγ activation in the pancreas to insulin sensitization by TZD remains to be determined [79]. Of note, tissue-specific knockout models are essential to elucidate the relative involvement of PPARγ to insulin sensitization in different tissues. Using this kind of mouse models, skeletal muscle and liver showed to play a role in the insulin sensitization induced by TZD, despite the relatively low expression of PPARγ in these tissues (reviewed by [79]). Finally, macrophages are also relevant targets of PPARγ agonists; they are recruited to adipose tissue, and their inflammatory potential can be modulated by PPARγ activation. However, the contribution of macrophage selective-PPARγ activation to control insulin resistance remains unclear since controversial results have been obtained from in vivo studies [80, 83, 84].

TZDs, such as rosiglitazone and pioglitazone, are very efficacious as insulin sensitizers and have been widely used for the treatment of type 2 diabetes mellitus (often as a second-line oral drug in combination with the metformin) [79]. They are also efficient drugs for treating hyperlipidaemia, reducing circulating FA. Unfortunately, TZDs have shown undesirable side effects in human, including weight gain, fluid retention, cardiovascular complications, bone fractures, bladder cancer, and hepatotoxicity, which have restricted their clinical use. It has been postulated that PPARγ over activation (inducing excessive fat storage in adipose tissue) or PPARγ unwanted activation in other tissues might be responsible for side effects. The molecular basis of PPARγ activation offers the opportunity to design *novel* agonists, capable of driving a partial activation of this receptor. Desirable agonists would induce a differential regulation of PPARγ target genes, leading to the beneficial effects on glucose and lipid homeostasis, but avoiding adverse side effects.

Acknowledgements The authors would like to thank Mauricio Castellano for his assistance with the preparation of figures. This work was supported by the Agencia Nacional de Investigación e Innovación (ANII) and Programa de Desarrollo de Ciencias Básicas (PEDECIBA, Uruguay).

References

1. Issemann I, Green S (1990) Activation of a member of the steroid hormone receptor superfamily by peroxisome proliferators. Nature 347:645–650
2. Desvergne B, Wahli W (1999) Peroxisome Proliferator-activated receptors: nuclear control of metabolism. Endocr Rev 20:649–688
3. Willson TM, Brown PJ, Sternbach DD, Henke BR (2000) The PPARs: from orphan receptors to drug discovery. J Med Chem 43:527–550
4. Sher T, Yi HF, McBride OW, Gonzalez FJ (1993) cDNA cloning, chromosomal mapping, and functional characterization of the human peroxisome proliferator activated receptor. Biochemistry 32:5598–5604
5. Yoshikawa T, Brkanac Z, Dupont BR, Xing G-Q, Leach RJ, Detera-Wadleigh SD (1996) Assignment of the human nuclear hormone receptor, NUC1 (PPARD), to chromosome 6p21.1–p21.2. Genomics 35:637–638
6. Greene ME, Blumberg B, McBride OW, Yi HF, Kronquist K, Kwan K, et al (1995) Isolation of the human peroxisome proliferator activated receptor gamma cDNA: expression in hematopoietic cells and chromosomal mapping. Gene Expr 4:281–299
7. Pawlak M, Lefebvre P, Staels B (2012) General molecular biology and architecture of nuclear receptors. Curr Top Med Chem. Inserm. 12:486–504
8. Mangelsdorf DJ, Borgmeyer U, Heyman RA, Zhou JY, Ong ES, Oro AE et al (1992) Characterization of three RXR genes that mediate the action of 9-cis retinoic acid. Genes Dev. Cold Spring Harbor Laboratory Press. 6:329–344
9. Kliewer SA, Umesono K, Noonan DJ, Heyman RA, Evans RM (1992) Convergence of 9-cis retinoic acid and peroxisome proliferator signalling pathways through heterodimer formation of their receptors. Nature 358:771–774
10. Palmer CN, Hsu MH, Griffin HJ, Johnson EF (1995) Novel sequence determinants in peroxisome proliferator signaling. J Biol Chem 270:16114–16121
11. Juge-Aubry C, Pernin A, Favez T, Burger AG, Wahli W, Meier CA et al (1997) DNA binding properties of peroxisome proliferator-activated receptor subtypes on various natural peroxisome proliferator response elements. Importance of the 5′-flanking region. J Biol Chem 272:25252–25259
12. IJpenberg A, Jeannin E, Wahli W, Desvergne B (1997) Polarity and specific sequence requirements of Peroxisome Proliferator-activated receptor (PPAR)/Retinoid X receptor Heterodimer binding to DNA: a functional analysis of the malic enzyme gene. J Biol Chem 272:20108–20117
13. Dna PR, Hsu M-H, Palmer CNA, Song W, Griffin KJ, Johnson EF (1998) A carboxyl-terminal extension of the zinc finger domain contributes to the specificity and polarity of peroxisome proliferator-activated receptor DNA Binding. J Biol Chem 273:27988–27997
14. Lonard DM, O'malley BW (2007) Nuclear receptor coregulators: judges, juries, and executioners of cellular regulation. Mol Cell. Elsevier 27:691–700
15. Guan H-P, Ishizuka T, Chui PC, Lehrke M, Lazar MA (2005) Corepressors selectively control the transcriptional activity of PPAR in adipocytes. Genes Dev 19:453–461

16. McInerney EM, Rose DW, Flynn SE, Westin S, Mullen TM, Krones A et al (1998) Determinants of coactivator LXXLL motif specificity in nuclear receptor transcriptional activation. Genes Dev 12:3357–3368
17. DiRenzo J, Söderstrom M, Kurokawa R, Ogliastro MH, Ricote M, Ingrey S et al (1997) Peroxisome proliferator-activated receptors and retinoic acid receptors differentially control the interactions of retinoid X receptor heterodimers with ligands, coactivators, and corepressors. Mol Cell Biol 17:2166–2176
18. Kalkhoven E (2004) CBP and p300: HATs for different occasions. Biochem Pharmacol 68:1145–1155
19. Hu X, Lazar MA (1999) The CoRNR motif controls the recruitment of corepressors by nuclear hormone receptors. Nature 402:93–96
20. Koppen A, Kalkhoven E (2010) Brown vs white adipocytes: the PPARγ coregulator story. FEBS Lett 584:3250–3259
21. Frkic RL, Marshall AC, Blayo A, Pukala TL, Theodore M, Griffin PR et al (2018) Tumble and trap mechanism of the activation Helix. iScience:69–79
22. Chrisman IM, Nemetchek MD, de Vera IMS, Shang J, Heidari Z, Long Y et al (2018) Defining a conformational ensemble that directs activation of PPARγ. Nat Commun 9:1794
23. Delerive P, De Bosscher K, Besnard S, Vanden Berghe W, Peters JM, Gonzalez FJ et al (1999) Peroxisome proliferator-activated receptor α negatively regulates the vascular inflammatory gene response by negative cross-talk with transcription factors NF-κB and AP-1. J Biol Chem 274:32048–32054
24. Gervois P, Vu-Dac N, Kleemann R, Kockx M, Dubois G, Laine B et al (2001) Negative regulation of human fibrinogen gene expression by Peroxisome Proliferator-activated receptor α agonists via inhibition of CCAAT box/enhancer-binding protein β. J Biol Chem 276:33471–33477
25. Lee CH, Chawla A, Urbiztondo N, Liao D, Boisvert WA, Evans RM (2003) Transcriptional repression of Atherogenic inflammation: modulation by PPARδ. Science 302:453–457
26. Fan Y, Wang Y, Tang Z, Zhang H, Qin X, Zhu Y et al (2007) Suppression of pro-inflammatory adhesion molecules by PPAR- in human vascular endothelial cells. Arterioscler Thromb Vasc Biol 28:315–321
27. Brunmeir R, Xu F (2018) Functional regulation of PPARs through post-translational modifications. Int J Mol Sci 19:1738
28. Chandra V, Huang P, Hamuro Y, Raghuram S, Wang Y, Burris TP et al (2008) Structure of the intact PPAR-gamma-RXR- nuclear receptor complex on DNA. Nature 456:350–356
29. Zoete V, Grosdidier A, Michielin O (2007) Peroxisome proliferator-activated receptor structures: ligand specificity, molecular switch and interactions with regulators. Biochim Biophys Acta Mol Cell Biol Lipids 1771:915–925
30. Kroker AJ, Bruning JB (2015) Review of the structural and dynamic mechanisms of PPAR γ partial agonism. PPAR Res. Hindawi 2015:1–15
31. Nolte RT, Wisely GB, Westin S, Cobb JE, Lambert MH, Kurokawa R et al (1998) Ligand binding and co-activator assembly of the peroxisome proliferator- activated receptor-γ. Nature 395:137–143
32. Bruning JB, Chalmers MJ, Prasad S, Busby SA, Kamenecka TM, He Y et al (2007) Partial agonists activate PPARγ using a Helix 12 independent mechanism. Structure 15:1258–1271
33. Brzozowski AM, Pike ACW, Dauter Z, Hubbard RE, Bonn T, Engström O et al (1997) Molecular basis of agonism and antagonism in the oestrogen receptor. Nature 389:753–758
34. Shiau AK, Barstad D, Loria PM, Cheng L, Kushner PJ, Agard DA et al (1998) The structural basis of estrogen receptor/coactivator recognition and the antagonism of this interaction by tamoxifen. Cell. Elsevier 95:927–937
35. Xu HE, Stanley TB, Montana VG, Lambert MH, Shearer BG, Cobb JE et al (2002) Structural basis for antagonist-mediated recruitment of nuclear corepressors by PPARα. Nature 415:813–817
36. Kliewer SASASA, Sundseth SSSSS, Jones SASAA, Brown PJJPJ, Wisely GBB, Koble CSSCS et al (1997) Fatty acids and eicosanoids regulate gene expression through direct interactions with peroxisome proliferator-activated receptors alpha and gamma. Proc Natl Acad Sci U S A 94:4318–4323
37. Forman BM, Chen J, Evans RM (1997) Hypolipidemic drugs, polyunsaturated fatty acids, and eicosanoids are ligands for peroxisome proliferator-activated receptors and. Proc Natl Acad Sci 94:4312–4317
38. Krey G, Braissant O, L'Horset F, Kalkhoven E, Perroud M, Parker MG et al (1997) Fatty acids, Eicosanoids, and Hypolipidemic agents identified as Ligands of Peroxisome Proliferator-activated receptors by Coactivator-dependent receptor Ligand assay. Mol Endocrinol 11:779–791
39. Schopfer FJ, Lin Y, Baker PRS, Cui T, Garcia-Barrio M, Zhang J et al (2005) Nitrolinoleic acid: an endogenous peroxisome proliferator-activated receptor gamma ligand. Proc Natl Acad Sci U S A 102:2340–2345
40. Pochetti G, Godio C, Mitro N, Caruso D, Galmozzi A, Scurati S et al (2007) Insights into the mechanism of partial agonism. J Biol Chem 282:17314–17324
41. Poulsen L la C, Siersbæk M, Mandrup S (2012) PPARs: fatty acid sensors controlling metabolism. Semin Cell Dev Biol. Elsevier Ltd 23:631–639
42. Dubois V, Eeckhoute J, Lefebvre P, Staels B (2017) Distinct but complementary contributions of PPAR isotypes to energy homeostasis. J Clin Invest 127:1202–1214
43. Cuaranta-Monroy I, Kiss M, Simandi Z, Nagy L (2015) Genomewide effects of peroxisome proliferator-activated receptor gamma in macrophages and dendritic cells – revealing complex-

ity through systems biology. Eur J Clin Investig 45:964–975
44. Campolongo P, Roozendaal B, Trezza V, Cuomo V, Astarita G, Fu J et al (2009) Fat-induced satiety factor oleoylethanolamide enhances memory consolidation. Proc Natl Acad Sci 106:8027–8031
45. LoVerme J, La Rana G, Russo R, Calignano A, Piomelli D (2005) The search for the palmitoylethanolamide receptor. Life Sci 77:1685–1698
46. Fang X, Dillon JS, Hu S, Harmon SD, Yao J, Anjaiah S et al (2007) 20-Carboxy-arachidonic acid is a dual activator of peroxisome proliferator-activated receptors α and γ. Prostaglandins Other Lipid Mediat 82:175–184
47. Narala VR, Adapala RK, Suresh MV, Brock TG, Peters-Golden M, Reddy RC (2010) Leukotriene B_4 is a physiologically relevant endogenous peroxisome proliferator-activated receptor-α agonist. J Biol Chem 285:22067–22074
48. Pawlak M, Lefebvre P, Staels B (2015) Molecular mechanism of PPARα action and its impact on lipid metabolism, inflammation and fibrosis in non-alcoholic fatty liver disease. J Hepatol 62:720–733
49. Badman MK, Pissios P, Kennedy AR, Koukos G, Flier JS, Maratos-Flier E (2007) Hepatic fibroblast growth factor 21 is regulated by PPARα and is a key mediator of hepatic lipid metabolism in Ketotic states. Cell Metab 5:426–437
50. Peeters A, Baes M (2010) Role of PPARα in hepatic carbohydrate metabolism. PPAR Res 2010
51. Shah A, Rader DJ, Millar JS (2010) The effect of PPAR-α agonism on apolipoprotein metabolism in humans. Atherosclerosis 210:35–40
52. Remick J, Weintraub H, Setton R, Offenbacher J, Fisher E, Schwartzbard A (2008) Fibrate therapy. Cardiol Rev 16:129–141
53. Wierzbicki AS (2009) Fibrates in the treatment of cardiovascular risk and atherogenic dyslipidaemia. Curr Opin Cardiol 24:372–379
54. Wierzbicki AS (2010) Fibrates: no ACCORD on their use in the treatment of dyslipidaemia. Curr Opin Lipidol 21:352–358
55. Wang D, Liu B, Tao W, Hao Z, Liu M (2015) Fibrates for secondary prevention of cardiovascular disease and stroke. Cochrane Database Syst Rev:CD009580
56. Xu HE, Lambert MH, Montana VG, Parks DJ, Blanchard SG, Brown PJ et al (1999) Molecular recognition of fatty acids by peroxisome proliferator-activated receptors. Mol Cell. Elsevier 3:397–403
57. Naruhn S, Meissner W, Adhikary T, Kaddatz K, Klein T, Watzer B et al (2010) 15-hydroxyeicosatetraenoic acid is a preferential peroxisome proliferator-activated receptor beta/delta agonist. Mol Pharmacol 77:171–184
58. Ritzenthaler JD, Roman J, Han S (2009) PPARβ/δ agonist increases the expression of PGE2 receptor subtype EP4 in human lung carcinoma cells. Meth Mol Biol. (Clifton, NJ):309–323
59. Jiménez R, Sánchez M, Zarzuelo MJ, Romero M, Quintela AM, López-Sepúlveda R et al (2010) Endothelium-dependent vasodilator effects of peroxisome proliferator-activated receptor beta agonists via the phosphatidyl-inositol-3 kinase-Akt pathway. J Pharmacol Exp Ther 332:554–561
60. Kuroda T, Hirota H, Fujio Y, Sugiyama S, Masaki M, Hiramoto Y et al (2007) Carbacyclin induces carnitine palmitoyltransferase-1 in cardiomyocytes via peroxisome proliferator-activated receptor (PPAR) δ independent of the IP receptor signaling pathway. J Mol Cell Cardiol 43:54–62
61. Cheng L, Ding G, Qin Q, Huang Y, Lewis W, He N et al (2004) Cardiomyocyte-restricted peroxisome proliferator-activated receptor-δ deletion perturbs myocardial fatty acid oxidation and leads to cardiomyopathy. Nat Med 10:1245–1250
62. Wang P, Liu J, Li Y, Wu S, Luo J, Yang H et al (2010) Peroxisome Proliferator-activated receptor δ is an essential transcriptional regulator for mitochondrial protection and biogenesis in adult heart. Circ Res 106:911–919
63. Burkart EM, Sambandam N, Han X, Gross RW, Courtois M, Gierasch CM et al (2007) Nuclear receptors PPARβ/δ and PPARα direct distinct metabolic regulatory programs in the mouse heart. J Clin Invest 117:3930–3939
64. Liu J, Wang P, Luo J, Huang Y, He L, Yang H et al (2011) Peroxisome proliferator-activated receptor β/δ activation in adult hearts facilitates mitochondrial function and cardiac performance under pressure-overload condition. Hypertension. (Dallas, Tex 1979) 57:223–230
65. Chen W, Gao R, Xie X, Zheng Z, Li H, Li S et al (2015) A metabolomic study of the PPARδ agonist GW501516 for enhancing running endurance in Kunming mice. Sci Rep 5:9884
66. Nahlé Z, Hsieh M, Pietka T, Coburn CT, Grimaldi PA, Zhang MQ et al (2008) CD36-dependent regulation of muscle FoxO1 and PDK4 in the PPAR delta/beta-mediated adaptation to metabolic stress. J Biol Chem 283:14317–14326
67. Palomer X, Barroso E, Pizarro-Delgado J, Peña L, Botteri G, Zarei M et al (2018) PPARβ/δ: a key therapeutic target in metabolic disorders. Int J Mol Sci 19:913
68. Müller R (2017) PPARβ/δ in human cancer. Biochimie 136:90–99
69. Peters JM, Gonzalez FJ, Müller R (2015) Establishing the role of PPARβ/δ in carcinogenesis. Trends Endocrinol Metab 26:595–607
70. Werman A, Hollenberg A, Solanes G, Bjørbæk C, Vidal-Puig AJ, Flier JS (1997) Ligand-independent activation domain in the N terminus of peroxisome proliferator-activated receptor γ (PPARγ). Differential activity of PPARγ1 and −2 isoforms and influence of insulin. J Biol Chem 272:20230–20235
71. Vidal-Puig A, Jimenez-Liñan M, Lowell BB, Hamann A, Hu E, Spiegelman B et al (1996) Regulation of PPAR gamma gene expression by nutrition and obesity in rodents. J Clin Invest 97:2553–2561

72. Loviscach M, Rehman N, Carter L, Mudaliar S, Mohadeen P, Ciaraldi TP et al (2000) Distribution of peroxisome proliferator-activated receptors (PPARs) in human skeletal muscle and adipose tissue: relation to insulin action. Diabetologia 43:304–311
73. Schopfer FJ, Cole MP, Groeger AL, Chen CS, Khoo NKH, Woodcock SR et al (2010) Covalent peroxisome proliferator-activated receptor γ adduction by nitro-fatty acids: selective ligand activity and anti-diabetic signaling actions. J Biol Chem 285:12321–12333
74. Schopfer FJ, Cipollina C, Freeman BA (2011) Formation and signaling actions of electrophilic lipids [Internet]. Chem Rev:5997–6021
75. Tontonoz P, Hu E, Spiegelman BM (1994) Stimulation of adipogenesis in fibroblasts by PPARγ2, a lipid-activated transcription factor. Cell 79:1147–1156
76. Zhang J, Fu M, Cui T, Xiong C, Xu K, Zhong W et al (2004) Selective disruption of PPARgamma 2 impairs the development of adipose tissue and insulin sensitivity. Proc Natl Acad Sci U S A 101:10703–10708
77. Hegele RA, Cao H, Frankowski C, Mathews ST, Leff T (2002) PPARG F388L, a transactivation-deficient mutant, in familial partial lipodystrophy. Diabetes Am Diabetes Assoc 51:3586–3590
78. Savage DB, Tan GD, Acerini CL, Jebb SA, Agostini M, Gurnell M et al (2003) Human metabolic syndrome resulting from dominant-negative mutations in the nuclear receptor peroxisome proliferator-activated receptor-γ. Diabetes 52:910–917
79. Ahmadian M, Suh JM, Hah N, Liddle C, Atkins AR, Downes M et al (2013) PPARγ signaling and metabolism: the good, the bad and the future. Nat Med 19:557–566
80. Sugii S, Olson P, Sears DD, Saberi M, Atkins AR, Barish GD et al (2009) PPAR activation in adipocytes is sufficient for systemic insulin sensitization. Proc Natl Acad Sci U S A 106:22504–22509
81. Yamauchi T, Kamon J, Waki H, Terauchi Y, Kubota N, Hara K et al (2001) The fat-derived hormone adiponectin reverses insulin resistance associated with both lipoatrophy and obesity. Nat Med 7:941–946
82. Berg AH, Combs TP, Du X, Brownlee M, Scherer PE (2001) The adipocyte-secreted protein Acrp30 enhances hepatic insulin action. Nat Med 7:947–953
83. Hevener AL, Olefsky JM, Reichart D, Nguyen MTA, Bandyopadyhay G, Leung H-Y et al (2007) Macrophage PPAR gamma is required for normal skeletal muscle and hepatic insulin sensitivity and full antidiabetic effects of thiazolidinediones. J Clin Invest 117:1658–1669
84. Odegaard JI, Ricardo-Gonzalez RR, Goforth MH, Morel CR, Subramanian V, Mukundan L et al (2007) Macrophage-specific PPARgamma controls alternative activation and improves insulin resistance. Nature 447:1116–1120
85. Ferry G, Bruneau V, Beauverger P, Goussard M, Rodriguez M, Lamamy V et al (2001) Binding of prostaglandins to human PPARgamma: tool assessment and new natural ligands. Eur J Pharmacol 417:77–89
86. Kliewer SA, Sundseth SS, Jones SA, Brown PJ, Wisely GB, Koble CS et al (1997) Fatty acids and eicosanoids regulate gene expression through direct interactions with peroxisome proliferator-activated receptors alpha and gamma. Proc Natl Acad Sci U S A 94:4318–4323
87. Palmer CN, Wolf CRR (1998) Cis Parinaric acid is a ligand for the human peroxisome proliferator activated receptor gamma – development of a novel spectrophotometric assay for the discovery of PPAR gamma ligands. FEBS Lett. Wiley-Blackwell 431:476–480
88. Kliewer SA, Lenhard JM, Willson TM, Patel I, Morris DC, Lehmann JM (1995) A prostaglandin J2 metabolite binds peroxisome proliferator-activated receptor γ and promotes adipocyte differentiation. Cell 83:813–819
89. Nagy L, Tontonoz P, Alvarez JGA, Chen H, Evans RM (1998) Oxidized LDL regulates macrophage gene expression through ligand activation of PPARγ. Cell 93:229–240
90. McIntyre TM, Pontsler AV, Silva AR, St. Hilaire A, Xu Y, Hinshaw JC et al (2003) Identification of an intracellular receptor for lysophosphatidic acid (LPA): LPA is a transcellular PPAR agonist. Proc Natl Acad Sci 100:131–136
91. Lehmann JM, Moore LB, Smith-Oliver TA, Wilkison WO, Willson TM, Kliewer SA (1995) An antidiabetic thiazolidinedione is a high affinity ligand for peroxisome proliferator-activated receptor gamma (PPAR gamma). J Biol Chem:12953–12956
92. Henke BR, Blanchard SG, Brackeen MF, Brown KK, Cobb JE, Collins JL et al (1998) N-(2-benzoylphenyl)-L-tyrosine PPARγ agonists. 1. Discovery of a novel series of potent antihyperglycemic and antihyperlipidemic agents. J Med Chem 41:5020–5036
93. Motani A, Wang Z, Weiszmann J, McGee LR, Lee G, Liu Q et al (2009) INT131: a selective modulator of PPARγ. J Mol Biol 386:1301–1311
94. Chen X, Osborne C (2005) Pharmacological profile of a novel, non-TZD PPARg agonist. Diabetes Obes Metab 7:536–546
95. Kurosaki E, Nakano R, Shimaya A, Yoshida S, Ida M, Suzuki T et al (2003) Differential effects of YM440 a hypoglycemic agent on binding to a peroxisome proliferator-activated receptor γ and its transactivation. Biochem Pharmacol 65:795–805
96. Leesnitzer LM, Parks DJ, Bledsoe RK, Cobb JE, Collins JL, Consler TG et al (2002) Functional consequences of cysteine modification in the ligand binding sites of peroxisome proliferator activated receptors by GW9662. Biochemistry 41:6640–6650
97. Hopkins CR, O'Neil SV, Laufersweiler MC, Wang Y, Pokross M, Mekel M et al (2006) Design and synthesis of novel N-sulfonyl-2-indole carboxamides as

potent PPAR-γ binding agents with potential application to the treatment of osteoporosis. Bioorg Med Chem Lett 16:5659–5663

98. Bernardes A, Batista FAH, de Oliveira Neto M, Figueira ACM, Webb P, Saidemberg D et al (2012) Low-resolution molecular models reveal the oligomeric state of the PPAR and the conformational organization of its domains in solution. PLoS One 7:e31852

99. Xu HE, Lambert MH, Montana VG, Plunket KD, Moore LB, Collins JL et al (2001) Structural determinants of ligand binding selectivity between the peroxisome proliferator-activated receptors. Proc Natl Acad Sci U S A 98:13919–13924

100. Artis DR, Lin JJ, Zhang C, Wang W, Mehra U, Perreault M et al (2009) Scaffold-based discovery of indeglitazar, a PPAR pan-active anti-diabetic agent. Proc Natl Acad Sci 106:262–267

4 Diminishing Inflammation by Reducing Oxidant Generation: Nitrated Fatty Acid-Mediated Inactivation of Xanthine Oxidoreductase

Eric E. Kelley

Abstract

Inhibition of xanthine oxidoreductase (XOR) has proven beneficial in a plethora of inflammatory disease processes due to a net reduction in pro-inflammatory oxidants and secondary nitrating species. Electrophilic nitrated fatty acid derivatives, such as nitro-oleic acid (OA-NO_2) are also noted to display a broad spectrum of anti-inflammatory effects via interaction with critical signaling pathways. An alternative process in which nitrated fatty acids may extend anti-inflammatory actions is via inactivation of XOR, a process that is more effective than allo/oxypurinol-mediated inhibition. Herein, we describe the molecular aspects of nitrated fatty acid-associated inactivation of XOR, identify specificity via structure function relationships and discuss XOR as a crucial component of the anti-inflammatory portfolio of nitrated fatty acids.

Keywords

Xanthine oxidoreductase · Nitro-fatty acids · Reactive oxygen species · Allopurinol

E. E. Kelley (✉)
Department of Physiology and Pharmacology, West Virginia University, Morgantown, WV, USA
e-mail: eric.kelley@hsc.wvu.edu

4.1 Xanthine Oxidoreductase (XOR)

Xanthine oxidoreductase is a complex molybdo-flavin protein that catalyzes the terminal two steps in purine degradation in primates; oxidation of hypoxanthine to xanthine and oxidation of xanthine to uric acid. The human XOR gene (*xdh*) resides on chromosome band 2p23.1 and is transcribed as xanthine dehydrogenase (XDH). The enzyme is a homodimer of ~295 kD with each subunit consisting of four redox centers: a molybdenum cofactor (Mo-co), one FAD and two Fe/S clusters. The Mo-co is comprised of a pterin derivative with a cyclized dithiolene side chain and one Mo atom pentacoordinated with the dithiolene, two oxygen atoms and a sulfur atom. The Mo-co is the site of purine oxidation whereas NAD^+ and O_2 reduction occur at the FAD, Fig. 4.1. The two Fe/S clusters serve as a conduit for electron flow between the Mo-co and the FAD. These Fe/S clusters are both of the ferredoxin type, but are not identical and thus are independently distinguishable by their electron paramagnetic resonance spectra [1–4].

Post-translation modification of XDH via limited proteolysis or oxidation of critical cysteine residues, results in conformational changes that confer alteration of the electrostatic and steriochemical environment in the vicinity of the FAD cofactor ultimately establishing oxidase capacity [1, 5]. When induced by cysteine oxidation,

A. Trostchansky, H. Rubbo (eds.), *Bioactive Lipids in Health and Disease*, Advances in Experimental Medicine and Biology 1127, https://doi.org/10.1007/978-3-030-11488-6_4

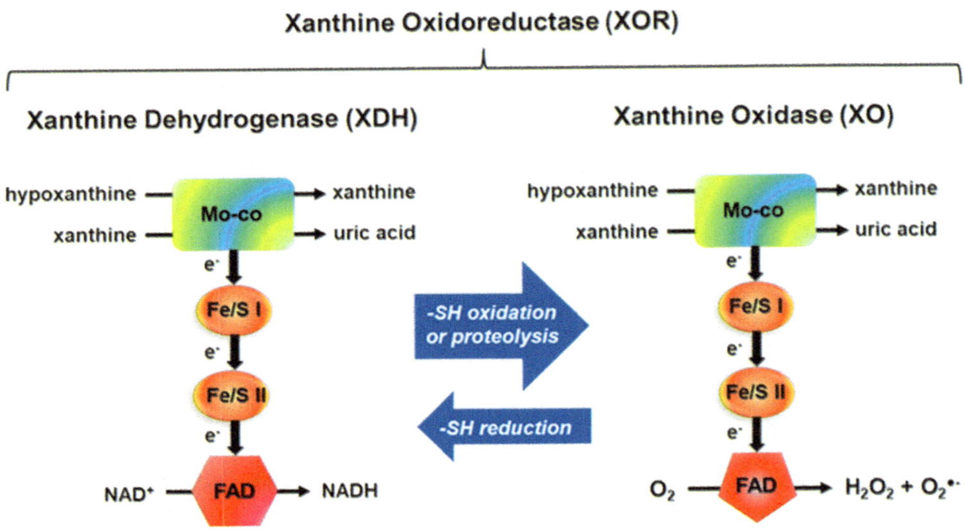

Fig. 4.1 Xanthine Oxidoreductase (XOR). For xanthine dehydrogenase (**XDH**), hypoxanthine is oxidized to xanthine and xanthine is oxidized to uric acid at the Mo-co. Electrons derived from the purine oxidation at the Mo-co are transferred via 2 Fe/S centers to the FAD where NAD^+ is reduced to NADH. Upon oxidation of critical cysteine residues (reversible) and/or limited proteolysis (irreversible), XDH is converted to xanthine oxidase (**XO**). For XO, hypoxanthine and xanthine are oxidized to uric acid at the Mo-co and electrons are transferred to the FAD where O_2 is reduced to generate $O_2^{\bullet-}$ and H_2O_2

conversion from XDH to XO is reversible upon endogenous reduction of these residues or by *in vitro* application of reducing agents such as dithiothreitol (DTT), Fig. 4.1. However, conversion to XO by partial proteolysis of XDH is irreversible [1]. While the post-translational modification resulting in XO activity has become synonymous with the conversion from a housekeeping enzyme to a source of deleterious reactive oxygen species (ROS), it is important to recognize that XDH can also reduce O_2 and thus generate ROS [6]. Although NAD^+ is the preferred electron acceptor for XDH, *in vivo* conditions where the concentration of NAD^+ concentration is decreased can result in utilization of O_2 as an electron acceptor. Such conditions include ischemia/hypoxia where O_2-dependent alterations in cellular respiration lead to decreased mitochondrial NADH oxidation and concomitant diminution of NAD^+ concentration [7]. Therefore, exclusively associating XO with oxidant generation while referring to XDH as a housekeeping enzyme is ill-advised; especially under conditions as described above where up-regulation of intracellular XDH, in the absence of post-translational modification to XO, may lead to unwanted amplification in intracellular oxidative stress. This is important as a plethora of reports over the past 35 years have demonstrated involvement of XOR-derived oxidants in the pathology of an array of inflammatory disease processes including heart failure, chronic obstructive pulmonary disease (COPD), pulmonary hypertension, sickle cell disease and diabetes [8–14].

It is unfortunate that XO is mainly referred to as a source of $O_2^{\bullet-}$ with rare mention of XO-catalyzed H_2O_2 production. It is often assumed that H_2O_2 formation results either from reaction of XO-derived $O_2^{\bullet-}$ with superoxide dismutase (SOD) ($2O_2^{\bullet-} \rightarrow H_2O_2$) or spontaneous dismutation of $O_2^{\bullet-}$. This assumption leads to a significant misrepresentation of XO-derived oxidant identity. For example, studies conducted decades ago revealed that 100% O_2 and pH 10 are requisite for 100% $O_2^{\bullet-}$ production from XO; a setting completely incompatible with cell physiology [15]. This same study revealed that under more physiologically-relevant conditions (21% O_2 and pH 7.0) XO generates ~25% $O_2^{\bullet-}$ and

~75% H_2O_2 [15]. Evaluation of the relative proportion of H_2O_2 versus $O_2^{\bullet-}$ produced by XO under various O_2 tensions (1–21%) at normal pH revealed XO-catalyzed H_2O_2 approaches 90–95% of total electron flux though the enzyme under clinically-relevant hypoxic conditions (1–2% O_2) [16]. It was noted that the focal point for transition towards enhanced H_2O_2 formation was ~2% O_2 or ~26 μM saturated O_2; a value similar to the K_m O_2 of ~27 μM at the FAD of XO [15, 17, 18] and this proclivity for XO-catalyzed H_2O_2 formation is further amplified by acidic pH. Therefore, under ischemia/hypoxia, where both O_2 tension and pH are reduced, H_2O_2 formation is favored suggesting that XO activity may significantly affect numerous signaling events where H_2O_2 is reported to be contributory [19, 20].

Up-regulation of cellular XDH can result in release to the circulation where it is rapidly (<1 min) converted to XO. While having a net negative charge at physiological pH, XO contains pockets of cationic amino acid motifs on the surface of the protein result in its high affinity (K_d = 6 nM) for negatively charged GAGs comprising the glycocaylx of vascular endothelium [21–25]. Evidence of the presence of GAG-associated XO is provided by both animal models and clinical studies of cardiovascular disease whereby intravenous administration of heparin results elevation of plasma XO activity [21, 26, 27]. This sequestration of proteins by GAGs (1) substantially amplifies local concentration, (2) diminishes rotational and translational mobility and (3) alters kinetic properties. For example, when compared to XO in free in solution, GAG-immobilized XO demonstrates an increased K_m for xanthine (6.5 vs. 21.2 μM) and a fivefold increase K_i for allo/oxypurinol; a consequence with significant implications for pharmacological intervention [24]. Similar alterations in XO kinetics were observed when XO was bound to the milk fat globule membrane where immobilization of the enzyme in this setting induced a twofold increase in the K_m for xanthine while enhancing affinity for NADH by increasing the K_m threefold at the FAD [28]. In addition to affecting kinetic properties at the Mo-co, binding of XO to GAGs confers alterations to the FAD resulting in reduction of $O_2^{\bullet-}$ production and thus elevation of H_2O_2 formation [24]. Combined, XO-GAG interaction results in: (1) diminished affinity for hypoxanthine/xanthine, (2) resistance to inhibition by the pyrazalopyrimidine-based inhibitors allo/oxypurinol and (3) diminished $O_2^{\bullet-}$ production and thus enhanced H_2O_2 generation. This vascular milieu where XO is sequestered on the surface of the endothelium is primed for prolonged enhancement of oxidant formation that is partially resistant to inhibition by the most commonly prescribed clinical agents.

4.2 Shortcomings Associated with Allo/Oxypurinol Inhibition of XOR

The overwhelming majority of studies addressing XOR-related enzymology and pathobiology have utilized the pyrazaloprymidine-based inhibitor allopurinol. Allopurinol (1*H*-pyrazolo[3,4-d] pyrimidin-4(2H)-one) was approved by the FDA for treatment of gout in 1966 and remains the anchor therapy for hyperuricemia [29]. While inhibition of XO-derived uric acid formation to address symptoms of gout has been accomplished successfully for over 50 years by clinical administration of allopurinol, only partial reduction of symptoms and restoration of function has been observed when applied to address vascular inflammatory disease. This phenomenon may be explained by examination of allopurinol reaction with the Mo-cofactor of XO. Allopurinol is a classic "suicide inhibitor" as its binding to and reduction of the Mo-cofactor induces self-oxidation to form the active, tight-binding competitive inhibitor, oxypurinol (1,2-dihydropyrazolo[4,3-e]pyrimidine-4,6-dione) [30], Fig. 4.2. Reduction of the Mo-cofactor by allopurinol ultimately leads to electron transfer to the FAD resulting in reduction of O_2 [31, 32]. It is equally important to note that oxypurinol binding and resultant inhibition requires the Mo-cofactor to be reduced [31, 32]. This is accomplished by initial reaction of allopurinol or, in the case of treatment with pure oxypurinol, XO substrates such as xanthine must

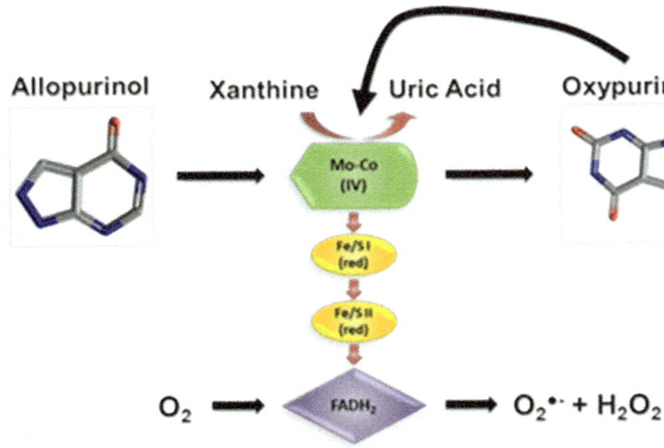

Fig. 4.2 Inhibition of XOR by Allopurinol. Allopurinol is oxidized at the Mo-co to oxypurinol. Oxypurinol binds tightly only to the reduced Mo-co (IV) and competitively prevents hypoxanthine oxidation to xanthine and ultimately oxidation of xanthine to uric acid. However, this process induces enzyme turn-over and unwanted ROS generation

provide these initial electrons. In either case, both allopurinol and/or oxypurinol require enzyme turn-over resulting in ROS formation before inhibition is attained. This undesirable action of allo/oxypurinol combined with the reduced capacity to inhibit endothelial GAG-associated XO and propensity to affect alternative purine catabolic pathways [33] may lead to significant misinterpretation of ROS-driven vascular pathology where XO is contributory. These limitations affirm the need for alternative inhibitors with superior specificity, reactivity in the absence of enzyme turn-over and resistance to GAG-immobilization alterations in kinetics. While some progress has been made with the FDA approval of febuxostat (Uloric®), an XOR-specific inhibitor that is more potent that allopurinol [32], there remains much room for alternative approaches to address XOR as a target for a variety of inflammatory disease processes.

4.3 Nitrated Fatty Acids

Nitration of unsaturated fatty acids represents the convergence of lipid-mediated, NO and ROS signaling. Our current understanding suggests that oxidative inflammatory conditions that include the generation of •NO-derived species as well as dietary conditions where low pH and nitrite are present induce the nitration of unsaturated fatty acids [34]. Two fatty acid nitroalkene derivatives, nitro-oleic acid (OA-NO_2; 9- or 10-nitro-9-*cis*-octadecenoic acids) and nitrolinoleic acid (LNO_2; 9-, 10-, 12-, or 13-nitro-octadecadienoic acids) have been reported to assume broad-spectrum anti-inflammatory signaling characteristics. For example, these nitrated fatty acids activate peroxisome proliferator-activated receptor-γ, inhibit inflammatory cytokine secretion from macrophages, activate stress signaling, impede platelet activation, and diminish obesity-associated metabolic and pulmonary dysfunction [35–40]. A key mechanism driving nitroalkene-mediated signaling is post-translational modification of proteins [41]. The alkenyl nitro configuration imparts electrophilic reactivity on the β-carbon adjacent to the nitro-bonded carbon, facilitating Michael addition reactions with nucleophiles such as cysteine and histidine residues of proteins [41]. For example, reaction of OA-NO_2 with the catalytic Cys-149 of glyceraldehyde-3-phosphate dehydrogenase (GAPDH) results in inhibition of enzymatic activity [41]. In this same vein, inhibitors and electrophilic substrates of XOR react at the molybdenum cofactor while long-chain alkene modifications augment the effectiveness of XOR inhibitors [42, 43]. This being the case, it was logical to assume nitrated fatty acids may also exert some degree of their anti-inflammatory influence via inhibition of XOR-derived oxidants.

4.4 OA-NO_2 and LNO_2 Inactivate XOR

Both nitro-oleic acid (OA-NO_2) and nitrolinoleic acid (LNO_2) are effective inhibitors of XO-catalyzed uric acid formation and $O_2^{\bullet-}$ generation with OA-NO_2 demonstrating slightly greater potency [44]. For example, the IC_{50} for OA-NO_2 inhibition of uric acid formation is 0.6 μM compared to 2.4 μM for allopurinol. This same superiority (fourfold) in potency is also realized for $O_2^{\bullet-}$ generation. This action of OA-NO_2 is realized under physiologic pH and is independent of enzyme turnover. In addition, structure-function studies demonstrate inhibitory effects on XOR are dependent specifically on the nitro group assuming the 9 or 10 position of the fatty acid chain. Furthermore, biochemical analysis whereby alternative electron acceptors were employed to identify the cofactor affected by OA-NO_2 reaction and revealed the Mo-co as the responsive domain. For example, oxygen consumption and spectrophotometric studies revealed OA-NO_2-exposed XOR maintains functional Fe/S centers as well as an operative FAD site whereas uric acid and $O_2^{\bullet-}$ generation are inhibited. While the exact site of OA-NO_2 reaction with XOR is yet to be determined, it is assumed that this reaction directly or indirectly affects the Mo-co. In fact, detailed kinetic analysis revealed that the inhibition of XOR by OA-NO_2 was actually an inactivation process which was apparent by incapacity to reactivate the enzyme.

4.5 XOR-Endothelium Interaction Does not Affect Inactivation by OA-NO_2

As mentioned above, XOR bound and immobilized on endothelial GAGs is resistant to inhibition by allo/oxypurinol at concentrations well above those achieved clinically. As such, alternative approaches for XOR inhibition that remain unaffected by XOR-GAG interaction would greatly elevate effectiveness towards reducing XOR-derived ROS formation and thus provide better tools for evaluating contributory roles for XOR in a vast array of disease processes. To this end, it has been observed that the inactivation capacity of OA-NO_2 for XOR is not affected by XOR binding to either heparin-Sepharose beads or bovine aortic endothelial cells (BAEC) [44]. This effect is portrayed in cartoon format in Fig. 4.3. While clinically-achievable concentrations of allo/oxypurinol have been successful in reducing uric acid levels enough to alleviate the symptoms of gout, they: (1) do not diminish XO-derived ROS generation below 50% and (2)

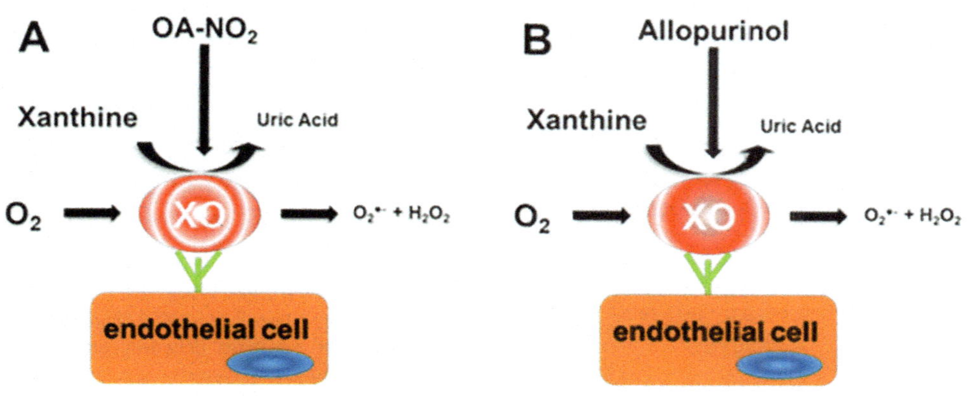

Fig. 4.3 Inhibition of Endothelial-Bound XOR by OA-NO_2 is Superior to Allopurinol. OA-NO_2 demonstrates near complete inhibition of XOR at single digit μM levels (**a**) whereas allopurinol fails to inhibit over 50% at concentrations (400 μM) above those attained clinically (**b**). The size of the font is representative of the amount of product formation while the white-out color of the Mo-co is representative of inhibition

generate unwanted ROS by inducing enzyme turnover during the inhibition process. These shortcomings may have resulted in a significant underestimation of contributory roles for XOR in inflammatory disease processes, especially vascular dysfunction, by failing to demonstrate benefit. It is in the vascular microenvironment that elevation of XOR activity may contribute critically to loss of homeostasis and supported organ dysfunction. It is the identification of novel and potent XOR inhibitors such as OA-NO_2 that will serve to push the field forward by more successfully ablating XOR-derived oxidant generation.

While nitrated fatty acids have demonstrated a variety of pathway-specific effects on inflammatory disease, a commonly overlooked aspect of their benefit has been the capacity of OA-NO_2 and potentially other derivatives to inactivate XOR. Since XOR is a critical source of oxidant generation and elevation in oxidant production is inextricably linked to inflammation, then it is logical to assume that, at least to some degree, salutary outcomes associated with treatment with nitrated fatty acid derivatives may be potentiated by XOR inactivation. As the field progresses, it will be interesting to see if this is the case.

References

1. Enroth C, Eger BT, Okamoto K, Nishino T, Nishino T, Pai EF (2000) Crystal structures of bovine milk xanthine dehydrogenase and xanthine oxidase: structure-based mechanism of conversion. Proc Natl Acad Sci USA 97:10723–10728
2. Nishino T, Okamoto K (2000) The role of the [2Fe-2s] cluster centers in xanthine oxidoreductase. J Inorg Biochem 82:43–49
3. Iwasaki T, Okamoto K, Nishino T, Mizushima J, Hori H (2000) Sequence motif-specific assignment of two [2Fe-2S] clusters in rat xanthine oxidoreductase studied by site-directed mutagenesis. J Biochem 127:771–778
4. Hunt J, Massey V, Dunham WR, Sands RH (1993) Redox potentials of milk xanthine dehydrogenase. Room temperature measurement of the FAD and 2Fe/2S center potentials. J Biol Chem 268:18685–18691
5. Kuwabara Y, Nishino T, Okamoto K, Matsumura T, Eger BT, Pai EF, Nishino T (2003) Unique amino acids cluster for switching from the dehydrogenase to oxidase form of xanthine oxidoreductase. Proc Natl Acad Sci USA 100:8170–8175
6. Harris CM, Massey V (1997) The Reaction of Reduced Xanthine Dehydrogenase with Molecular Oxygen. Reaction kinetics and measurement of superoxide radical. J Biol Chem 272:8370–8379
7. Zhou L, Stanley WC, Saidel GM, Yu X, Cabrera ME (2005) Regulation of lactate production at the onset of ischaemia is independent of mitochondrial NADH/NAD+: insights from in silico studies. J Physiol 569:925–937
8. Butler R, Morris AD, Belch JJ, Hill A, Struthers AD (2000) Allopurinol normalizes endothelial dysfunction in type 2 diabetics with mild hypertension. Hypertension 35:746–751
9. Desco MC, Asensi M, Marquez R, Martinez-Valls J, Vento M, Pallardo FV, Sastre J, Vina J (2002) Xanthine oxidase is involved in free radical production in type 1 diabetes: protection by allopurinol. Diabetes 51:1118–1124
10. Farquharson CA, Butler R, Hill A, Belch JJ, Struthers AD (2002) Allopurinol improves endothelial dysfunction in chronic heart failure. Circulation 106:221–226
11. Aslan M, Ryan TM, Adler B, Townes TM, Parks DA, Thompson JA, Tousson A, Gladwin MT, Patel RP, Tarpey MM, Batinic-Haberle I, White CR, Freeman BA (2001) Oxygen radical inhibition of nitric oxide-dependent vascular function in sickle cell disease. Proc Natl Acad Sci USA 98:15215–15220
12. Heunks LMA, Vina J, van Herwaarden CLA, Folgering HTM, Gimeno A, Dekhuijzen PNR (1999) Xanthine oxidase is involved in exercise-induced oxidative stress in chronic obstructive pulmonary disease. Am J Physiol Regul Integr Comp Physiol 277:R1697–R1704
13. Ahmed MI, Gladden JD, Litovsky SH, Lloyd SG, Gupta H, Inusah S, Denney T Jr, Powell P, McGiffin DC, Dell'Italia LJ (2010) Increased oxidative stress and cardiomyocyte myofibrillar degeneration in patients with chronic isolated mitral regurgitation and ejection fraction >60%. J Am Coll Cardiol 55:671–679
14. Rajesh M, Mukhopadhyay P, Batkai S, Mukhopadhyay B, Patel V, Hasko G, Szabo C, Mabley JG, Liaudet L, Pacher P (2009) Xanthine oxidase inhibitor allopurinol attenuates the development of diabetic cardiomyopathy. J Cell Mol Med 13:2330–2341
15. Fridovich I (1970) Quantitative aspects of the production of superoxide anion radical by milk xanthine oxidase. J Biol Chem 245:4053–4057
16. Kelley EE, Khoo NK, Hundley NJ, Malik UZ, Freeman BA, Tarpey MM (2010) Hydrogen peroxide is the major oxidant product of xanthine oxidase. Free Radic Biol Med 48:493–498
17. Harris CM, Massey V (1997) The oxidative half-reaction of xanthine dehydrogenase with NAD; reaction kinetics and steady-state mechanism. J Biol Chem 272:28335–28341
18. Olson JS, Ballow DP, Palmer G, Massey V (1974) The reaction of xanthine oxidase with molecular oxygen. J Biol Chem 249:4350–4362

19. Stone JR, Yang S (2006) Hydrogen peroxide: a signaling messenger. Antioxid Redox Signal 8:243–270
20. Veal EA, Day AM, Morgan BA (2007) Hydrogen peroxide sensing and signaling. Mol Cell 26:1–14
21. Houston M, Estevez A, Chumley P, Aslan M, Marklund S, Parks DA, Freeman BA (1999) Binding of xanthine oxidase to vascular endothelium. Kinetic characterization and oxidative impairment of nitric oxide-dependent signaling. J Biol Chem 274:4985–4994
22. Adachi T, Fukushima T, Usami Y, Hirano K (1993) Binding of human xanthine oxidase to sulphated glycosaminoglycans on the endothelial-cell surface. Biochem J 289:523–527
23. Fukushima T, Adachi T, Hirano K (1995) The heparin-binding site of human xanthine oxidase. Biol Pharm Bull 18:156–158
24. Kelley EE, Trostchansky A, Rubbo H, Freeman BA, Radi R, Tarpey MM (2004) Binding of xanthine oxidase to glycosaminoglycans limits inhibition by oxypurinol. J Biol Chem 279:37231–37234
25. Radi R, Rubbo H, Bush K, Freeman BA (1997) Xanthine oxidase binding to glycosaminoglycans: kinetics and superoxide dismutase interactions of immobilized xanthine oxidase-heparin complexes. Arch Biochem Biophys 339:125–135
26. Granell S, Gironella M, Bulbena O, Panes J, Mauri M, Sabater L, Aparisi L, Gelpi E, Closa D (2003) Heparin mobilizes xanthine oxidase and induces lung inflammation in acute pancreatitis. Crit Care Med 31:525–530
27. Landmesser U, Spiekermann S, Dikalov S, Tatge H, Wilke R, Kohler C, Harrison DG, Hornig B, Drexler H (2002) Vascular oxidative stress and endothelial dysfunction in patients with chronic heart failure: role of xanthine-oxidase and extracellular superoxide dismutase. Circulation 106:3073–3078
28. Briley MS, Eisenthal R (1974) Association of xanthine oxidase with the bovine milk-fat-globule membrane. Biochem J 143:149–157
29. Pacher P, Nivorozhkin A, Szabo C (2006) Therapeutic effects of xanthine oxidase inhibitors: renaissance half a century after the discovery of allopurinol. Pharmacol Rev 58:87–114
30. Massey V, Komai H, Palmer G, Elion GB (1970) On the mechanism of inactivation of xanthine oxidase by allopurinol and other Pyrazolo[3,4-d]pyrimidines. J Biol Chem 245:2837–2844
31. Galbusera C, Orth P, Fedida D, Spector T (2006) Superoxide radical production by allopurinol and xanthine oxidase. Biochem Pharmacol 71:1747–1752
32. Malik UZ, Hundley NJ, Romero G, Radi R, Freeman BA, Tarpey MM, Kelley EE (2011) Febuxostat inhibition of endothelial-bound XO: Implications for targeting vascular ROS production. Free Radic Biol Med 51:179–184
33. Takano Y, Hase-Aoki K, Horiuchi H, Zhao L, Kasahara Y, Kondo S, Becker MA (2005) Selectivity of febuxostat, a novel non-purine inhibitor of xanthine oxidase/xanthine dehydrogenase. Life Sci 76:1835–1847
34. Schopfer FJ, Vitturi DA, Jorkasky DK, Freeman BA (2018) Nitro-fatty acids: New drug candidates for chronic inflammatory and fibrotic diseases. Nitric Oxide 79:31–37
35. Kelley EE, Baust J, Bonacci G, Golin-Bisello F, Devlin JE, St Croix CM, Watkins SC, Gor S, Cantu-Medellin N, Weidert ER, Frisbee JC, Gladwin MT, Champion HC, Freeman BA, Khoo NK (2014) Fatty acid nitroalkenes ameliorate glucose intolerance and pulmonary hypertension in high fat diet-induced obesity. Cardiovasc Res 101:352–363
36. Deen AJ, Sihvola V, Harkonen J, Patinen T, Adinolfi S, Levonen AL (2018) Regulation of stress signaling pathways by nitro-fatty acids. Nitric Oxide 78:140–145
37. Schopfer FJ, Baker PR, Giles G, Chumley P, Batthyany C, Crawford J, Patel RP, Hogg N, Branchaud BP, Lancaster JR Jr, Freeman BA (2005) Fatty acid transduction of nitric oxide signaling. Nitrolinoleic acid is a hydrophobically stabilized nitric oxide donor. J Biol Chem 280:19289–19297
38. Schopfer FJ, Lin Y, Baker PR, Cui T, Garcia-Barrio M, Zhang J, Chen K, Chen YE, Freeman BA (2005) Nitrolinoleic acid: an endogenous peroxisome proliferator-activated receptor gamma ligand. Proc Natl Acad Sci USA 102:2340–2345
39. Coles B, Bloodsworth A, Clark SR, Lewis MJ, Cross AR, Freeman BA, O'Donnell VB (2002) Nitrolinoleate inhibits superoxide generation, degranulation, and integrin expression by human neutrophils: novel antiinflammatory properties of nitric oxide-derived reactive species in vascular cells. Circ Res 91:375–381
40. Coles B, Bloodsworth A, Eiserich JP, Coffey MJ, McLoughlin RM, Giddings JC, Lewis MJ, Haslam RJ, Freeman BA, O'Donnell VB (2002) Nitrolinoleate inhibits platelet activation by attenuating calcium mobilization and inducing phosphorylation of vasodilator-stimulated phosphoprotein through elevation of cAMP. J Biol Chem 277:5832–5840
41. Batthyany C, Schopfer FJ, Baker PR, Duran R, Baker LM, Huang Y, Cervenansky C, Branchaud BP, Freeman BA (2006) Reversible post-translational modification of proteins by nitrated fatty acids in vivo. J Biol Chem 281:20450–20463
42. Masuoka N, Kubo I (2004) Characterization of xanthine oxidase inhibition by anacardic acids. Biochim Biophys Acta 1688:245–249
43. Masuoka N, Nihei K, Kubo I (2006) Xanthine oxidase inhibitory activity of alkyl gallates. Mol Nutr Food Res 50:725–731
44. Kelley EE, Batthyany CI, Hundley NJ, Woodcock SR, Bonacci G, Del Rio JM, Schopfer FJ, Lancaster JR Jr, Freeman BA, Tarpey MM (2008) Nitro-oleic acid, a novel and irreversible inhibitor of xanthine oxidoreductase. J Biol Chem 283:36176–36184

5 Unfolded Protein Response: Cause or Consequence of Lipid and Lipoprotein Metabolism Disturbances?

Bruno Araújo Serra Pinto, Lucas Martins França, Francisco Rafael Martins Laurindo, and Antonio Marcus de Andrade Paes

Abstract

The liver plays a capital role in the control of whole body energy homeostasis through the metabolization of dietary carbohydrates and lipids. However, under excess macronutrient uptake, those pathways overcharge nucleus-to-endoplasmic reticulum (ER) traffic pathways, leading to luminal overload of unfolded proteins which activates a series of adaptive signaling pathways known as unfolded protein response (UPR). The UPR is a central network mechanism for cellular stress adaptation, however far from a global nonspecific all-or-nothing response. Such a complex signaling network is able to display considerable specificity of responses, with activation of specific signaling branches trimmed for distinct types of stimuli. This makes the UPR a fundamental mechanism underlying metabolic processes and diseases, especially those related to lipid and carbohydrate metabolism. Thus, for a better understanding of the role of UPR on the physiopathology of lipid metabolism disorders, the concepts discussed along this chapter will demonstrate how several metabolic derangements activate UPR components and, in turn, how UPR triggers several metabolic adaptations through its component signaling proteins. This dual role of UPR on lipid metabolism will certainly foment the pursuit of an answer for the question: is UPR cause or consequence of lipid and lipoprotein metabolism disturbances?

Keywords

Unfolded protein response · ER stress · PERK · IRE-1 α · ATF-6

5.1 Introduction

In mammals, the liver plays a capital role in the control of whole body energy homeostasis through the metabolization of excess dietary carbohydrates and lipids to glycogen or fatty acids. Lipid-derived fatty acids are promptly esterified to triacylglycerols (TAG), meanwhile conversion of carbohydrates into TAG demands longer and distinct processes. Glucose is broken down to generate acetyl CoA, which is subsequently chained to build up fatty acids, whereas fructose is directly converted to fatty acids.

B. A. S. Pinto · L. M. França · A. M. d. A. Paes (✉)
Laboratory of Experimental Physiology, Department of Physiological Sciences, Federal University of Maranhão, São Luís, Brazil
e-mail: marcuspaes@ufma.br

F. R. M. Laurindo
Laboratory of Vascular Biology, Heart Institute of the School of Medicine, University of São Paulo, São Paulo, Brazil

A. Trostchansky, H. Rubbo (eds.), *Bioactive Lipids in Health and Disease*, Advances in Experimental Medicine and Biology 1127, https://doi.org/10.1007/978-3-030-11488-6_5

Through such distinct routes, carbohydrates and lipids are ultimately converted to TAG and secreted via very-low density lipoproteins (VLDL) for long-term energy storage in the white adipose tissue (WAT), leading to adipocyte hypertrophy [95]. Enzymes involved in glycolytic and lipogenic pathways are dynamically regulated at both transcriptional and posttranslational levels by various factors such as substrate concentrations and hormones. However, under excess macronutrient uptake, those pathways overcharge nucleus-to-endoplasmic reticulum (ER) traffic pathways, leading to disturbance of ER homeostasis [55].

Disruption of ER homeostasis results in transient luminal overload of un/misfolded proteins which elicits a series of adaptive signaling pathways, collectively known as the unfolded protein response (UPR), towards reestablishment of normal ER function. The UPR is critically required for quality control processes in the ER for secretory proteins in exocrine and endocrine tissues, for instance the hepatic synthesis, assembly and secretion of lipoproteins [105]. Primarily, both glycolytic and lipogenic pathways are regulated by insulin-mediated activation of two well-characterized lipogenic transcription factors, sterol regulatory element binding protein-1c (SREBP-1c) and carbohydrate response element-binding protein (ChREBP) [52, 119]. These studies have followed from the inaugural report that the X-box binding protein 1 (XBP1), a key regulator of the UPR, was required for the normal fatty acid synthesis in the liver [58]. However, this issue remains controversial, as the role of XPB1 and other UPR transducers on lipid metabolism has been settled down [38], and some other findings have contrariwise proposed anti-lipogenic roles for them [35].

This chapter will concisely describe the classical functions of the UPR in maintaining ER protein homeostasis to afterwards discuss its importance for hepatic lipid accumulation, dyslipidemias and other metabolic diseases. Notwithstanding, previous reports have also proposed that hepatic lipid accumulation [106], as well as aberrant lipid compositions of the ER membrane [26] are indeed major triggers of UPR. Thus, we will discuss insights on this two-way road in pursuit of an answer for the question: *is UPR cause or consequence of lipid and lipoprotein metabolism disturbances?*

5.2 The Unfolded Protein Response

The ER is an eukaryote-exclusive organelle formed from nucleus membrane invaginations, which originates sheet-like cisternae and a polygonal array of tubules connecting the perinuclear space to the other cytosolic organelles. It is morphologically divided into two functionally distinct structures, rough- and smooth-ER, defined according to the respective presence or absence of ribosomes anchored to the membrane cytosolic surface. Smooth-ER functions are mostly related to lipid and steroid synthesis, membrane biogenesis, and calcium storage, whereas rough-ER is the primary site for protein folding [7, 92]. Protein folding quality control is rigorously maintained by the surveillance of transmembrane and luminal chaperones, which promote nascent protein folding, mature protein trafficking and secretion, while in parallel driving incorrectly folded proteins to retrograde traffic to the cytosol and proteasomal degradation [2].

The chaperoning core is mainly composed by calnexin (CNX) and calreticulin (CRT), which present an extended arm-like domain for association with ERP57, a thiol oxidoreductase belonging to protein disulfide isomerase (PDI) family. Nascent proteins are catched up by CNX/CRT/ERP57 complex to allow the action of foldase and isomerase enzymes toward tertiary structure building. Additionally, the 78 kDa glucose-regulated protein (GRP78 or BiP) also plays a role on modulation of nascent protein influx, sealing of inactive translocation channels, protein folding, oligomerization and disaggregation. Should the chaperoning process fail, un/misfolded proteins undergo ER-associated degradation (ERAD), which includes ubiquitin labeling, ER-to-cytosol translocation and proteasomal lysis. All these processes collectively support the maintainance of ER protein processing homeostasis, often designated as ER proteostasis [2, 34].

However, despite such a rigorous quality control, ER homeostasis is challenged in a number of conditions such as hypoxia, ischemia, sleep deprivation, prolonged fasting, excess carbohydrate and lipid cellular uptake, abnormal intracellular calcium fluctuations, as well as oxidative stress. These conditions compromise the quality of protein folding, trafficking and secretion, leading to accumulation of un/misfolded proteins inside ER lumen, a condition designated by the general term *ER stress*. To reestablish ER homeostasis, virtually all cell types activate UPR-driven adaptive signaling pathways, which aim to: (1) attenuate global protein synthesis to prevent ER overload and rescue the quality of protein folding; (2) upregulate the synthesis of chaperones and antioxidant proteins; (3) upregulate the synthesis of ERAD constituents; (4) degrade aberrant mRNAs; and (5) activate autophagy pathways [105].

UPR adaptive pathways are triggered by a set of three ER transmembrane proteins, namely: inositol-requiring enzyme-1 (IRE-1), protein kinase RNA-like ER kinase (PERK) and activating transcription factor 6 (ATF6) (Fig. 5.1). A common feature of these sensors is the presence of a luminal domain for unfolded protein detection and a cytosolic domain for transcriptional and translational signal transmission [87]. Under homeostasis, GRP78 associates to all the three sensor proteins and covers their luminal sensing domain. However, upon ER homeostasis disruption, GRP78 dissociates to assist protein folding, allowing the sensor proteins to undergo autophosphorylation, di- and oligomerization to activate their respective signaling pathways [6, 81].

IRE-1 is a type I ER transmembrane protein found in two isoforms. IRE-1α is ubiquitously expressed from yeast to humans and essential for embryonic development, whereas IRE-1β is specifically expressed in intestinal epithelial cells of mammals [39, 100]. Both isoforms possess Ser/Thr kinase and unique endoribonuclease activities, which are responsible for catalyzing an unconventional splicing of a 26-nucleotide intron of unspliced *Xbp1* mRNA (*Xpb1u*) to generate the spliced form (*Xbp1s*), the translation of which results in a potent transcription factor considered a master regulator of ER capacity [21, 33]. Into the nucleus, XBP1s upregulates expression of genes encoding chaperones, ERAD components, and phospholipid synthesis machinery, which are required for ER membrane expansion during ER stress [56]. The function of XBP1u protein translated from the *Xbp1u* mRNA is poorly known, but it has been shown to heterodimerize with XBP1s protein to suppress its function under certain circumstances [113]. Of note, the function of XBP1 in hepatic lipogenesis (see next section) is unrelated to its function in the UPR, but nevertheless requires splicing by IRE1α [58]. Independently from XBP1, IRE-1α also degrades certain mRNAs through regulated IRE-1-dependent decay (RIDD) [40].

PERK is also a type I ER transmembrane protein structurally and functionally related to IRE-1α, whose activation leads to phosphorylation of Ser_{51} residue on the α-subunit of eukaryotic initiation factor 2 (eIF2α) and subsequent translation of activating transcription factor 4 (ATF4) [29]. Phosphorylation of eIF2α by PERK inhibits the assembly of the 80S ribosome and results in a general inhibition of protein synthesis, whereas ATF4 promotes the expression of pro-survival genes related to protein folding (mostly chaperones), redox balance, aminoacid metabolism and autophagy [1]. In parallel, PERK also phosphorylates and activates the nuclear factor erythroid 2-related factor 2 (NRF2), inducing its dissociation from Kelch-like ECH-associated protein 1 (KEAP1) [14] and migration to the nucleus, where it binds to antioxidant responsive element (ARE) promoter sequences to regulate the transcription of antioxidant and phase II detoxifying genes [70]. Interestingly, a recent study showed a well-conserved mechanism by which local reactive oxygen species generated at the ER rapidly oxidizes a single Cys residue within the IRE-1α kinase active site, inhibiting the IRE-1α/XBP1 axis and directing IRE-1α to play a different role in which it activates a p38/NRF2 branch, thereby promoting oxidative stress resistance [41].

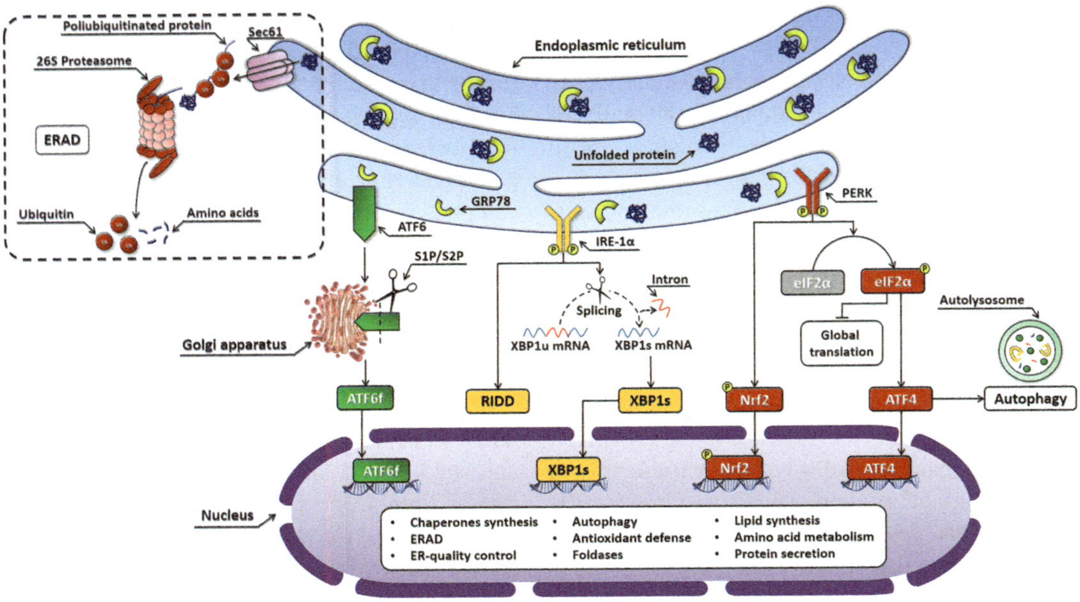

Fig. 5.1 The UPR signalling pathways. The impairment of protein folding into endoplasmic reticulum (ER) lumen promotes the accumulation of unfolded proteins, a condition designated as ER stress. As an attempt to reestablish ER homeostasis, a cascade of adaptive processes is concurrently activated. The first line of defense is the correct chaperoning of aberrant proteins by several transmembrane and luminal chaperones, especially the 78 kDa glucose-regulated protein (GRP78 or BiP). Unsolved proteins are forwarded to ER-associated degradation (ERAD), a mechanism by which the unfolded proteins are retro-translocated to cytosol (Sec61 channel), poliubiquitinated and lysed by 26S proteasomes. In parallel, the unfolded protein response (UPR) is triggered into ER lumen with the activation of three Fig. 5.1 (continued) trans membrane effectors: inositol-requiring enzyme-1 (IRE-1), protein kinase RNA-like ER kinase (PERK) and activating transcription factor 6 (ATF6). The activation of these three effectors is initiated after dissociation of GRP78 coupled to the luminal domain of those proteins to assist the protein folding. Upon release from GRP78, ATF6 is trafficked to the Golgi apparatus and cleaved by the site-1 and site-2 proteases (S1P and S2P, respectively), to release a soluble cytosolic fragment (ATF6f) that enters the nucleus to induce the expression of target genes. Likewise, IRE-1 dimerize and autophosphorylate triggering its RNase activity, which processes the splicing of mRNA encoding unspliced X box-binding protein 1 (XBP1u) to produce an active transcription factor, the spliced XBP1 (XBP1s), which migrates to nucleus. In addition, IRE-1 also degrades aberrant mRNAs through regulated IRE1-dependent decay (RIDD). At last, upon activation, PERK phosphorylates the eukaryotic translation initiator factor 2α (eIF2α) which attenuates the global protein translation and encodes the transcription factor ATF4. Independent of eIF2α, PERK may also phosphorylate the nuclear factor erythroid 2-related factor 2 (NRF2). Together, the transcription factors ATF6f, XBP1s, NRF2 and ATF4 upregulates the expression of adaptive genes encoding chaperones, ERAD components, improve the ER-quality control, autophagy, antioxidant defense, foldases, lipid synthesis, amino acid metabolism and protein secretion, aimed to guaranteeing cellular homeostasis

In contrast to IRE-1α and PERK, ATF6 (isoforms α and β) is a 90 kDa type II transmembrane protein encoding a basic leucine zipper (bZIP) transcription factor in its cytosolic domain. Under ER stress, ATF6 interacts with coat protein II (COPII) and undergoes ER-to-Golgi traffic, where it is sequentially cleaved by the site-1 and site-2 proteases (S1P and S2P, respectively) to release the 50 kDa active ATF6 transcription factor (ATF6f). ATF6f is translocated to the nucleus to upregulate the expression of genes encoding ERAD components and *Xbp1* [91]. Studies have demonstrated that ATF6 is more selective than the other two UPR-sensing proteins, being mostly activated at circumstances of decreased N-glycosylation and redox alterations [63].

As aforementioned, the UPR is activated to restore ER homeostasis on acute ER stress.

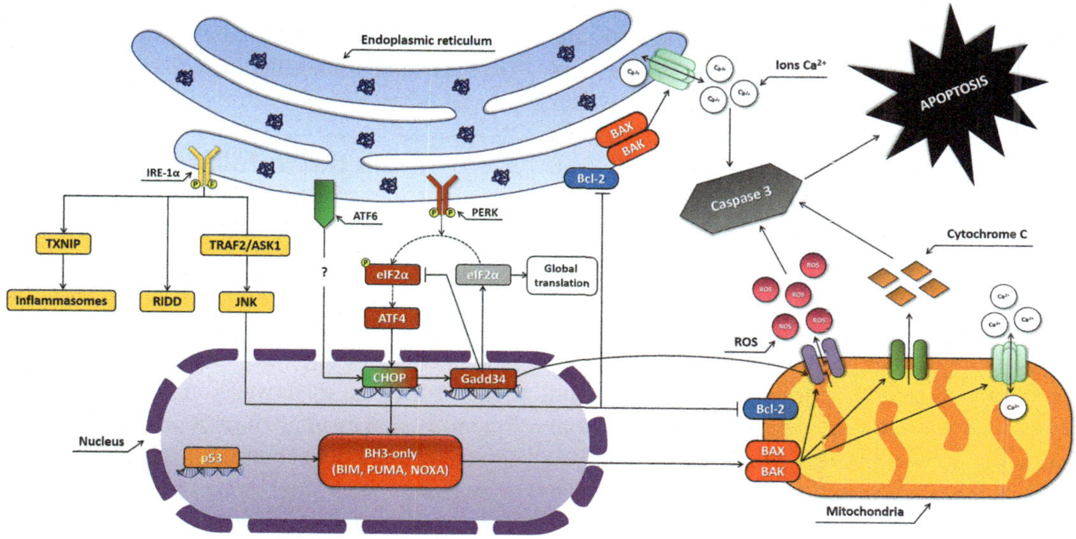

Fig. 5.2 Apoptotic pathways associated to ER stress. Chronic ER stress promotes transition from UPR-adaptive signaling to pro-apoptotic pattern. In this context, IRE-1α recruiting the adaptor protein TNFR-associated factor 2 (TRAF2), which results in the activation of the apoptosis signal regulating kinase 1 (ASK1) and its downstream target c-jun n-terminal kinase (JNK), which suppresses the anti-apoptotic factor BCL-2 both in ER and mitochondria. Furthermore, IRE-1α promotes the degradation of mRNAs encoding for key folding mediators through RIDD, as well as induces the expression of pro-oxidant thioredoxin-interacting protein (TXNIP), which activates inflammasomes. Through ATF4 and ATF6 signaling, the pro-apoptotic transcription factor C/EBP-homologous protein (CHOP) is overexpressed and then promotes BCL-2 suppression and upregulates the BH3-only apoptotic factors (Bcl-2-like protein 11 – BIM; p53 upregulated modulator of apoptosis – PUMA; and NOXA) as well as the Bcl-2-associated X protein (BAX) and Bcl-2 homologous antagonist killer (BAK). Farther, CHOP increase the expression of growth arrest and DNA damage-inducible 34 protein (GADD34), which dephosphorylate eIF2a and favor the mitochondrial reactive oxygen species (ROS) production. In its turn, the tumor protein p53 (p53) also contributes to BH3-only expressions. On mitochondria, this apoptotic signaling leads to an oxidative environment ROS-mediated, allows cytochrome C release and disrupt the calcium homeostasis, which together induce the route of apoptosis Caspase-regulated, which irreversibly conduct the cell to death by apoptosis

However, prolonged UPR activation induces apoptosis, mainly through ATF4- and ATF6-mediated activation of C/EBP homologous protein (CHOP) (Fig. 5.2). CHOP is a member of the C/EBP family of bZIP transcription factors widely recognized as a key marker of ER stress-mediated apoptotic pathway. Downstream effects associated with CHOP include: (1) suppression of the anti-apoptotic factor B-cell lymphoma 2 (BCL-2) [67]; (2) upregulation of apoptotic factors, such as Bcl-2-like protein 11 (BIM) [84], p53 upregulated modulator of apoptosis (PUMA) [9], Bcl-2-associated X protein (BAX) and Bcl-2 homologous antagonist killer (BAK) [122]; (3) upregulation of the tribbles homolog 3 (TRB3), a pseudokinase modulator of signal transduction in apoptotic cascades [73]; and (4) upregulation of the growth arrest and DNA damage-inducible 34 protein (GADD34), a promoter of eIF2α dephosphorylation [64]. Additionally, CHOP also promotes higher mitochondria-derived reactive oxygen species (ROS) generation, resulting in oxidation of critical thiols in the ryanodine-sensitive Ca^{2+} channels and impairment of Ca^{2+} homeostasis [85]. Furthermore, CHOP inhibits GRP78 expression, favoring ER stress recrudescence [121].

Besides CHOP, other pathways are also involved in ER stress-related apoptosis, such as IRE-1α-mediated activation of c-jun n-terminal kinase (JNK), which suppresses BCL-2 and further upregulates BIM expression [59, 112]. Likewise, the tumor protein p53 (p53) increases PUMA and NOXA expression, which are important members of BH3-only apoptotic family [93] (Fig. 5.2). The hyperactivation of IRE-1α still induces the expression of pro-oxidant thioredoxin-interacting protein (TXNIP), which activates inflammasomes besides its Caspase-1-dependent pro-death pathway [11, 60]. The upregulation of all such ER pro-apoptotic factors ultimately converges on mitochondrial disruption and later cell death. The overexpression of BH3-only proteins (BIM, PUMA, and NOXA) incapacitates the mitochondrial protecting proteins (*e.g.*, BCL-2 anti-apoptotic family) and directly activates the pro-apoptotic BAX and BAK. BAX/BAK damage outer mitochondrial membrane integrity and allow cytochrome C release into the cytoplasm, leading to activation of downstream effector Caspases (*e.g.*, Caspase-3), irreversibly driving the cell to apoptosis and/or necrosis [44, 97].

Mechanisms responsible for the transition from adaptive to apoptotic UPR signaling are not yet fully established. Indeed, studies conducted both *in vitro* and *in vivo* have demonstrated that adapting factors, as well as ERAD components are concomitantly expressed with cell death-related UPR components [36, 82]. IRE-1α and PERK seem to play an important role in this transition. During the adaptive response, both IRE-1α and PERK branches are upregulated. However, upon persistent ER stress, IRE-1α expression is attenuated in response to negative modulators, whereas PERK and its associated pro-apoptotic factors (CHOP and GADD34) are kept upregulated [62, 82, 108].

Besides the well-established roles of UPR and ER stress in cell survival, evidence indicates their pivotal function in the regulation of cell physiology and metabolism, as well as initiation and progression of several diseases in humans [86]. UPR components are engaged in modulation of several physiological processes, such as cell differentiation [28, 45], secretory activity [30, 42, 57], innate immunity [66], cognition and neurogenesis [13, 32], as well as glucose [104] and lipid metabolism [107]. Moreover, as demonstrated in many studies, failure of adaptive UPR signaling and activation of pro-apoptotic ER stress pathways, are directly related to developmental diseases, including genetic disorders [43, 71], cancer [101], metabolic dysfunctions [31, 75, 88] and neurodegenerative diseases [37].

5.3 From the Unfolded Protein Response to Lipid Metabolism Disturbances

The ER constitutes the core of protein and lipid synthesis, membrane biogenesis and cellular calcium storage, besides posttranslational modifications such as glycosylation, hydroxylation, lipidation, and disulfide formation, thus playing a capital role in the control of membrane lipid composition and lipid homeostasis. Metabolic perturbations such as advanced glycation of proteins and lipids, S-nitrosylation, oxidative and carbonyl stress, mostly related to excess dietary carbohydrates and lipid uptake, lead to UPR activation as a mean of restoring ER homeostasis [27, 89]. Interestingly, ER stress-induced hepatic steatosis is exacerbated in animals with liver-specific knockout of ER chaperones such as GRP78 [88]. On the other hand, overexpression of GRP78 was found to inhibit *de novo* lipogenesis by reducing SREBP-1c activation, thereby alleviating hepatic steatosis in diabetic mice [48]. Thus, it is not unexpected that the UPR is directly linked to metabolic disturbances, such as type 2 diabetes, obesity and atherogenic dyslipidemia [38, 105].

IRE-1α/XBP1 is the most well-conserved branch of UPR [39]. The primary evidence of its role on lipogenesis came from studies on *Xbp1Δ* mice bearing an inducible, conditional disruption of the Xbp1 gene in the liver. *Xbp1Δ* mice displayed a decreased rate of plasma TAG accumulation with no lipid retention in the liver. Assessment of lipogenic gene expression levels revealed downregulation of those encoding stearyl coenzyme A (CoA) desaturase 1 (*Scd1*), diacylglycerol acetyltransferase 2 (*Dgat2*), and acetyl CoA carboxylase 2 (*Acc2*), while the expression of ChREBP and SREBP family–regulated genes were unaltered. Accordingly, exposure to high-fructose diet markedly increased transcriptional levels of *Scd1*, *Acc1*, and *Acc2* in wild-type but not *Xbp1Δ* livers [58]. The demonstration that glucose itself, but not insulin, increased XBP1 protein levels in the liver was suggestive of a carbohydrate-driven mechanism for XBP1 roles on lipogenesis [23]. Nevertheless, silencing of *Xbp1* in the liver transiently decreased plasma TG and cholesterol levels in both C57BL/6 and apoE-deficient mice, supporting that ablation of XBP1 can efficiently reduce even very elevated plasma lipid levels [94].

IRE-1α plays an essential role in maintaining ER homeostasis through initiating unconventional splicing of *Xbp1 [33]*. Studies on a hepatocyte-specific *Ire1α*-null mouse model showed that IRE1α is essential to minimize lipid accumulation in response to acute ER stress, whereas under normal physiological conditions *Ire-1α* deletion only resulted in modest increase of cellular TAG levels and mild steatosis [118]. Importantly, *Ire-1α* deletion also promoted early activation of PERK/eIF2α branch, leading to selective upregulation of ER stress-inducible pro-apoptotic factors ATF4, CHOP, and ATF3. These data suggest a limited ability of *Ire-1α*-null hepatocytes to adapt to prolonged ER stress [118]. Afterwards, it was demonstrated that the same hepatocyte-specific *Ire-1α*-null mice, once challenged by a 12-week exposure to high-fructose diet, displayed substantial hepatic steatosis, although no differences were detected in transcriptional levels of *Chrebp*, *Srebp-1c*, as well as fatty acid oxidation key genes [107]. The absence of lipogenic alterations led the authors to investigate the role of IRE-1α/XBP1 axis on TAG-rich VLDL particle assembly and secretion, unveiling a specific regulatory mechanism by which IRE-1α/XBP1 modulate PDI translational levels and increase the activity of PDI heterodimer microsomal TAG-transfer protein (MTP), a cofactor absolutely required for VLDL biogenesis [107]. The relevance of this mechanism for hepatic steatosis and hypertriglyceridemia was subsequently demonstrated in monosodium L-glutamate-induced obese rats, a non-dietary but rather neuroendocrine obesity animal model [20] (Fig. 5.3).

On the other hand, early findings had demonstrated that XBP1s translocation to the nucleus is regulated by p85, a regulatory subunit of phosphatidyl inositol 3-kinase (PI3K), allowing efficient chaperone response during metabolic overload through insulin receptor signaling in the liver. Such mechanism was found to be defective in obese and insulin-resistant *ob/ob* mice, advocating against a lipogenic role for XBP1s on the onset and progression of hepatic steatosis [77]. To address this discrepancy, Herrema et al. [35] investigated the role of XBP1s in the development of non-alcoholic fatty liver disease (NAFLD). Overexpression of XBP1s induced by tail vein injection of adXBP1s led to downregulation of key lipogenic genes and decreased TAG content in the liver of both high-fat diet fed C57Bl/6J and *ob/ob* mice, whereas circulating TAG levels were significantly increased. The anti-lipogenic activity of XBP1s was independent of its transcriptional activity, leading the authors to suggest that XBP1s exerts anti-lipogenic effects through a protein-protein interaction [35]. This assumption is corroborated by the supportive role of IRE-1α/XBP1 axis on MTP-mediated VLDL particle assembly and secretion [107].

PERK/eIF2α axis has also been shown to regulate lipogenesis and hepatic steatosis. PERK and eIF2α phosphorylation induced by antipsychotic drugs led to increased lipid accumulation in hepatocytes through activation of SREBP-1c and SREBP-2 [53]. Impairment of eIF2α phosphorylation by hepatic overexpression of GADD34 reduced high-fat diet-induced hepatic steatosis [74]. Corroborating evidence came from studies with *Atf4*$^{-/-}$ mice fed a high-carbohydrate diet, which showed marked decrease of hepatic SCD1 expression and consequent lower TAG accumulation in comparison to wild-type mice [61]. Similarly, hepatic lipid accumulation induced by high-fructose diet was also attenuated in *Atf4*$^{-/-}$ mice due to downregulation of SREBP-1c, ACC and fatty acid synthase (FAS) [109]. As the main linker between PERK/eIF2α axis and apoptosis, CHOP also seems to be involved in lipid metabolism [88], although the exact molecular and cellular mechanisms remain to be elucidated [27]. Recently, we demonstrated that 60-day exposure of Swiss mice to high-

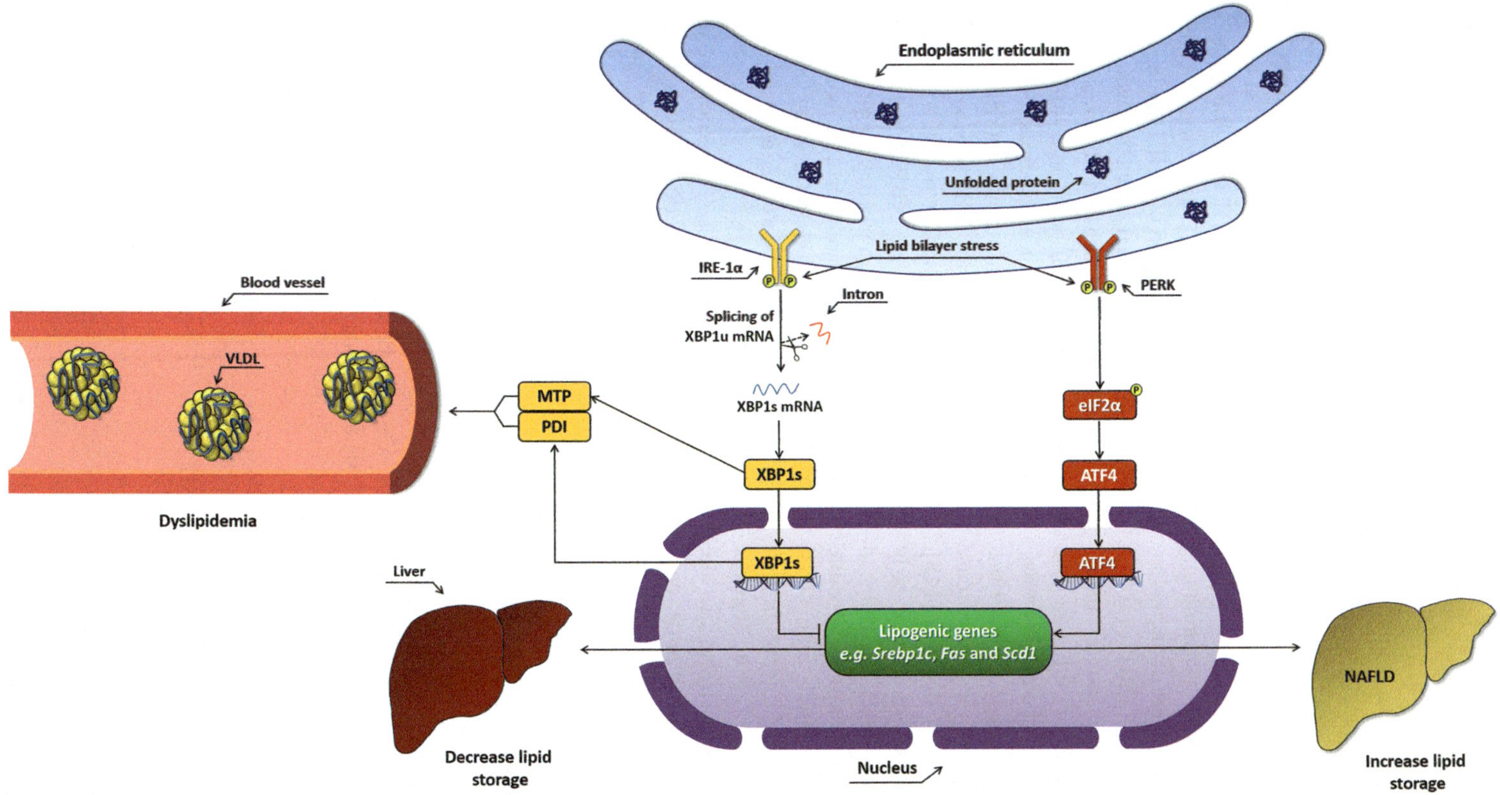

Fig. 5.3 UPR sensors and hepatic lipid metabolism. The accumulation of unfolded proteins or lipid bilayer stress in the endoplasmic reticulum (ER) induce activation of IRE-1α and PERK. Each sensing pathway play different role on lipid metabolism. IRE-1α /XBP1s pathway decreases the hepatic lipid storage both by inhibiting lipogenic genes expression and by activating microsomal triacylglycerol-transfer protein (MTP) and increasing the expression of protein disulfide isomerase (PDI), which result in increased VLDL secretion and dyslipidemia. On the other hand, PERK/eIF2α pathway elevates hepatic lipid storage through activation of transcription factor 4 (ATF4), which induces the expression of lipogenic genes, contributing to the development of nonalcoholic fatty liver disease (NAFLD)

sucrose diet induced switch from an ER-driven adaptive pattern to an apoptotic pattern mediated by translationally upregulated levels of BAK instead of CHOP [19].

Despite the early report that cleaved ATF6 translocates into the nucleus to form a suppressive complex with SREBP2 and histone deacetylase 1 (HDAC1) to downregulate *de novo* lipogenesis gene expression upon glucose deprivation [115], ATF6 remains as the least investigated UPR branch regarding the relationship between hepatic ER stress and lipogenesis. *Atf6α*-deleted mice presented persistent hepatic dysfunction and steatosis in response to pharmacological ER stress, effects at least partially related to the failure of ATF6α-mediated induction of genes encoding protein chaperone, trafficking and ERAD functions [110]. Moreover, high-fat diet feeding of *Atf6α*$^{-/-}$ mice induced greater hepatic steatosis and glucose intolerance in association with upregulation of SREBP-1c [102]. Such effects may be, at least partially, related to the previously described capacity of ATF6 to augment the acute hepatic inflammation mediated by cAMP-responsive element-binding protein H (CREBH), which was posteriorly characterized as a key metabolic regulator of hepatic lipogenesis, fatty acid oxidation, and lipolysis under metabolic stress [116, 120].

A major consequence of ER stress contributing to metabolic diseases is the occurrence of inflammatory response, which may be activated by all UPR-sensing proteins [25]. PERK/eIf2α attenuates translation of both IκB kinase and nuclear factor-κB (NF-κB) expression. However, the shorter half-life of IκB results in an increase of NF-κB/IκB ratio and inflammation triggering [98]. ATF6 also activates NF-κB signaling via AKT phosphorylation [111]. IRE1α forms a complex with tumour necrosis factor (TNF) receptor-associated factor 2 (TRAF2) to induce phosphorylation of JNK and upregulation of pro-inflammatory genes through activator protein 1 (AP1) [100]. ATF4 increases the expression of interleukin 6 (IL6) by direct binding to the cytokine promoter [46]. In addition, XBP1s can also directly binds to the same promoter to stimulate the expression of TNF and IL6 [66], driving NAFLD progression toward non-alcoholic steatohepatitis.

5.4 From Lipotoxicity to Unfolded Protein Response

As aforementioned, excess dietary carbohydrates and lipids are ultimately converted to TAG in the liver and secreted via VLDL for long-term energy storage in WAT, leading to adipocyte hypertrophy [95]. However, when there is exaggerated hypertrophy, adipocyte loses its capacity to keep storing more fat and starts to secrete it back into circulation in the form of free fatty acids (FFA). Excess circulating FFA overload non-adipose tissues, including liver, pancreas, muscle and heart, causing cell dysfunction because of a lipid-specific toxicity, so-called lipotoxicity. Lipotoxicity contributes to the development of several metabolic diseases through both classic mechanisms, such as oxidative stress and inflammation, and, as shown more recently, by UPR activation [22, 38, 106]. Lipotoxicity-derived activation of UPR has been demonstrated in experimental models of obesity induced by both dietary and genetic manipulation [75, 106]. ER stress is also observed in the liver of obese patients suffering from steatosis and steatohepatitis [24, 83]. Interestingly, weight loss after gastric bypass surgery decreased the lipid accumulation and UPR markers levels in the liver and other tissues from those individuals [24].

Distinct lipids and their byproducts are involved in lipotoxicity process. However, the literature highlights saturated fatty acids as the main activator of UPR in several tissues, particularly in the liver [27]. For instance, when L02 immortal hepatic cells and HepG2 hepatoma cells were exposure with saturated fatty acids there was less cellular viability with increased PERK phosphorylation and upregulation of downstream genes, such as ATF4 and CHOP. Moreover, knock-down of PERK in those hepatocyte lines reduced palmitate-induced cell death [8]. It is well-established that UPR induction contributes to ROS overproduction [96]. However, ER stress-induced ROS generation can be mediated by saturated fatty acids. Egnatchik et al. [18] showed that palmitate was also able to compromise the capability of ER to maintain Ca^{2+} stores in primary hepatocytes, resulting in stimulation of mitochondrial oxidative metabolism, ROS generation and UPR activation. In addition, increase levels of phosphatidylcholine within ER membrane from *ob/ob* mice hepatocytes inhibited sarco/endoplasmic reticulum calcium ATPase (SERCA) activity, depleting ER Ca^{2+} stores and causing ER stress [4].

In pancreatic β-cells, saturated fatty acids were shown to be more potent UPR inducers than unsaturated fatty acids [15]. Saturated fatty acids-activated UPR in pancreatic β-cells has been characterized by increased splicing of *Xbp1*, as well as increased transcriptional levels of *Atf4* and *Chop* [49, 51, 54]. Palmitate-induced lipotoxicity promotes changes in ER membrane rigidity and fluidity subsequent to alterations of its phospholipid composition [68], a mechanism shown to activate UPR sensors, such as IRE-1α and PERK, via their transmembrane domains [3, 103]. Noteworthy, chemical chaperones, such as 4-phenylbutyric acid (4-PBA) [12] or tauroursodeoxycholic acid (TUDCA) [10], were able to reduce ER stress in pancreatic β-cells exposed to palmitate. High-density lipoprotein (HDL) has also been described as a potential inhibitor of ER stress in pancreatic β-cells. Treatment of these cells with HDL attenuated ER stress-mediated apoptosis induced by thapsigargin, a SERCA inhibitor. Moreover, HDL still prevented palmitate-induced UPR activation and β-cell death through restoration of ER capacity to perform protein folding and trafficking [80].

Lipid metabolism is very important to the heart since, unlike other tissues, adult cardiomyocytes use fatty acids as main energy source [27]. However, TAG accumulation in the heart is regarded as a hallmark of cardiac lipotoxicity that can lead to heart failure [65, 72]. Lipotoxicity has been suggested as a mediator linking ischemia to cardiomyocyte ER stress [5]. This hypothesis was corroborated by Perman et al. [78], who demonstrated that induction of hypoxia and ischemia in HL-1 cardiomyocytes and mice hearts, respectively, lead to ER stress in a lipid-dependent manner mediated by higher VLDL receptor expression. Moreover, palmitate also triggered UPR in AC16 cells, a human cardiomyocyte cell line [76]. In parallel, it has been recently shown that Nox4, an integral UPR component, activates an ATF4-mediated pathway that switches cardiac substrate metabolism from glucose oxidation to fatty acid oxidation as a manner to make the heart capable of resisting pathological remodeling in the face of chronic stress [69]. On its turn, the role of lipotoxicity-induced ER stress in skeletal muscle cells is paradoxical. Palmitate induces UPR and reduces viability of cultured human myotubes [79]. Also, high-fat diet fed mice markedly increased the transcriptional levels of *Grp78*, *Xbp1s*, and *Atf4* in skeletal muscle [16]. In human studies, TUDCA [50] and 4-PBA [117] improved skeletal muscle insulin sensitivity in obese individuals, however TUDCA did not change ER stress markers when compared with placebo [50]. Moreover, skeletal muscle samples from high-fat diet fed humans showed increased lipid content but no UPR marker expression [17].

Despite the variable extent of lipotoxicity-induced UPR in different tissues, alteration of phospholipid composition within ER membrane seems to be a common mechanism in many tissues, such as liver [4] and pancreas [68]. These changes in the integrity and fluidity of the membranes is part of a phenomenon called lipid bilayer stress, which encompasses any increase of lipid saturation degrees, which include increased phosphatidylcholine/ phosphatidylethanolamine ratio or sterol levels [26]. It has been shown that IRE-1α and PERK lacking their luminal sensing domains, and thus unable to be activated by misfolded proteins, still trigger UPR upon lipid bilayer stress [103]. Thus, lipid bilayer stress has been emerged as an UPR activating mechanism independently from misfolded proteins and potentially responsible for lipotoxicity-induced ER stress [38].

5.5 Closing Remarks and Perspectives

The UPR is a central network mechanism for cellular stress adaptation, however far from a global nonspecific all-or-nothing response. Such a complex signaling network is able to display considerable specificity of responses, with activation of specific signaling branches trimmed for distinct types of stimuli. In this context, hijacking of these mechanistic cascades for other physiological functions has been increasingly evident, generating a frontier zone in which the UPR serves in part to adapt cells to stress, while in parallel exerting several functions including in particular metabolic regulation. This makes the UPR a fundamental mechanism underlying metabolic processes and diseases, especially those related to lipid and carbohydrate metabolism. The concepts discussed along this chapter indicate that several metabolic derangements activate UPR components and, in turn, the UPR triggers several metabolic adaptations through its component signaling proteins (Fig. 5.1). Such interplay occurs at distinct levels in each cell type such as liver, pancreatic β-cells, heart, adipose tissue, macrophages and others, and integrate either to the adaptive (Fig. 5.1) or to the pro-apoptotic (Fig. 5.2) branches of the UPR.

Understanding the precise roles of UPR signaling in metabolic disease is likely to yield new diagnostic and therapeutic options. At this time, these developments are in progress and comprise experimental observations on metabolic effects of chemical chaperones or genetic deletion of specific UPR components, as discussed in this chapter. Further progress is expected as novel and more specific tools for pharmacological modulation of the UPR become available, given the effects of agents able to antagonize distinct UPR branches [47, 90] Recently, targeting of IRE-1α with a series of novel specific small molecule inhibitors was shown to prevent atherosclerosis progression [99]. However, as with all other forms of therapy, collateral effects may arise, *e.g.*, diabetes and insulin resistance upon PERK loss-of-function [114]. Given the rapid advances in the understanding of molecular mechanisms and implications of the UPR, one should expect a rapid progress in those developments.

References

1. Ameri K, Harris AL (2008) Activating transcription factor 4. Int J Biochem Cell Biol 40:14–21
2. Araki K, Nagata K (2011) Protein folding and quality control in the ER. Cold Spring Harb Perspect Biol 3:a007526
3. Ariyama H, Kono N, Matsuda S, Inoue T, Arai H (2010) Decrease in membrane phospholipid unsaturation induces unfolded protein response. J Biol Chem 285:22027–22035
4. Arruda AP, Hotamisligil GS (2015) Calcium homeostasis and organelle function in the pathogenesis of obesity and diabetes. Cell Metab 22:381–397
5. Azfer A, Niu J, Rogers LM, Adamski FM, Kolattukudy PE (2006) Activation of endoplasmic reticulum stress response during the development of ischemic heart disease. Am J Physiol Heart Circ Physiol 291:H1411–H1420
6. Bertolotti A, Zhang Y, Hendershot LM, Harding HP, Ron D (2000) Dynamic interaction of BiP and ER stress transducers in the unfolded-protein response. Nat Cell Biol 2:326
7. Bravo R, Parra V, Gatica D, Rodriguez AE, Torrealba N, Paredes F, Wang ZV, Zorzano A, Hill JA, Jaimovich E (2013) Endoplasmic reticulum and the unfolded protein response: dynamics and metabolic integration. Int Rev Cell Mol Biol 301:215
8. Cao J, Dai DL, Yao L, Yu HH, Ning B, Zhang Q, Chen J, Cheng WH, Shen W, Yang ZX (2012) Saturated fatty acid induction of endoplasmic reticulum stress and apoptosis in human liver cells via the PERK/ATF4/CHOP signaling pathway. Mol Cell Biochem 364:115–129
9. Cazanave SC, Elmi NA, Akazawa Y, Bronk SF, Mott JL, Gores GJ (2010) CHOP and AP-1 cooperatively mediate PUMA expression during lipoapoptosis. Am J Physiol Gastrointest Liver Physiol 299:G236–G243
10. Chen YY, Sun LQ, Wang BA, Zou XM, Mu YM, Lu JM (2013) Palmitate induces autophagy in pancreatic beta-cells via endoplasmic reticulum stress and its downstream JNK pathway. Int J Mol Med 32:1401–1406
11. Chen D, Dixon BJ, Doycheva DM, Li B, Zhang Y, Hu Q, He Y, Guo Z, Nowrangi D, Flores J (2018) IRE1α inhibition decreased TXNIP/NLRP3 inflammasome activation through miR-17-5p after neonatal hypoxic–ischemic brain injury in rats. J Neuroinflammation 15:32
12. Choi SE, Lee YJ, Jang HJ, Lee KW, Kim YS, Jun HS, Kang SS, Chun J, Kang Y (2008) A chemical chaperone 4-PBA ameliorates palmitate-induced inhibition of glucose-stimulated insulin secretion (GSIS). Arch Biochem Biophys 475:109–114
13. Costa-Mattioli M, Gobert D, Harding H, Herdy B, Azzi M, Bruno M, Bidinosti M, Mamou CB, Marcinkiewicz E, Yoshida M (2005) Translational control of hippocampal synaptic plasticity and memory by the eIF2α kinase GCN2. Nature 436:1166–1173
14. Cullinan SB, Diehl JA (2006) Coordination of ER and oxidative stress signaling: the PERK/Nrf2 signaling pathway. Int J Biochem Cell Biol 38:317–332
15. Cunha DA, Hekerman P, Ladriere L, Bazarra-Castro A, Ortis F, Wakeham MC, Moore F, Rasschaert J, Cardozo AK, Bellomo E, Overbergh L, Mathieu C,

Lupi R, Hai T, Herchuelz A, Marchetti P, Rutter GA, Eizirik DL, Cnop M (2008) Initiation and execution of lipotoxic ER stress in pancreatic beta-cells. J Cell Sci 121:2308–2318
16. Deldicque L, Cani PD, Philp A, Raymackers JM, Meakin PJ, Ashford ML, Delzenne NM, Francaux M, Baar K (2010) The unfolded protein response is activated in skeletal muscle by high-fat feeding: potential role in the downregulation of protein synthesis. Am J Physiol Endocrinol Metab 299:E695–E705
17. Deldicque L, Van Proeyen K, Francaux M, Hespel P (2011) The unfolded protein response in human skeletal muscle is not involved in the onset of glucose tolerance impairment induced by a fat-rich diet. Eur J Appl Physiol 111:1553–1558
18. Egnatchik RA, Leamy AK, Jacobson DA, Shiota M, Young JD (2014) ER calcium release promotes mitochondrial dysfunction and hepatic cell lipotoxicity in response to palmitate overload. Mol Metab 3:544–553
19. Flister KFT, Serra BA, França LM, Coêlho CFF, Dos Santos PC, Vale CC, Kajihara D, Debbas V, Laurindo FRM, De Andrade Paes AM (2018) Long-term exposure to high-sucrose diet downregulates hepatic endoplasmic reticulum-stress adaptive pathways and potentiates de novo lipogenesis in weaned male mice. J Nutr Biochem 62:155–166
20. Franca LM, Freitas LN, Chagas VT, Coelho CF, Barroso WA, Costa GC, Silva LA, Debbas V, Laurindo FR, Paes AM (2014) Mechanisms underlying hypertriglyceridemia in rats with monosodium L-glutamate-induced obesity: evidence of XBP-1/PDI/MTP axis activation. Biochem Biophys Res Commun 443:725–730
21. Gardner BM, Walter P (2011) Unfolded proteins are Ire1-activating ligands that directly induce the unfolded protein response. Science 333:1891–1894
22. Ghemrawi R, Battaglia-Hsu SF, Arnold C (2018) Endoplasmic reticulum stress in metabolic disorders. Cell 7:63
23. Glimcher LH, Lee AH (2009) From sugar to fat: how the transcription factor XBP1 regulates hepatic lipogenesis. Ann N Y Acad Sci 1173(Suppl 1):E2–E9
24. Gregor MF, Yang L, Fabbrini E, Mohammed BS, Eagon JC, Hotamisligil GS, Klein S (2009) Endoplasmic reticulum stress is reduced in tissues of obese subjects after weight loss. Diabetes 58:693–700
25. Grootjans J, Kaser A, Kaufman RJ, Blumberg RS (2016) The unfolded protein response in immunity and inflammation. Nat Rev Immunol 16:469–484
26. Halbleib K, Pesek K, Covino R, Hofbauer HF, Wunnicke D, Hanelt I, Hummer G, Ernst R (2017) Activation of the unfolded protein response by lipid Bilayer stress. Mol Cell 67:673–684. e678
27. Han J, Kaufman RJ (2016) The role of ER stress in lipid metabolism and lipotoxicity. J Lipid Res 57:1329–1338
28. Han J, Murthy R, Wood B, Song B, Wang S, Sun B, Malhi H, Kaufman R (2013) ER stress signalling through eIF2α and CHOP, but not IRE1α, attenuates adipogenesis in mice. Diabetologia 56:911–924
29. Harding HP, Novoa I, Zhang Y, Zeng H, Wek R, Schapira M, Ron D (2000) Regulated translation initiation controls stress-induced gene expression in mammalian cells. Mol Cell 6:1099–1108
30. Harding HP, Zeng H, Zhang Y, Jungries R, Chung P, Plesken H, Sabatini DD, Ron D (2001) Diabetes mellitus and exocrine pancreatic dysfunction in perk−/− mice reveals a role for translational control in secretory cell survival. Mol Cell 7:1153–1163
31. Hasnain SZ, Borg DJ, Harcourt BE, Tong H, Sheng YH, Ng CP, Das I, Wang R, Chen AC, Loudovaris T (2014) Glycemic control in diabetes is restored by therapeutic manipulation of cytokines that regulate beta cell stress. Nat Med 20:1417
32. Hayashi A, Kasahara T, Iwamoto K, Ishiwata M, Kametani M, Kakiuchi C, Furuichi T, Kato T (2007) The role of brain-derived neurotrophic factor (BDNF)-induced XBP1 splicing during brain development. J Biol Chem 282:34525–34534
33. He Y, Sun S, Sha H, Liu Z, Yang L, Xue Z, Chen H, Qi L (2010) Emerging roles for XBP1, a sUPeR transcription factor. Gene Expr 15:13–25
34. Hebert DN, Molinari M (2007) In and out of the ER: protein folding, quality control, degradation, and related human diseases. Physiol Rev 87:1377–1408
35. Herrema H, Zhou Y, Zhang D, Lee J, Salazar Hernandez MA, Shulman GI, Ozcan U (2016) XBP1s is an anti-lipogenic protein. J Biol Chem 291:17394–17404
36. Hetz C (2012) The unfolded protein response: controlling cell fate decisions under ER stress and beyond. Nat Rev Mol Cell Biol 13:89–102
37. Hetz C, Mollereau B (2014) Disturbance of endoplasmic reticulum proteostasis in neurodegenerative diseases. Nat Rev Neurosci 15:233–249
38. Ho N, Xu C, Thibault G (2018) From the unfolded protein response to metabolic diseases - lipids under the spotlight. J Cell Sci 131
39. Hollien J (2013) Evolution of the unfolded protein response. Biochim Biophys Acta 1833:2458–2463
40. Hollien J, Lin JH, Li H, Stevens N, Walter P, Weissman JS (2009) Regulated Ire1-dependent decay of messenger RNAs in mammalian cells. J Cell Biol 186:323–331
41. Hourihan JM, Moronetti Mazzeo LE, Fernandez-Cardenas LP, Blackwell TK (2016) Cysteine Sulfenylation directs IRE-1 to activate the SKN-1/Nrf2 antioxidant response. Mol Cell 63:553–566
42. Huh WJ, Esen E, Geahlen JH, Bredemeyer AJ, Lee AH, Shi G, Konieczny SF, Glimcher LH, Mills JC (2010) XBP1 controls maturation of gastric zymogenic cells by induction of MIST1 and expansion of the rough endoplasmic reticulum. Gastroenterology 139:2038–2049
43. Imai Y, Soda M, Hatakeyama S, Akagi T, Hashikawa T, Nakayama KI, Takahashi R (2002) CHIP is associated with Parkin, a gene responsible for familial

Parkinson's disease, and enhances its ubiquitin ligase activity. Mol Cell 10:55–67

44. Iurlaro R, Munoz-Pinedo C (2016) Cell death induced by endoplasmic reticulum stress. FEBS J 283:2640–2652
45. Iwakoshi NN, Lee A-H, Vallabhajosyula P, Otipoby KL, Rajewsky K, Glimcher LH (2003) Plasma cell differentiation and the unfolded protein response intersect at the transcription factor XBP-1. Nat Immunol 4:321–329
46. Iwasaki Y, Suganami T, Hachiya R, Shirakawa I, Kim-Saijo M, Tanaka M, Hamaguchi M, Takai-Igarashi T, Nakai M, Miyamoto Y, Ogawa Y (2014) Activating transcription factor 4 links metabolic stress to interleukin-6 expression in macrophages. Diabetes 63:152–161
47. Jung TW, Choi KM (2016) Pharmacological modulators of endoplasmic reticulum stress in metabolic diseases. Int J Mol Sci 17:192
48. Kammoun HL, Chabanon H, Hainault I, Luquet S, Magnan C, Koike T, Ferre P, Foufelle F (2009) GRP78 expression inhibits insulin and ER stress-induced SREBP-1c activation and reduces hepatic steatosis in mice. J Clin Invest 119:1201–1215
49. Karaskov E, Scott C, Zhang L, Teodoro T, Ravazzola M, Volchuk A (2006) Chronic palmitate but not oleate exposure induces endoplasmic reticulum stress, which may contribute to INS-1 pancreatic beta-cell apoptosis. Endocrinology 147:3398–3407
50. Kars M, Yang L, Gregor MF, Mohammed BS, Pietka TA, Finck BN, Patterson BW, Horton JD, Mittendorfer B, Hotamisligil GS, Klein S (2010) Tauroursodeoxycholic acid may improve liver and muscle but not adipose tissue insulin sensitivity in obese men and women. Diabetes 59:1899–1905
51. Kharroubi I, Ladriere L, Cardozo AK, Dogusan Z, Cnop M, Eizirik DL (2004) Free fatty acids and cytokines induce pancreatic beta-cell apoptosis by different mechanisms: role of nuclear factor-kappaB and endoplasmic reticulum stress. Endocrinology 145:5087–5096
52. Kim MS, Krawczyk SA, Doridot L, Fowler AJ, Wang JX, Trauger SA, Noh HL, Kang HJ, Meissen JK, Blatnik M, Kim JK, Lai M, Herman MA (2016) ChREBP regulates fructose-induced glucose production independently of insulin signaling. J Clin Invest 126:4372–4386
53. Lauressergues E, Bert E, Duriez P, Hum D, Majd Z, Staels B, Cussac D (2012) Does endoplasmic reticulum stress participate in APD-induced hepatic metabolic dysregulation? Neuropharmacology 62:784–796
54. Laybutt DR, Preston AM, Akerfeldt MC, Kench JG, Busch AK, Biankin AV, Biden TJ (2007) Endoplasmic reticulum stress contributes to beta cell apoptosis in type 2 diabetes. Diabetologia 50:752–763
55. Lee AH, Glimcher LH (2009) Intersection of the unfolded protein response and hepatic lipid metabolism. Cell Mol Life Sci 66:2835–2850
56. Lee A-H, Iwakoshi NN, Glimcher LH (2003) XBP-1 regulates a subset of endoplasmic reticulum resident chaperone genes in the unfolded protein response. Mol Cell Biol 23:7448–7459
57. Lee AH, Chu GC, Iwakoshi NN, Glimcher LH (2005) XBP-1 is required for biogenesis of cellular secretory machinery of exocrine glands. EMBO J 24:4368–4380
58. Lee AH, Scapa EF, Cohen DE, Glimcher LH (2008) Regulation of hepatic lipogenesis by the transcription factor XBP1. Science 320:1492–1496
59. Lei K, Davis RJ (2003) JNK phosphorylation of Bim-related members of the Bcl2 family induces Bax-dependent apoptosis. Proc Natl Acad Sci 100:2432–2437
60. Lerner AG, Upton J-P, Praveen P, Ghosh R, Nakagawa Y, Igbaria A, Shen S, Nguyen V, Backes BJ, Heiman M (2012) IRE1α induces thioredoxin-interacting protein to activate the NLRP3 inflammasome and promote programmed cell death under irremediable ER stress. Cell Metab 16:250–264
61. Li H, Meng Q, Xiao F, Chen S, Du Y, Yu J, Wang C, Guo F (2011) ATF4 deficiency protects mice from high-carbohydrate-diet-induced liver steatosis. Biochem J 438:283–289
62. Lin JH, Li H, Zhang Y, Ron D, Walter P (2009) Divergent effects of PERK and IRE1 signaling on cell viability. PLoS One 4:e4170
63. Maiuolo J, Bulotta S, Verderio C, Benfante R, Borgese N (2011) Selective activation of the transcription factor ATF6 mediates endoplasmic reticulum proliferation triggered by a membrane protein. Proc Natl Acad Sci 108:7832–7837
64. Marciniak SJ, Yun CY, Oyadomari S, Novoa I, Zhang Y, Jungreis R, Nagata K, Harding HP, Ron D (2004) CHOP induces death by promoting protein synthesis and oxidation in the stressed endoplasmic reticulum. Genes Dev 18:3066–3077
65. Marfella R, Di Filippo C, Portoghese M, Barbieri M, Ferraraccio F, Siniscalchi M, Cacciapuoti F, Rossi F, D'amico M, Paolisso G (2009) Myocardial lipid accumulation in patients with pressure-overloaded heart and metabolic syndrome. J Lipid Res 50:2314–2323
66. Martinon F, Chen X, Lee A-H, Glimcher LH (2010) TLR activation of the transcription factor XBP1 regulates innate immune responses in macrophages. Nat Immunol 11:411–418
67. Mccullough KD, Martindale JL, Klotz LO, Aw TY, Holbrook NJ (2001) Gadd153 sensitizes cells to endoplasmic reticulum stress by down-regulating Bcl2 and perturbing the cellular redox state. Mol Cell Biol 21:1249–1259
68. Moffitt JH, Fielding BA, Evershed R, Berstan R, Currie JM, Clark A (2005) Adverse physicochemical properties of tripalmitin in beta cells lead to morphological changes and lipotoxicity in vitro. Diabetologia 48:1819–1829
69. Nabeebaccus AA, Zoccarato A, Hafstad AD, Santos CX, Aasum E, Brewer AC, Zhang M, Beretta M,

Yin X, West JA, Schroder K, Griffin JL, Eykyn TR, Abel ED, Mayr M, Shah AM (2017) Nox4 reprograms cardiac substrate metabolism via protein O-GlcNAcylation to enhance stress adaptation. JCI Insight 2
70. Nguyen T, Nioi P, Pickett CB (2009) The Nrf2-antioxidant response element signaling pathway and its activation by oxidative stress. J Biol Chem 284:13291–13295
71. Nichols WC, Seligsohn U, Zivelin A, Terry VH, Hertel CE, Wheatley MA, Moussalli MJ, Hauri H-P, Ciavarella N, Kaufman RJ (1998) Mutations in the ER–Golgi intermediate compartment protein ERGIC-53 cause combined deficiency of coagulation factors V and VIII. Cell 93:61–70
72. Nielsen LB, Perko M, Arendrup H, Andersen CB (2002) Microsomal triglyceride transfer protein gene expression and triglyceride accumulation in hypoxic human hearts. Arterioscler Thromb Vasc Biol 22:1489–1494
73. Ohoka N, Yoshii S, Hattori T, Onozaki K, Hayashi H (2005) TRB3, a novel ER stress-inducible gene, is induced via ATF4-CHOP pathway and is involved in cell death. EMBO J 24:1243–1255
74. Oyadomari S, Harding HP, Zhang Y, Oyadomari M, Ron D (2008) Dephosphorylation of translation initiation factor 2alpha enhances glucose tolerance and attenuates hepatosteatosis in mice. Cell Metab 7:520–532
75. Özcan U, Cao Q, Yilmaz E, Lee A-H, Iwakoshi NN, Özdelen E, Tuncman G, Görgün C, Glimcher LH, Hotamisligil GS (2004) Endoplasmic reticulum stress links obesity, insulin action, and type 2 diabetes. Science 306:457–461
76. Palomer X, Capdevila-Busquets E, Botteri G, Salvado L, Barroso E, Davidson MM, Michalik L, Wahli W, Vazquez-Carrera M (2014) PPARbeta/delta attenuates palmitate-induced endoplasmic reticulum stress and induces autophagic markers in human cardiac cells. Int J Cardiol 174:110–118
77. Park SW, Zhou Y, Lee J, Lu A, Sun C, Chung J, Ueki K, Ozcan U (2010) The regulatory subunits of PI3K, p85alpha and p85beta, interact with XBP-1 and increase its nuclear translocation. Nat Med 16:429–437
78. Perman JC, Bostrom P, Lindbom M, Lidberg U, Stahlman M, Hagg D, Lindskog H, Scharin Tang M, Omerovic E, Mattsson Hulten L, Jeppsson A, Petursson P, Herlitz J, Olivecrona G, Strickland DK, Ekroos K, Olofsson SO, Boren J (2011) The VLDL receptor promotes lipotoxicity and increases mortality in mice following an acute myocardial infarction. J Clin Invest 121:2625–2640
79. Peter A, Weigert C, Staiger H, Machicao F, Schick F, Machann J, Stefan N, Thamer C, Haring HU, Schleicher E (2009) Individual stearoyl-coa desaturase 1 expression modulates endoplasmic reticulum stress and inflammation in human myotubes and is associated with skeletal muscle lipid storage and insulin sensitivity in vivo. Diabetes 58:1757–1765
80. Petremand J, Puyal J, Chatton JY, Duprez J, Allagnat F, Frias M, James RW, Waeber G, Jonas JC, Widmann C (2012) HDLs protect pancreatic beta-cells against ER stress by restoring protein folding and trafficking. Diabetes 61:1100–1111
81. Pfaffenbach KT, Lee AS (2011) The critical role of GRP78 in physiologic and pathologic stress. Curr Opin Cell Biol 23:150–156
82. Pinto BA, Melo TM, Flister KF, Franca LM, Kajihara D, Tanaka LY, Laurindo FR, Paes AM (2016) Early and sustained exposure to high-sucrose diet triggers hippocampal ER stress in young rats. Metab Brain Dis 31:917–927
83. Puri P, Mirshahi F, Cheung O, Natarajan R, Maher JW, Kellum JM, Sanyal AJ (2008) Activation and dysregulation of the unfolded protein response in nonalcoholic fatty liver disease. Gastroenterology 134:568–576
84. Puthalakath H, O'reilly LA, Gunn P, Lee L, Kelly PN, Huntington ND, Hughes PD, Michalak EM, Mckimm-Breschkin J, Motoyama N, Gotoh T, Akira S, Bouillet P, Strasser A (2007) ER stress triggers apoptosis by activating BH3-only protein Bim. Cell 129:1337–1349
85. Rozpedek W, Markiewicz L, Alan Diehl J, Pytel D, Majsterek I (2015) Unfolded protein response and PERK Kinase as a new therapeutic target in the pathogenesis of Alzheimer's disease. Curr Med Chem 22:3169–3184
86. Rutkowski DT, Hegde RS (2010) Regulation of basal cellular physiology by the homeostatic unfolded protein response. J Cell Biol 189:783–794
87. Rutkowski DT, Kaufman RJ (2004) A trip to the ER: coping with stress. Trends Cell Biol 14:20–28
88. Rutkowski DT, Wu J, Back SH, Callaghan MU, Ferris SP, Iqbal J, Clark R, Miao H, Hassler JR, Fornek J, Katze MG, Hussain MM, Song B, Swathirajan J, Wang J, Yau GD, Kaufman RJ (2008) UPR pathways combine to prevent hepatic steatosis caused by ER stress-mediated suppression of transcriptional master regulators. Dev Cell 15:829–840
89. Santos CX, Tanaka LY, Wosniak J, Laurindo FR (2009) Mechanisms and implications of reactive oxygen species generation during the unfolded protein response: roles of endoplasmic reticulum oxidoreductases, mitochondrial electron transport, and NADPH oxidase. Antioxid Redox Signal 11:2409–2427
90. Sarvani C, Sireesh D, Ramkumar KM (2017) Unraveling the role of ER stress inhibitors in the context of metabolic diseases. Pharmacol Res 119:412–421
91. Shen J, Prywes R (2005) ER stress signaling by regulated proteolysis of ATF6. Methods 35:382–389
92. Shibata Y, Voeltz GK, Rapoport TA (2006) Rough sheets and smooth tubules. Cell 126:435–439
93. Shibue T, Suzuki S, Okamoto H, Yoshida H, Ohba Y, Takaoka A, Taniguchi T (2006) Differential contribution of Puma and Noxa in dual regula-

tion of p53-mediated apoptotic pathways. EMBO J 25:4952–4962
94. So JS, Hur KY, Tarrio M, Ruda V, Frank-Kamenetsky M, Fitzgerald K, Koteliansky V, Lichtman AH, Iwawaki T, Glimcher LH, Lee AH (2012) Silencing of lipid metabolism genes through IRE1alpha-mediated mRNA decay lowers plasma lipids in mice. Cell Metab 16:487–499
95. Softic S, Gupta MK, Wang GX, Fujisaka S, O'neill BT, Rao TN, Willoughby J, Harbison C, Fitzgerald K, Ilkayeva O, Newgard CB, Cohen DE, Kahn CR (2017) Divergent effects of glucose and fructose on hepatic lipogenesis and insulin signaling. J Clin Invest 127:4059–4074
96. Song B, Scheuner D, Ron D, Pennathur S, Kaufman RJ (2008) Chop deletion reduces oxidative stress, improves beta cell function, and promotes cell survival in multiple mouse models of diabetes. J Clin Invest 118:3378–3389
97. Tait SW, Green DR (2013) Mitochondrial regulation of cell death. Cold Spring Harb Perspect Biol 5:a008706
98. Tam AB, Mercado EL, Hoffmann A, Niwa M (2012) ER stress activates NF-kappaB by integrating functions of basal IKK activity, IRE1 and PERK. PLoS One 7:e45078
99. Tufanli O, Telkoparan Akillilar P, Acosta-Alvear D, Kocaturk B, Onat UI, Hamid SM, Cimen I, Walter P, Weber C, Erbay E (2017) Targeting IRE1 with small molecules counteracts progression of atherosclerosis. Proc Natl Acad Sci U S A 114:E1395–E1404
100. Urano F, Wang X, Bertolotti A, Zhang Y, Chung P, Harding HP, Ron D (2000) Coupling of stress in the ER to activation of JNK protein kinases by transmembrane protein kinase IRE1. Science 287:664–666
101. Urra H, Dufey E, Avril T, Chevet E, Hetz C (2016) Endoplasmic reticulum stress and the hallmarks of cancer. Trends Cancer 2:252–262
102. Usui M, Yamaguchi S, Tanji Y, Tominaga R, Ishigaki Y, Fukumoto M, Katagiri H, Mori K, Oka Y, Ishihara H (2012) Atf6alpha-null mice are glucose intolerant due to pancreatic beta-cell failure on a high-fat diet but partially resistant to diet-induced insulin resistance. Metabolism 61:1118–1128
103. Volmer R, Van Der Ploeg K, Ron D (2013) Membrane lipid saturation activates endoplasmic reticulum unfolded protein response transducers through their transmembrane domains. Proc Natl Acad Sci U S A 110:4628–4633
104. Wagner M, Moore DD (2011) Endoplasmic reticulum stress and glucose homeostasis. Curr Opini Clin Nutr Metab Care 14:367–373
105. Wang M, Kaufman RJ (2016) Protein misfolding in the endoplasmic reticulum as a conduit to human disease. Nature 529:326–335
106. Wang D, Wei Y, Pagliassotti MJ (2006) Saturated fatty acids promote endoplasmic reticulum stress and liver injury in rats with hepatic steatosis. Endocrinology 147:943–951
107. Wang S, Chen Z, Lam V, Han J, Hassler J, Finck BN, Davidson NO, Kaufman RJ (2012) IRE1alpha-XBP1s induces PDI expression to increase MTP activity for hepatic VLDL assembly and lipid homeostasis. Cell Metab 16:473–486
108. Woehlbier U, Hetz C (2011) Modulating stress responses by the UPRosome: a matter of life and death. Trends Biochem Sci 36:329–337
109. Xiao G, Zhang T, Yu S, Lee S, Calabuig-Navarro V, Yamauchi J, Ringquist S, Dong HH (2013) ATF4 protein deficiency protects against high fructose-induced hypertriglyceridemia in mice. J Biol Chem 288:25350–25361
110. Yamamoto K, Takahara K, Oyadomari S, Okada T, Sato T, Harada A, Mori K (2010) Induction of liver steatosis and lipid droplet formation in ATF6alpha-knockout mice burdened with pharmacological endoplasmic reticulum stress. Mol Biol Cell 21:2975–2986
111. Yamazaki H, Hiramatsu N, Hayakawa K, Tagawa Y, Okamura M, Ogata R, Huang T, Nakajima S, Yao J, Paton AW, Paton JC, Kitamura M (2009) Activation of the Akt-NF-kappaB pathway by subtilase cytotoxin through the ATF6 branch of the unfolded protein response. J Immunol 183:1480–1487
112. Yang J, Yao S (2015) JNK-Bcl-2/Bcl-xL-Bax/Bak pathway mediates the crosstalk between matrine-induced autophagy and apoptosis via interplay with beclin 1. Int J Mol Sci 16:25744–25758
113. Yoshida H, Oku M, Suzuki M, Mori K (2006) pXBP1(U) encoded in XBP1 pre-mRNA negatively regulates unfolded protein response activator pXBP1(S) in mammalian ER stress response. J Cell Biol 172:565–575
114. Yu Q, Zhao B, Gui J, Katlinski KV, Brice A, Gao Y, Li C, Kushner JA, Koumenis C, Diehl JA, Fuchs SY (2015) Type I interferons mediate pancreatic toxicities of PERK inhibition. Proc Natl Acad Sci U S A 112:15420–15425
115. Zeng L, Lu M, Mori K, Luo S, Lee AS, Zhu Y, Shyy JY (2004) ATF6 modulates SREBP2-mediated lipogenesis. EMBO J 23:950–958
116. Zhang K, Shen X, Wu J, Sakaki K, Saunders T, Rutkowski DT, Back SH, Kaufman RJ (2006) Endoplasmic reticulum stress activates cleavage of CREBH to induce a systemic inflammatory response. Cell 124:587–599
117. Zhang H, Wang Y, Li J, Yu J, Pu J, Li L, Zhang H, Zhang S, Peng G, Yang F, Liu P (2011a) Proteome of skeletal muscle lipid droplet reveals association with mitochondria and apolipoprotein a-I. J Proteome Res 10:4757–4768
118. Zhang K, Wang S, Malhotra J, Hassler JR, Back SH, Wang G, Chang L, Xu W, Miao H, Leonardi R, Chen YE, Jackowski S, Kaufman RJ (2011b) The unfolded protein response transducer IRE1alpha prevents ER stress-induced hepatic steatosis. EMBO J 30:1357–1375
119. Zhang C, Chen X, Zhu RM, Zhang Y, Yu T, Wang H, Zhao H, Zhao M, Ji YL, Chen YH, Meng XH, Wei

W, Xu DX (2012a) Endoplasmic reticulum stress is involved in hepatic SREBP-1c activation and lipid accumulation in fructose-fed mice. Toxicol Lett 212:229–240
120. Zhang C, Wang G, Zheng Z, Maddipati KR, Zhang X, Dyson G, Williams P, Duncan SA, Kaufman RJ, Zhang K (2012b) Endoplasmic reticulum-tethered transcription factor cAMP responsive element-binding protein, hepatocyte specific, regulates hepatic lipogenesis, fatty acid oxidation, and lipolysis upon metabolic stress in mice. Hepatology 55:1070–1082
121. Zhao Y, Yan Y, Zhao Z, Li S, Yin J (2015) The dynamic changes of endoplasmic reticulum stress pathway markers GRP78 and CHOP in the hippocampus of diabetic mice. Brain Res Bull 111:27–35
122. Zong W-X, Li C, Hatzivassiliou G, Lindsten T, Yu Q-C, Yuan J, Thompson CB (2003) Bax and Bak can localize to the endoplasmic reticulum to initiate apoptosis. J Cell Biol 162:59–69

Part III

Bioactive Lipids in Inflammatory and Cardiovascular Diseases

Arachidonic Acid and Nitroarachidonic: Effects on NADPH Oxidase Activity

6

Lucía Gonzalez-Perilli, Carolina Prolo, and María Noel Álvarez

Abstract

Arachidonic acid (AA) is a polyunsaturated fatty acid that participates in the inflammatory response mainly through bioactive-lipids formation in macrophages and also in the phagocytic NADPH oxidase 2 (NOX2) activation. NOX2 is the enzyme responsible for a huge superoxide formation in macrophages, essential to eliminate pathogens inside the phagosome. The oxidase is an enzymatic complex comprised of a membrane–bound flavocytochrome b_{558} (gp91^{phox}/p22^{phox}), three cytosolic subunits (p47^{phox}, p40^{phox} and p67^{phox}) and a Rac-GTPase. The enzyme becomes active when macrophages are exposed to appropriate stimuli that trigger the phosphorylation of cytosolic subunits and its migration to plasmatic membrane to form the active complex. It is proposed that AA stimulates NOX2 activity through AA interaction with different components of the NADPH oxidase complex. In inflammatory conditions, there is an increase in reactive oxygen and nitrogen species that results in the production of nitrated derivatives of AA, such as nitroarachidonic acid (NO_2-AA). NO_2-AA is capable to inhibit NOX2 activity by interfering with p47^{phox} migration to the membrane without affecting phosphorylation of cytosolic proteins. Also, NO_2-AA is capable to interact with protein disulfide isomerase (PDI), which is involved on NOX2 active complex formation. It has been demonstrated that NO_2-AA forms a covalent adduct with PDI that could prevent the interaction with NOX2 and it would explain the inhibitory effects of the fatty acid upon NOX2. Together, current data indicate that AA is an important activator of NOX2 formed in the early events of the inflammatory response, leading to a massive production of oxidants that may, in turn, promote NO_2-AA formation and shutting down the oxidative burst. Hence, AA and its derivatives could have antagonistic roles on NOX2 activity regulation.

Keywords

Arachidonic acid · NADPH oxidase · Nitro-arachidonic acid · Macrophages · Reactive oxygen species

6.1 Introduction

Macrophages participate in a variety of processes including defense against pathogens, tissue homeostasis and remodeling. To correctly fulfill

L. Gonzalez-Perilli · C. Prolo · M. N. Álvarez (✉)
Departamento de Bioquímica and Center for Free Radical and Biomedical Research, Facultad de Medicina-Universidad de la República, Montevideo, Uruguay
e-mail: noelalv@fmed.edu.uy

A. Trostchansky, H. Rubbo (eds.), *Bioactive Lipids in Health and Disease*, Advances in Experimental Medicine and Biology 1127, https://doi.org/10.1007/978-3-030-11488-6_6

this heterogeneous functional repertoire, macrophages have a notable plasticity and adopt various activation states. Environmental signals induce changes that provide macrophages with enhanced antimicrobial activity or cells more susceptible to infections. In the literature macrophages have been classified in two categories, M1 and M2, where M1 represent classically pro-inflammatory activated macrophages and M2 alternatively-activated macrophages, nevertheless M2 designation includes different types of phenotypes with diverse functions. Precise regulation of macrophage activation is indispensable for the control of infection and preservation of tissue homeostasis [1].

Classically-activated effector macrophages are recruited during cell-mediated immune responses. The combination of interferon gamma (IFN-γ), secreted by innate and adaptive immune cells, interleukin 1 (IL-1) and tumor necrosis factor α (TNF-α), induced by Toll-Like Receptor (TLR) ligand, results in an enhanced microbicidal capacity with high levels of pro-inflammatory cytokines production, activation of arachidonic acid (AA) metabolism, nitric oxide synthase induction and NADPH oxidase priming [2, 3]. The macrophage-pathogen interaction through TLR triggers two central effector responses: cytosolic phospholipase A_2 (PLA_2) activation with AA release from membrane phospholipids and NADPH oxidase 2 (NOX2) assembly and activation with a huge production of superoxide anion inside the phagosome. Both responses generate diverse inflammatory mediators and apparently act in independent paths.

NOX2, the isoform of NADPH oxidase found primarily in professional phagocytic cells such as macrophages and neutrophils [4, 5], reduces molecular oxygen to superoxide ($O_2^{\bullet-}$) at expenses of NADPH oxidation inside phagocytic vacuole. Expression of inducible nitric oxide synthase (iNOS) in immunostimulated macrophages also leads to an overproduction of nitric oxide ($^{\bullet}NO$) that plays an important role as cytotoxic tool against tumor cells and invading pathogens. $^{\bullet}NO$ reacts with $O_2^{\bullet-}$ in a diffusion-controlled reaction to yield the strong oxidant peroxynitrite [6, 7]. Peroxynitrite is a short half-lived species that react with several biological targets including thiols and transition metal centers. An important target for peroxynitrite in biological milieu is carbon dioxide (CO_2) [8]. Its reaction produces nitrosoperoxycarboxylate, which decomposes generating two potent short-lived oxidants that can oxidize and nitrate proteins and nucleic acids, *i.e.* carbonate radical ($CO_3^{\bullet-}$) and nitrogen dioxide ($^{\bullet}NO_2$). Peroxynitrite dependent modification is not limited to proteins or DNA, fatty acid present in membranes are also susceptible to oxidation and nitration. There is evidence that nitro-fatty acids are formed *in vivo* and its concentration increases during inflammatory diseases [9]. Products of fatty acid nitration, such as nitroarachidonic acid (NO_2-AA), have the potential to act as biomarkers and signaling molecules [10–12].

In an inflammatory process, macrophages activation results therefore in the simultaneous formation of superoxide, nitric oxide, peroxynitrite and AA, by activation of different routes. In this chapter we explain the mechanisms of each of these processes as well as the interactions between them. The formation of AA helps and amplifies the activation of NOX2, increasing superoxide and peroxynitrite, leading to the occurrence of oxidized and nitrated products of AA. In particular, NO_2-AA is a potent inhibitor of both the assembly of NADPH oxidase and the induction of nitric oxide synthase [13], modulating inflammatory activation.

6.2 Arachidonic Acid Metabolism in Macrophage Activation

AA is a polyunsaturated fatty acid with four double bonds (20:4, $\Delta^{5,8,11,14}$) which is metabolized by different enzymatic pathways generating a variety of bioactive lipids, with antagonists effects [14]. In macrophages, PLA_2 activation releases AA from plasmatic membrane, which has an important role in the inflammatory response, involving prostaglandins synthesis and interaction with NOX2. Following receptor-mediated activation of phagocytes, phospholipase

Cβ catalyzes the hydrolysis of phosphatidylinositol biphosphate (PIP_2), releasing inositol triphosphate (IP_3) and diacylglycerol (DAG). IP_3 promotes the opening of calcium channels in the endoplasmic reticulum (ER) leading to a transient increase of cytosolic calcium concentration, which -along with DAG- triggers the activation of PKC. The calcium influx also promotes the release of AA from the plasma membrane by the action of the PLA_2 [5], which is a precursor for the generation of pro-inflammatory mediators [15].

In macrophages, free AA can be metabolized for cyclooxygenase (COX) and lipoxygenase (LOX) pathways [14–16]. The COX pathway involves COX-1 and COX-2, along with downstream enzymes that mediate the production of prostaglandins (PGE2, PGD2 and PGF2α), prostacyclin (PGI2), and thromboxanes (TXA2, TXB2). The COX-1, is an enzyme constitutively expressed in all tissues and COX-2 is an inducible isoform present in macrophages and it is up-regulated by inflammatory stimuli [17]. LOX pathway implies LOX-5, LOX-8, LOX-12, and LOX-15 enzymes which produce leukotrienes (LTB4, LTC4, LTD4 and LTE4), lipoxins (LXA4 and LXB4 formed upon LXA4 degradation) and 8–12- 15- hydroperoxyeicosatetraenoic acid (HPETE) [18, 19]. Most of these molecules (prostaglandins, leukotrienes) are pro-inflammatory mediators while lipoxins are involved in resolution of inflammation [14].

AA derivatives modulate the intensity and duration of inflammatory responses by means of pro and anti-inflammatory effects as well as coordinating important cellular responses [20]. The profile of bioactive lipids derived from AA at sites of inflammation is regulated by stimuli present in the immune response (IL-1, TNF-α, LPS, etc.) [17, 21], being specific to the enzymatic content of each cell type. For example, resting macrophages produce TXA2 in excess over PGE2, however, activation with bacterial lipopolysaccharide (LPS) switches the profile to favor PGE2 production [20].

The importance of AA in macrophage activation derives from its metabolism through COX-2, but also from NOX2 modulation through direct interactions with the enzyme, which is discussed further on this chapter [22–24].

6.3 NADPH Oxidase 2 Activation

NOX2, a phagocytic isoform of NADPH oxidase, is responsible for the NADPH-dependent univalent reduction of oxygen to superoxide anion in the phagosome promoting a toxic environment [5]. NOX2 is an enzymatic complex containing 6 subunits: two transmembrane proteins (NOX2 and $p22^{phox}$) that form the heterodimer flavocytochrome b_{558}, three cytosolic regulatory proteins: $p47^{phox}$, $p40^{phox}$ and $p67^{phox}$ and a Rac-GTPase [25]. The flavocytochrome b_{558} complex constitutes the catalytic nucleus of the enzyme allowing the electron transfer across the membrane, from NADPH to oxygen (O_2), through a series of cofactors (FAD and 2 haem groups) [26, 27]. During superoxide formation, NADPH provides two electrons that are translocated across the plasma membrane to reduce O_2 generating two protons in the cytoplasm, what could lead to a drop in cytosolic pH. For that, there are proton extrusion mechanisms to prevent this acidification, being voltage-gated proton channel H_V1 the most important path responsible for the proton extrusion during NOX enzyme catalysis [28, 29].

In resting macrophages, the NOX complex is disassembled and inactive. After activation, the cytosolic components undergo a series of covalent modifications and conformational changes that induce their translocation towards the membrane where they associate with NOX2 and $p22^{phox}$ forming the full complex and producing the radical $O_2^{\bullet-}$ [30] (Fig. 6.1). To prevent the high toxicity of reactive oxygen species under physiological conditions, the NADPH oxidase activity is highly regulated spatially and temporally, and NOX2 activation depends on specific protein interactions of the different subunits to form the enzymatic complex.

NOX2 activation mechanism involves phosphorylation of cytosolic subunits, lipid metabolism activation and nucleotide exchange on RAC protein [31]. This set of events is triggered upon microorganism recognition and phagocytosis,

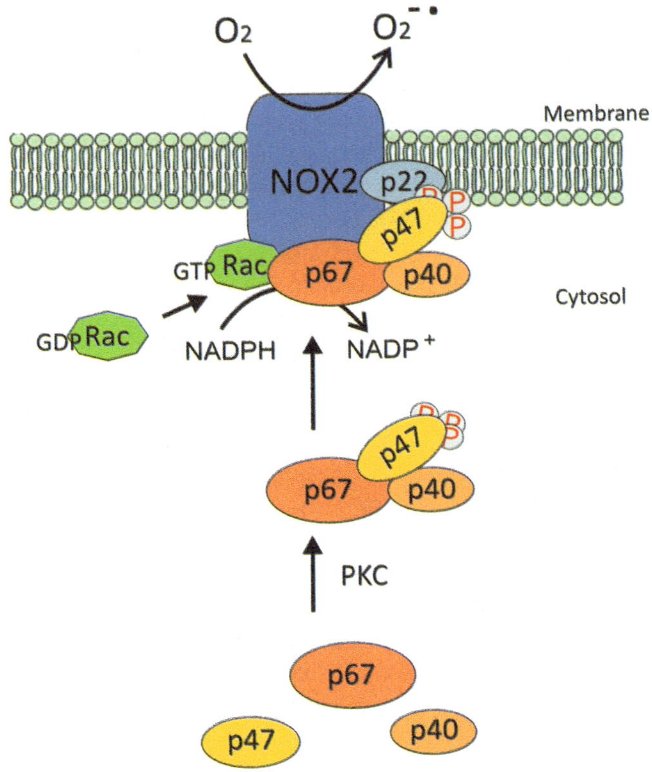

Fig. 6.1 Activation of phagocytic NADPH oxidase. Several agonists stimulate signaling pathways that lead to the activation of PKC, responsible for p47[phox] phosphorylation. Once phosphorylated, the cytosolic subunits interact and migrate to the membrane where they form the active complex with NOX2 and p22[phox]. In addition, activation induces the exchange of GDP by GTP on Rac, which also migrates to the membrane during the process of activation of the enzyme [80]

which leads to the activation of phospholipase Cβ and phosphatidylinositol biphosphate (PIP_2) hydrolysis. The second messengers formed through this mechanism (inositol triphosphate, IP3 and diacylglycerol, DAG) activate protein kinase C (PKC), a key modulator of NOX2 activity, which promotes phosphorylation of cytosolic subunits. Some pharmacological NOX2 activators, such as phorbol myristate (PMA) mimic DAG actions activating PKC.

p47[phox] remains inactive in resting cells, but after phosphorylation events it interacts with other cytosolic subunits, p40[phox] and p67[phox], and migrate to plasmatic membrane (Fig. 6.1). For that to occur, p47[phox] undergoes important conformational changes releasing the blockade of two SHC homology 3 (SH3) domains from the interaction with an *autoinhibitory region* (AIR) [32, 33]. PKC is the main kinase responsible for AIR phosphorylation and disruption of the inhibited conformation [34–36], although several kinases may modify p47[phox] in the same way [37–40] (Figs. 6.1 and 6.2). Once exposed, SH3 domains interact with a *proline rich region* (PRR) present in the membrane-bound subunit p22[phox] initiating full NOX2 complex assembly (Fig. 6.2) [41, 42]. p47[phox] translocation to plasmatic membrane may depend on the interaction of its *phox homology* domain (PX) with a citoesqueleton protein, moesin [43–47]. The same domain mediates p47[phox] anchorage within the lipid bilayer involving the interaction with phosphorylated phosphatidylinositols [48]. The other subunits also migrate to plasmatic membrane and phosphorylation of p40[phox] and p67[phox] seems to be necessary for NOX2 activation [49, 50]. A PRR present in p47[phox] interacts with a SH3 domain of p67[phox] [51], while p40[phox] and p67[phox] interact with each other through PB1 domains (Fig. 6.2) [52, 53]. Importantly, the translocation of p40[phox] and p67[phox] takes place only after p47[phox] activation and migration [54].

The other important event in NOX2 assembly is activation of the GTPase protein Rac. This protein remains bound to GDP and a *Rho dissociation inhibitor* in resting cells. Stimuli for NOX2

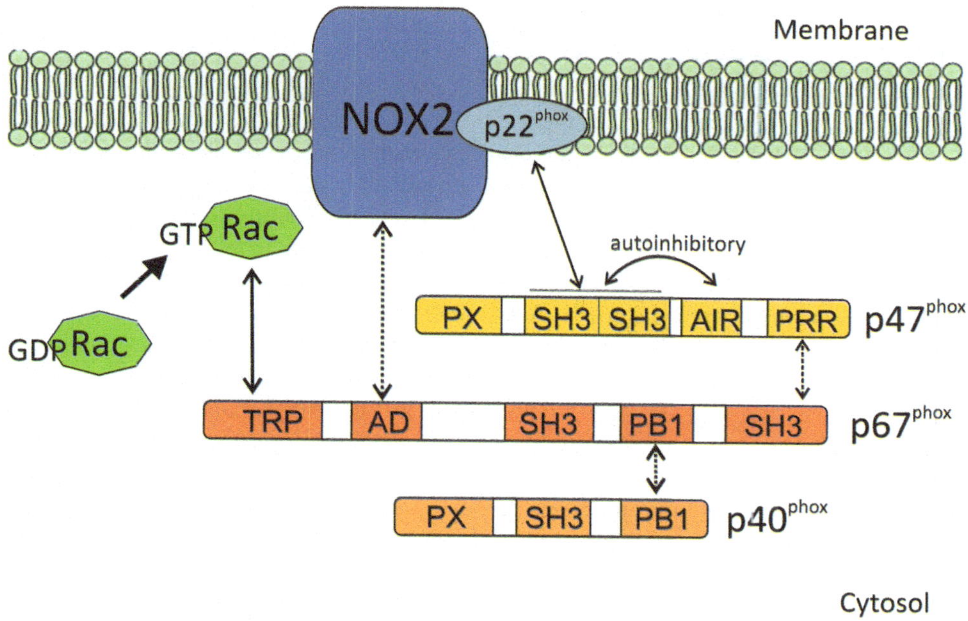

Fig. 6.2 Interaction of the NOX2 subunits. p47^{phox} has one PX domain, two SH3 domains, one AIR domain and one PRR domain. The AIR domain interacts with the SH3 domains of the same p47^{phox} in the resting state (autoinhibition). During activation, a conformational change in p47^{phox} exposes the SH3 domains in p47^{phox} to interact with p22^{phox}. Likewise, the PRR domain of p47^{phox} interacts with the SH3 domain of p67^{phox}. In addition, p67^{phox} forms a complex with p40^{phox} by interaction between PB1 of each protein. Rac-GTP is in contact with the TPR region of p67^{phox}. Finally, when the activation domain (AD) of p67^{phox} is activated, it interacts with NOX2, which is necessary for the regulation of $O_2^{\bullet -}$ production

activation promote GDP exchange on Rac protein contributing to full complex assembling. Rac-GTP (but not Rac-GDP) interacts with a *tetratricopeptide* region (TPR) of p67^{phox} [55, 56] (Fig. 6.2).

The whole process of NOX2 assembly leads to an important event, which is the interaction of the activation domain (AD) on p67^{phox} with the flavin region of flavocytochrome b_{558}, triggering the electron transfer from NADPH to FAD, the rate limiting step of superoxide formation [57, 58].

6.4 Activation of NOX2 by Arachidonic Acid

Among the well described actions of AA in the inflammatory response, it has been shown to be involved in the activation of NOX2 [22–24] in phagocytic cells [59] and in cell-free system [23, 60]. Reports have demonstrated that in the presence of PLA_2 inhibitors, superoxide production is blocked in phagocytic cells, and in the same line, PLA_2-deficient phagocytic cells do not produce superoxide [61–63]. Those observations implicate the requirement for PLA_2 and AA release for the activation of the assembled NOX2 in neutrophils.

The proposed activation mechanism involves AA interaction with different components of the NADPH oxidase complex. It has been demonstrated that a conformational change occurs in flavocytochrome b_{558} upon interaction with AA leading to its activation [64–66]. Besides, there is evidence of an enhanced NADPH oxidase stimulation due to AA-mediated activation of the Hv1 channel of NOX2, modulating the membrane potential and efflux of the H^+ ions generated along with the superoxide by NADPH oxidase [67, 68]. On the other hand, *in vitro* experiments reported an interaction of AA with regulatory proteins [69]. The most studied case is interaction with p47^{phox}. AA can alter the intramolecular bonds in p47^{phox} between the polybasic domain and the SH3 tandem – mimicking the phosphorylation events – and unmasking the SH3 regions that interact with p22^{phox} [70] (Fig. 6.2), a crucial event in activation of the NOX2 [41, 42]. Furthermore, AA elicits GDP-to-GTP exchange on Rac necessary for NOX2 activation [71] and induces the interaction between the p67^{phox}-Rac-GTP complex and NOX2 on the C-terminal cytosolic region, which also is required for superoxide production [71].

These observations suggest that AA acts as a signaling intermediate for the NADPH oxidase activation by functioning at multiple steps in stimulated cells. Actually, AA is widely used as an anionic amphiphile (like SDS or LDS) to activate NADPH oxidase cell free system [72, 73]. Interestingly, it has been shown that trans-AA isomers are unable to activate the NADPH oxidase complex [74]. However, other derivatives of AA may also have different effects on NOX2 activity.

6.5 Inhibition of NOX2 by Nitroarachidonic Acid

In an inflammatory environment, nitrogen-derived species (peroxynitrite and peroxynitrite-derived radicals) [75, 76] could react with unsaturated fatty acids present in biological membranes to form nitro-, nitrite- and nitroepoxy-derivatives, generating a complex mixture of oxidized and nitrated lipid products. A nitrated fatty acid of relevance in inflammation is the nitrated derivative of AA, the nitroarachidonic acid (NO_2-AA) [13].

In recent years, several important biological effects of NO_2-AA have been demonstrated [13, 77–79]. It exerts anti-inflammatory protective actions, such as the negatively regulation of pro-inflammatory cytokines secretion induced by LPS, the decreased expression of iNOS and the irreversible inhibition of COX-1 and -2 in addition to the activation of the Nrf2 pathway, contributing to the physiological shutdown of the inflammatory response in macrophages [11, 13, 79–82].

NO_2-AA has been reported to inhibit NOX2-dependent superoxide production in activated macrophages in a cell line and in primary macrophages in a dose -and time- dependent manner [83]. The inhibition power of NO_2-AA (IC_{50} of 4.1 μM) is similar to that of other common NOX2 inhibitors (Table 6.1). The observed *in vitro* effects were confirmed in an *in vivo* inflammatory model, in which subcutaneous injection of NO_2-AA in mice led to a decrease in NOX2 activity in thioglycolate-elicited macrophages [83]. NO_2-AA treatment did not promote any changes in the phosphorylation levels of the cytosolic subunits; however, it led to a reduction in cytosolic subunits migration towards the membrane, and therefore to an inhibition of the formation of the active complex [83]. The exact mechanism of migration inhibition is not elucidated yet. Since nitroalkenes have a nitro group bound to a double bond, which gives it electrophilic properties [84–86], a possible mechanism of inhibition of NOX2 could involve the formation of a covalent Michael adduct between NO_2-AA and critical nucleophilic residues (*e.g.* histidines or cysteines) of some component involved in the activation of NOX2, preventing then the formation of the active complex.

Recently, some works explore the role of protein disulfide isomerase (PDI) activity on NOX2 active complex formation [98–101]. PDI is a highly conserved redox chaperone, which is found mainly in the endoplasmic reticulum, where it assists protein folding through thiol oxidation and disulfide bonds formation with cysteine residues at the active site of the enzyme [102]. PDI exhibits a positive effect on the activity of NOX2 by controlling the migration of p47^{phox} to the membrane and interacting with its cytosolic components [98]. Given the role that it seems to play in the activation of NOX2 and the presence of critical cysteine residues for enzymatic activity, PDI is a potential target for NO_2-AA, which could explain the inhibitory effects on NOX2 without affecting the components of the enzymatic complex. Both the reductase and chaperone activities of PDI are inhibited by NO_2-AA in

Table 6.1 Most relevant reported inhibitors of NADPH oxidase

Inhibitor	Description	Mechanism	IC_{50}, Ki cell free assay	IC_{50} intact cells	Comments	Refs
Diphenylene-iodonium (DPI)	Iodonium compound	Generate phenyl radical that reacts with flavins	5.6 μM[a]	0.9 μM[b]	Non-specific. Inhibits mitochondrial complex I, xanthine oxidase, NOS and others.	[87, 88]
Nox2 ds-tat	Peptide containing a fragment analogue to B-loop of Nox2	Blocks the interaction between p47*phox* and Nox2	0.74 μM[c]	ND	No inhibition of Nox1 and Nox4	[89, 90]
Nitro-arachidonic acid	Nitrated form of arachidonic acid	Blocks NADPH oxidase assembly	ND	4.1 μM[d]	Reacts with nucleophiles	[83]
VAS2870	Triazolo-pyrimidine	NK	10.6 μM[e]	0.08 μM[f]	Also inhibits Nox1 and Nox4 and exerts off-target thiol alkylation	[91–94]
ML171	2-acetyl-phenotiazine	NK	5 μM[g]	>10 μM[h]	Potent inhibitor of Nox1	[95]
GKT137831	Pyrazolo-pyrimidine	NK	1.7 μM[i]	ND	Inhibits other Nox isoforms (1 and 4)	[96, 97]

[a]K_i, 6.2 μg of neutrophil membrane protein
[b]Rat peritoneal macrophages stimulated with PMA
[c]IC_{50}, COS-Nox2 cells, 5×10^5 cells lysate
[d]J774A-1 macrophages, stimulated with PMA
[e]IC_{50}, 100 μg of VSMC membrane protein
[f]Human neutrophils stimulated with PMA
[g]IC_{50}, HEK293 Cells expressing NOX2
[h]Human neutrophils stimulated with fMLP
[i]K_i, Hepatic stellate cells expressing NOX2

a dose and time- dependent manner, being a reversible effect [103]. Interestingly, NO_2-AA forms a reversible covalent adduct with a cysteine residue present in the active site of PDI *in vitro*, which could explain the inhibitory effect [103]. The adduct formation between NO_2-AA and PDI could prevent its interaction NOX2 subunits affecting the correct assembly of the enzyme (Fig. 6.3).

Therefore, NO_2-AA is capable to inhibit NOX2 activity in activated macrophages by modulating the active complex formation in the plasmatic membrane without affecting the phosphorylation of the cytosolic subunits. This effect may be due to NO_2-AA-dependent covalent modification of PDI, preventing its interaction with p47^{phox}, which has been related to NOX2 modulation (Fig. 6.3).

6.6 Summary

AA participates in the inflammatory response mainly through bioactive-lipids formation but also activating the phagocytic NADPH oxidase 2. The latter depends on direct interaction of AA with different subunits of NOX, as well as with kinases that promote full NOX2 complex assembling.

It is possible to hypothesize that there is a balance of AA and NO_2-AA with antagonistic roles as modulators of NOX2 activity. AA is formed in the early events of cell activation and promotes or amplifies the activation of NOX2. The important increase in oxidant formation after these events, could lead to NO_2-AA formation, which in turn, would inhibit superoxide production. Therefore, AA and its derivatives may have an important role in modulating the NOX2 activity.

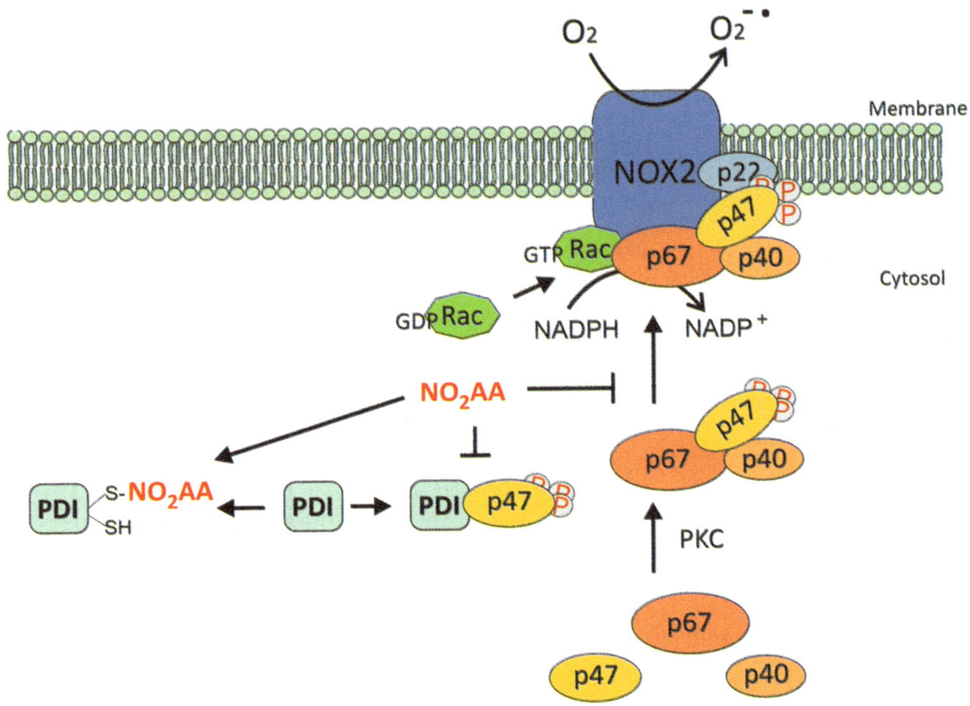

Fig. 6.3 **Effect of NO_2-AA on the PDI and NOX2.** NO_2-AA inhibit NOX2 activity mediating the prevention of the translocation of the cytosolic components to the membrane, through the formation of a covalent adduct with the PDI. This avoids the PDI-p47^{phox} interaction which is necessary for the formation of the NOX2 active complex

References

1. Gordon S, Taylor PR (2005) Monocyte and macrophage heterogeneity. Nat Rev Immunol 5(12):953–964
2. Murakami A, Ohigashi H (2007) Targeting NOX, INOS and COX-2 in inflammatory cells: chemoprevention using food phytochemicals. Int J Cancer 121(11):2357–2363
3. Qi HY, Shelhamer JH (2005) Toll-like receptor 4 signaling regulates cytosolic phospholipase A2 activation and lipid generation in lipopolysaccharide-stimulated macrophages. J Biol Chem 280(47):38969–38975
4. Bedard K, Krause KH (2007) The NOX family of ROS-generating NADPH oxidases: physiology and pathophysiology. Physiol Rev 87(1):245–313
5. Cathcart MK (2004) Regulation of superoxide anion production by NADPH oxidase in monocytes/macrophages: contributions to atherosclerosis. Arterioscler Thromb Vasc Biol 24(1):23–28
6. Kissner R et al (1998) Formation and properties of peroxynitrite as studied by laser flash photolysis, high-pressure stopped-flow technique, and pulse radiolysis volume 10, number 11, November 1997, pp 1285–1292. Chem Res Toxicol 11(5):557
7. Beckman JS et al (1990) Apparent hydroxyl radical production by peroxynitrite: implications for endothelial injury from nitric oxide and superoxide. Proc Natl Acad Sci U S A 87(4):1620–1624
8. Denicola A et al (1996) Peroxynitrite reaction with carbon dioxide/bicarbonate: kinetics and influence on peroxynitrite-mediated oxidations. Arch Biochem Biophys 333(1):49–58
9. Rubbo H, Trostchansky A, O'Donnell VB (2009) Peroxynitrite-mediated lipid oxidation and nitration: mechanisms and consequences. Arch Biochem Biophys 484(2):167–172
10. Kalyanaraman B (2004) Nitrated lipids: a class of cell-signaling molecules. Proc Natl Acad Sci U S A 101(32):11527–11528
11. Trostchansky A et al (2013) Nitro-fatty acids: formation, redox signaling, and therapeutic potential. Antioxid Redox Signal 19(11):1257–1265
12. Trostchansky A, Rubbo H (2008) Nitrated fatty acids: mechanisms of formation, chemical characterization, and biological properties. Free Radic Biol Med 44(11):1887–1896
13. Trostchansky A et al (2007) Synthesis, isomer characterization, and anti-inflammatory properties of nitroarachidonate. Biochemistry 46(15):4645–4653

14. Seeds MC, Bass DA (1999) Regulation and metabolism of arachidonic acid. Clin Rev Allergy Immunol 17(1–2):5–26
15. Tallima H, El Ridi R (2018) Arachidonic acid: physiological roles and potential health benefits – a review. J Adv Res 11:33–41
16. Kuehl FA Jr, Egan RW (1980) Prostaglandins, arachidonic acid, and inflammation. Science 210(4473):978–984
17. Smith WL, DeWitt DL, Garavito RM (2000) Cyclooxygenases: structural, cellular, and molecular biology. Annu Rev Biochem 69:145–182
18. Lewis RA, Austen KF, Soberman RJ (1990) Leukotrienes and other products of the 5-lipoxygenase pathway. Biochemistry and relation to pathobiology in human diseases. N Engl J Med 323(10):645–655
19. Rouzer CA, Marnett LJ (2003) Mechanism of free radical oxygenation of polyunsaturated fatty acids by cyclooxygenases. Chem Rev 103(6):2239–2304
20. Tilley SL, Coffman TM, Koller BH (2001) Mixed messages: modulation of inflammation and immune responses by prostaglandins and thromboxanes. J Clin Invest 108(1):15–23
21. Smith WL, Song I (2002) The enzymology of prostaglandin endoperoxide H synthases-1 and -2. Prostaglandins Other Lipid Mediat 68-69:115–128
22. Bromberg Y, Pick E (1981) Activation of macrophage adenylate cyclase by stimulants of the oxidative burst and by arachidonic acid—two distinct mechanisms. Cell Immunol 61(1):90–103
23. Bromberg Y, Pick E (1983) Unsaturated fatty acids as second messengers of superoxide generation by macrophages. Cell Immunol 79(2):240–252
24. Bromberg Y, Pick E (1984) Unsaturated fatty acids stimulate NADPH-dependent superoxide production by cell-free system derived from macrophages. Cell Immunol 88(1):213–221
25. Vignais PV (2002) The superoxide-generating NADPH oxidase: structural aspects and activation mechanism. Cell Mol Life Sci 59(9):1428–1459
26. Babior BM, Kipnes RS (1977) Superoxide-forming enzyme from human neutrophils: evidence for a flavin requirement. Blood 50(3):517–524
27. Doussiere J et al (1993) Critical assessment of the presence of an NADPH binding site on neutrophil cytochrome b558 by photoaffinity and immunochemical labeling. Biochemistry 32(34):8880–8887
28. DeCoursey TE (2016) The intimate and controversial relationship between voltage-gated proton channels and the phagocyte NADPH oxidase. Immunol Rev 273(1):194–218
29. Morgan D et al (2009) Voltage-gated proton channels maintain pH in human neutrophils during phagocytosis. Proc Natl Acad Sci U S A 106(42):18022–18027
30. Groemping Y, Rittinger K (2005) Activation and assembly of the NADPH oxidase: a structural perspective. Biochem J 386(Pt 3):401–416
31. Lambeth JD (2004) NOX enzymes and the biology of reactive oxygen. Nat Rev Immunol 4(3):181–189
32. Ago T et al (1999) Mechanism for phosphorylation-induced activation of the phagocyte NADPH oxidase protein p47(phox). Triple replacement of serines 303, 304, and 328 with aspartates disrupts the SH3 domain-mediated intramolecular interaction in p47(phox), thereby activating the oxidase. J Biol Chem 274(47):33644–33653
33. Yuzawa S et al (2004) A molecular mechanism for autoinhibition of the tandem SH3 domains of p47phox, the regulatory subunit of the phagocyte NADPH oxidase. Genes Cells 9(5):443–456
34. Fontayne A et al (2002) Phosphorylation of p47phox sites by PKC alpha, beta II, delta, and zeta: effect on binding to p22phox and on NADPH oxidase activation. Biochemistry 41(24):7743–7750
35. Kawakami N et al (1998) Indication of a protein kinase C-independent pathway for NADPH oxidase activation in human neutrophils. Arch Biochem Biophys 349(1):89–94
36. Reeves EP et al (1999) Direct interaction between p47phox and protein kinase C: evidence for targeting of protein kinase C by p47phox in neutrophils. Biochem J 344(Pt 3):859–866
37. Dewas C et al (2000) The mitogen-activated protein kinase extracellular signal-regulated kinase 1/2 pathway is involved in formyl-methionyl-leucyl-phenylalanine-induced p47phox phosphorylation in human neutrophils. J Immunol 165(9):5238–5244
38. El Benna J et al (1996) Phosphorylation of the respiratory burst oxidase subunit p47phox as determined by two-dimensional phosphopeptide mapping. Phosphorylation by protein kinase C, protein kinase A, and a mitogen-activated protein kinase. J Biol Chem 271(11):6374–6378
39. El Benna J et al (1996) Activation of p38 in stimulated human neutrophils: phosphorylation of the oxidase component p47phox by p38 and ERK but not by JNK. Arch Biochem Biophys 334(2):395–400
40. Hoyal CR et al (2003) Modulation of p47PHOX activity by site-specific phosphorylation: Akt-dependent activation of the NADPH oxidase. Proc Natl Acad Sci U S A 100(9):5130–5135
41. Leto TL, Adams AG, de Mendez I (1994) Assembly of the phagocyte NADPH oxidase: binding of Src homology 3 domains to proline-rich targets. Proc Natl Acad Sci U S A 91(22):10650–10654
42. Sumimoto H et al (1994) Role of Src homology 3 domains in assembly and activation of the phagocyte NADPH oxidase. Proc Natl Acad Sci U S A 91(12):5345–5349
43. El Benna J et al (1999) P40phox associates with the neutrophil Triton X-100-insoluble cytoskeletal fraction and PMA-activated membrane skeleton: a comparative study with P67phox and P47phox. J Leukoc Biol 66(6):1014–1020
44. el Benna J, Ruedi JM, Babior BM (1994) Cytosolic guanine nucleotide-binding protein Rac2 operates in vivo as a component of the neutrophil respiratory burst oxidase. Transfer of Rac2 and the cytosolic oxidase components p47phox and p67phox to the

submembranous actin cytoskeleton during oxidase activation. J Biol Chem 269(9):6729–6734
45. Nauseef WM et al (1991) Assembly of the neutrophil respiratory burst oxidase. Protein kinase C promotes cytoskeletal and membrane association of cytosolic oxidase components. J Biol Chem 266(9):5911–5917
46. Wientjes FB et al (2001) The NADPH oxidase components p47(phox) and p40(phox) bind to moesin through their PX domain. Biochem Biophys Res Commun 289(2):382–388
47. Woodman RC et al (1991) Respiratory burst oxidase and three of four oxidase-related polypeptides are associated with the cytoskeleton of human neutrophils. J Clin Invest 87(4):1345–1351
48. Kanai F et al (2001) The PX domains of p47phox and p40phox bind to lipid products of PI(3)K. Nat Cell Biol 3(7):675–678
49. Benna JE et al (1997) Phosphorylation of the respiratory burst oxidase subunit p67(phox) during human neutrophil activation. Regulation by protein kinase C-dependent and independent pathways. J Biol Chem 272(27):17204–17208
50. Bouin AP et al (1998) p40(phox) is phosphorylated on threonine 154 and serine 315 during activation of the phagocyte NADPH oxidase. Implication of a protein kinase c-type kinase in the phosphorylation process. J Biol Chem 273(46):30097–30103
51. Kami K et al (2002) Diverse recognition of non-PxxP peptide ligands by the SH3 domains from p67(phox), Grb2 and Pex13p. EMBO J 21(16):4268–4276
52. Ito T et al (2001) Novel modular domain PB1 recognizes PC motif to mediate functional protein-protein interactions. EMBO J 20(15):3938–3946
53. Nakamura R et al (1998) The PC motif: a novel and evolutionarily conserved sequence involved in interaction between p40phox and p67phox, SH3 domain-containing cytosolic factors of the phagocyte NADPH oxidase. Eur J Biochem 251(3):583–589
54. Dusi S, Donini M, Rossi F (1996) Mechanisms of NADPH oxidase activation: translocation of p40phox, Rac1 and Rac2 from the cytosol to the membranes in human neutrophils lacking p47phox or p67phox. Biochem J 314(Pt 2):409–412
55. Koga H et al (1999) Tetratricopeptide repeat (TPR) motifs of p67(phox) participate in interaction with the small GTPase Rac and activation of the phagocyte NADPH oxidase. J Biol Chem 274(35):25051–25060
56. Lapouge K et al (2000) Structure of the TPR domain of p67phox in complex with Rac.GTP. Mol Cell 6(4):899–907
57. Nisimoto Y et al (1999) The p67(phox) activation domain regulates electron flow from NADPH to flavin in flavocytochrome b(558). J Biol Chem 274(33):22999–23005
58. Han CH et al (1998) Regulation of the neutrophil respiratory burst oxidase. Identification of an activation domain in p67(phox). J Biol Chem 273(27):16663–16668
59. Curnutte JT et al (1984) Studies on the mechanism of superoxide release from human neutrophils stimulated with arachidonate. J Biol Chem 259(19):11851–11857
60. Henderson LM, Moule SK, Chappell JB (1993) The immediate activator of the NADPH oxidase is arachidonate not phosphorylation. Eur J Biochem 211(1–2):157–162
61. Dana R, Malech HL, Levy R (1994) The requirement for phospholipase A2 for activation of the assembled NADPH oxidase in human neutrophils. Biochem J 297(Pt 1):217–223
62. Henderson LM, Chappell JB, Jones OT (1989) Superoxide generation is inhibited by phospholipase A2 inhibitors. Role for phospholipase A2 in the activation of the NADPH oxidase. Biochem J 264(1):249–255
63. Sakata A et al (1987) Arachidonic acid acts as an intracellular activator of NADPH-oxidase in Fc gamma receptor-mediated superoxide generation in macrophages. J Immunol 138(12):4353–4359
64. Cross AR, Erickson RW, Curnutte JT (1999) Simultaneous presence of p47(phox) and flavocytochrome b-245 are required for the activation of NADPH oxidase by anionic amphiphiles. Evidence for an intermediate state of oxidase activation. J Biol Chem 274(22):15519–15525
65. Foubert TR et al (2002) Structural changes are induced in human neutrophil cytochrome b by NADPH oxidase activators, LDS, SDS, and arachidonate: intermolecular resonance energy transfer between trisulfopyrenyl-wheat germ agglutinin and cytochrome b(558). Biochim Biophys Acta 1567(1–2):221–231
66. Taylor RM et al (2004) Anionic amphiphile and phospholipid-induced conformational changes in human neutrophil flavocytochrome b observed by fluorescence resonance energy transfer. Biochim Biophys Acta 1663(1–2):201–213
67. Henderson LM (1998) Role of histidines identified by mutagenesis in the NADPH oxidase-associated H+ channel. J Biol Chem 273(50):33216–33223
68. Henderson LM et al (1997) The arachidonate-activatable, NADPH oxidase-associated H+ channel is contained within the multi-membrane-spanning N-terminal region of gp91-phox. Biochem J 325(Pt 3):701–705
69. Shiose A, Sumimoto H (2000) Arachidonic acid and phosphorylation synergistically induce a conformational change of p47phox to activate the phagocyte NADPH oxidase. J Biol Chem 275(18):13793–13801
70. Groemping Y et al (2003) Molecular basis of phosphorylation-induced activation of the NADPH oxidase. Cell 113(3):343–355
71. Matono R et al (2014) Arachidonic acid induces direct interaction of the p67(phox)-Rac complex with the phagocyte oxidase Nox2, leading to superoxide production. J Biol Chem 289(36):24874–24884
72. Molshanski-Mor S et al (2007) Cell-free assays: the reductionist approach to the study of NADPH oxidase assembly, or "all you wanted to know about

cell-free assays but did not dare to ask". Methods Mol Biol 412:385–428
73. Pick E (2014) Cell-free NADPH oxidase activation assays: "in vitro veritas". Methods Mol Biol 1124:339–403
74. Souabni H et al (2012) trans Arachidonic acid isomers inhibit NADPH-oxidase activity by direct interaction with enzyme components. Biochim Biophys Acta 1818(9):2314–2324
75. Alvarez MN et al (2011) Intraphagosomal peroxynitrite as a macrophage-derived cytotoxin against internalized Trypanosoma cruzi: consequences for oxidative killing and role of microbial peroxiredoxins in infectivity. J Biol Chem 286(8):6627–6640
76. Prolo C, Alvarez MN, Radi R (2014) Peroxynitrite, a potent macrophage-derived oxidizing cytotoxin to combat invading pathogens. Biofactors 40(2):215–225
77. Bonilla L et al (2013) Regulation of protein kinase C by nitroarachidonic acid: impact on human platelet activation. Arch Biochem Biophys 533(1–2):55–61
78. Ferreira AM et al (2008) Nitroalkenes: synthesis, characterization, and effects on macrophage activation. Methods Enzymol 441:33–51
79. Trostchansky A et al (2011) Nitroarachidonic acid, a novel peroxidase inhibitor of prostaglandin endoperoxide H synthases 1 and 2. J Biol Chem 286(15):12891–12900
80. Ferreira AM et al (2009) Macrophage activation induces formation of the anti-inflammatory lipid cholesteryl-nitrolinoleate. Biochem J 417(1):223–234
81. Ferreira AM et al (2012) Nitro-fatty acids as novel electrophilic ligands for peroxisome proliferator-activated receptors. Free Radic Biol Med 53(9):1654–1663
82. Trostchansky A, Rubbo H (2007) Lipid nitration and formation of lipid-protein adducts: biological insights. Amino Acids 32(4):517–522
83. Gonzalez-Perilli L et al (2013) Nitroarachidonic acid prevents NADPH oxidase assembly and superoxide radical production in activated macrophages. Free Radic Biol Med 58:126–133
84. Batthyany C et al (2006) Reversible post-translational modification of proteins by nitrated fatty acids in vivo. J Biol Chem 281(29):20450–20463
85. Schopfer FJ et al (2009) Detection and quantification of protein adduction by electrophilic fatty acids: mitochondrial generation of fatty acid nitroalkene derivatives. Free Radic Biol Med 46(9):1250–1259
86. Trostchansky A et al (2001) Formation of lipid-protein adducts in low-density lipoprotein by fluxes of peroxynitrite and its inhibition by nitric oxide. Arch Biochem Biophys 395(2):225–232
87. O'Donnell BV et al (1993) Studies on the inhibitory mechanism of iodonium compounds with special reference to neutrophil NADPH oxidase. Biochem J 290(Pt 1):41–49
88. Hancock JT, Jones OT (1987) The inhibition by diphenyleneiodonium and its analogues of superoxide generation by macrophages. Biochem J 242(1):103–107
89. Rey FE et al (2001) Novel competitive inhibitor of NAD(P)H oxidase assembly attenuates vascular O(2)(−) and systolic blood pressure in mice. Circ Res 89(5):408–414
90. Csanyi G et al (2011) Nox2 B-loop peptide, Nox2ds, specifically inhibits the NADPH oxidase Nox2. Free Radic Biol Med 51(6):1116–1125
91. ten Freyhaus H et al (2006) Novel Nox inhibitor VAS2870 attenuates PDGF-dependent smooth muscle cell chemotaxis, but not proliferation. Cardiovasc Res 71(2):331–341
92. Kleinschnitz C et al (2010) Post-stroke inhibition of induced NADPH oxidase type 4 prevents oxidative stress and neurodegeneration. PLoS Biol 8(9):e1000479
93. Sun QA et al (2012) Off-target thiol alkylation by the NADPH oxidase inhibitor 3-benzyl-7-(2-benzoxazolyl)thio-1,2,3-triazolo[4,5-d]pyrimidine (VAS2870). Free Radic Biol Med 52(9):1897–1902
94. Gatto GJ Jr et al (2013) NADPH oxidase-dependent and -independent mechanisms of reported inhibitors of reactive oxygen generation. J Enzyme Inhib Med Chem 28(1):95–104
95. Gianni D et al (2010) A novel and specific NADPH oxidase-1 (Nox1) small-molecule inhibitor blocks the formation of functional invadopodia in human colon cancer cells. ACS Chem Biol 5(10):981–993
96. Aoyama T et al (2012) Nicotinamide adenine dinucleotide phosphate oxidase in experimental liver fibrosis: GKT137831 as a novel potential therapeutic agent. Hepatology 56(5):2316–2327
97. Jha JC et al (2014) Genetic targeting or pharmacologic inhibition of NADPH oxidase nox4 provides renoprotection in long-term diabetic nephropathy. J Am Soc Nephrol 25(6):1237–1254
98. de A Paes AM et al (2011) Protein disulfide isomerase redox-dependent association with p47(phox): evidence for an organizer role in leukocyte NADPH oxidase activation. J Leukoc Biol 90(4):799–810
99. Janiszewski M et al (2005) Regulation of NAD(P)H oxidase by associated protein disulfide isomerase in vascular smooth muscle cells. J Biol Chem 280(49):40813–40819
100. Laurindo FR et al (2008) Novel role of protein disulfide isomerase in the regulation of NADPH oxidase activity: pathophysiological implications in vascular diseases. Antioxid Redox Signal 10(6):1101–1113
101. Santos CX et al (2009) Protein disulfide isomerase (PDI) associates with NADPH oxidase and is required for phagocytosis of Leishmania chagasi promastigotes by macrophages. J Leukoc Biol 86(4):989–998
102. Laurindo FR, Pescatore LA, Fernandes Dde C (2012) Protein disulfide isomerase in redox cell signaling and homeostasis. Free Radic Biol Med 52(9):1954–1969
103. Gonzalez-Perilli L et al (2017) Nitroarachidonic acid (NO2AA) inhibits protein disulfide isomerase (PDI) through reversible covalent adduct formation with critical cysteines. Biochim Biophys Acta 1861(5 Pt A):1131–1139

Lipid Metabolism and Signaling in Platelet Function

7

Antonio Marcus de Andrade Paes, Renato Simões Gaspar, Eduardo Fuentes, Sergio Wehinger, Iván Palomo, and Andrés Trostchansky

Abstract

Modern society has changed its diet composition, transitioning to a higher intake of saturated fat with a 50% increase of cardiovascular risk (CVD). Within the context of increased CVD, there is an induction of a prothrombotic phenotype mainly due to increased platelet reactivity as well as decreased platelet response to inhibitors. Platelets maintain haemostasis through both blood components and endothelial cells that secrete inhibitory or stimulatory molecules to regulate thrombus formation. There exist a correlation between platelets' polyunsaturated fatty acid (PUFA) and the increase in platelet reactivity. The aim of this chapter is to review the metabolism of the main PUFAs involved in platelet function associated with the role that their enzyme-derived oxidized metabolites exert in platelet function and fate. Finally, how lipid metabolism in the organism affect platelet aggregation and activation and the pharmacological modulation of these processes will also be discussed.

Keywords

Platelets · Arachidonic acid · Eicosanoids · Thrombosis · Dyslipidemias

7.1 Introduction

With increased industrialization, modern society has changed its diet composition, transitioning to a higher intake of saturated fat [1]. This coupled with an array of genetic mutations, prone humans to develop dyslipidemia, a disorder characterized by high levels of circulating lipids such as low-density lipoprotein (LDL), total cholesterol (TC) or triglycerides and/or decreased levels of high-density lipoprotein (HDL). The last survey conducted by the world health organization (WHO) in 2008 estimated the prevalence of dyslipidemia (characterized by TC > 5 mmol/L) to be 23.1% in

A. M. d. A. Paes
Laboratory of Experimental Physiology, Department of Physiological Sciences, Federal University of Maranhão, São Luís, Brazil

R. S. Gaspar
Institute of Cardiovascular and Metabolic Research, University of Reading, Reading, UK

E. Fuentes · S. Wehinger · I. Palomo
Thrombosis Research Center, Department of Clinical Biochemistry and Immunohaematology, Faculty of Health Sciences, Interdisciplinary Excellence Research Program on Healthy Aging (PIEI-ES), Universidad de Talca, Talca, Chile

A. Trostchansky (✉)
Departamento de Bioquímica and Centro de Investigaciones Biomédicas (CEINBIO), Facultad de Medicina, Universidad de la República, Montevideo, Uruguay
e-mail: trocha@fmed.edu.uy

A. Trostchansky, H. Rubbo (eds.), *Bioactive Lipids in Health and Disease*, Advances in Experimental Medicine and Biology 1127, https://doi.org/10.1007/978-3-030-11488-6_7

Africa, 53.7% in Europe, with a global prevalence of 38.9% [2]. Such high prevalence developed onto a public health debate, given that a rise in 2 mmol/L in TC has been shown to increase cardiovascular risk (CVD) by 50% [3], while a more recent meta-analysis questioned this assumption [4]. Within the context of increased CVD, it is well established that dyslipidemia induces a prothrombotic phenotype mainly due to increased platelet reactivity as well as decreased platelet response to inhibitors [5–8].

Platelets are fragments of megakaryocytes circulating within the bloodstream to block leakages – an evolutionary adaptation required for mammal survival. These cells maintain haemostasis through both blood components and endothelial cells that secrete inhibitory (e.g. nitric oxide) or stimulatory (e.g. thromboxane) molecules to regulate thrombus formation, which involves the exposure of sub-endothelial collagen and fibrinogen after endothelial injury. Platelets will then bind to both collagen and fibrinogen through glycoprotein (GP) VI and GPIb-V-IX receptors, respectively culminating in platelet adhesion, secretion of stimulatory molecules such as adenosine diphosphate (ADP) and aggregation to other platelets – a process deemed irreversible through binding to the platelet fibrinogen receptor αIIbβ3 [9–11]. While these processes are important to maintain haemostasis in physiological environments, it has been described an up-regulation of stimulatory signals as well as down-regulation of inhibitory ones in chronic diseases such as dyslipidemia and atherosclerosis [12]. Therefore, increased platelet activity has been perceived as a maladaptation implicated in chronic non-communicable diseases, opening up perspectives of new scientific discoveries – according to Prof Barry Coller, this is the "Golden Age of Platelet Research" [13].

The rise in platelet reactivity is due to the actions of different lipids on platelet function, which will be discussed in the following sections of this chapter. We will review the metabolism of the main polyunsaturated fatty acid (PUFA) involved in platelet function, arachidonic acid (AA), associated with the role that its enzyme-derived oxidized metabolites exert in platelet function and fate. Finally, how lipid metabolism in the organism affect platelet aggregation and activation and the pharmacological modulation of these processes will also be discussed.

7.2 Overview of Arachidonic Acid Metabolism

Arachidonic acid (AA, all-cis-5, 8, 11, 14-eicosatetraenoic acid) is an omega-6 PUFA. It is an essential constituent of cell membrane, necessary for membrane fluidity, flexibility and function in all cell types. The presence of its cis-four double bonds gives the compound a certain degree of flexibility to interact with proteins or to react against molecular oxygen, forming a range of bioactive oxygenated molecules via enzymatic and non-enzymatic mechanisms [14–18]. Esterified AA is usually localized in the glycerol backbone sn-2 position, constituting an important part of phospholipid-content in the membranes and cytosol of mammalian cells and tissues. In platelets, up to 25% of phospholipid fatty acids are AA [14–18], reaching levels near to 5 mM in resting platelets [19]. Arachidonic acid cellular concentration may influence both normal cellular functions and the development of platelet diseases. More important, it has been proposed that a low dietary intake of AA decrease the production of pro-inflammatory eicosanoids, which contributes to processes resolution in chronic inflammatory diseases [15, 17]. However, our body's AA needs are higher than the concentration found in human diet. Thus, our tissues depend on endogenous formation of AA from its precursor, linoleic acid (LA 18: 2n-6), in a process regulated by the action of enzymes such as desaturases and elongases. The activity of Δ6 and Δ5 desaturases (d-5-d) converts LA to gamma-linolenic acid (GLA, 18: 3), dihomo-GLA (DGLA, 20:3) and AA [15, 17].

Arachidonic acid is released from its esterified form by the action of phospholipase A_2 (PLA_2) complex, which mediates the one-step hydrolysis of the AA present at the sn-2 position on phospholipid backbone [14, 20–23]. Cytosolic calcium (Ca^{+2})-dependent group IV PLA_2 ($cPLA_2\alpha$) catalyzes that hydrolysis to generate a free fatty acid and a lysophospholipid. Secretory PLA_2 ($sPLA_2$), which is induced by the action of $cPLA_2$, controls magnitude and duration of free AA release, as well as paracrinely propagates the inflammatory response to neighboring cells [24–26]. Lastly, cytosolic Ca^{+2}-independent PLA_2 plays a role on cellular homeostasis through generation of SPM (specialized pro-resolvins mediators) and reacylation of free AA in membranes as well [16, 27]. Nevertheless, there are other phospholipases equally able to release AA: phospholipase C (PLC) and phospholipase D (PLD). PLC participates in the formation of diacylglycerol (DAG) by the action of diacylglycerol lipase and lipid products containing AA by the action of monoacylglycerol lipases [15, 20, 26, 27]. Similarly, phosphatidic acid or DAG are formed from posphatidylcholine by PLD. The former can be further catalyzed by phosphatidate phosphohydrolase to form DAG. Then, DAG-lipase hydrolyzes DAG to generate AA. Upon stimulation, Gq protein-coupled receptor activates phospholipase C that cleaves phosphatidylinositol 4,5 bisphosphate (PIP2) to inositol 1,4,5 triphosphate (IP3) and diacylglycerol (DAG). The presence of DAG and Ca^{2+} activates protein kinase C and IP3 opens ER Ca^{2+} channel [28–30]. The increase in intracellular Ca^{2+} ions levels guides the PLA_2 enzyme translocation to the perinuclear membrane where exerts the explained release of AA from the membrane to let the AA-metabolizing enzymes their substrate [14, 20–23].

Free AA can undergo four possible enzymatic pathways: prostaglandin endoperoxide H synthase (PGHS), lipoxygenase (LOX), cytochrome p450 (CYP 450) and anandamide pathway, which create bioactive 20-carbon oxygenated PUFA generally denominated eicosanoids. These bioactive lipids act as local hormones and/or signaling molecules produced both in response to basal metabolism as well as in inflammatory sites upon regulation by normal immune response stimuli (IL-1, TNF-alpha, lipopolysaccharides, etc.). The metabolic fate subsequent to AA release into the cytosol depends on characteristics of each tissue type, although on different tissues the same precursor often gives rise to products with antagonistic function, e.g. PGE_2 [14, 17]. Cyclooxygenase (COX), more precisely known as Prostaglandin Endoperoxide H synthase (PGHS), converts AA first into PGG_2, via a COX function and then to PGH_2 following a peroxidase reaction [31]. Two different PGHS isoforms (PGHS-1 and PGHS-2) are involved in the PGHS pathway to generate prostaglandin H_2 (PGH_2), an intermediate hub whose metabolization by downstream enzymes leads to different prostaglandins: PGE_2, PGD_2 and $PGF_2\alpha$, prostacyclin (PGI_2), or thromboxaneA_2 (TXA_2). LOX pathway consists of AA oxidation by the enzyme isoforms LOX-5, LOX-8, LOX-12, and LOX-15 to generate their products, leukotrienes (LTA_4, LTB_4, LTC_4, LTD_4 and LTE_4), lipoxins (LXA) and 8-, 12- or 15- hydroperoxyeicosatetraenoic (HpETE) acids or their corresponding reduced derivatives hydroxyeicosatetraenoic (HETE) acids. CYP 450 pathway involves two enzymes, CYP450 epoxygenase and CYP450 hydroxylase giving rise to epoxyeicosatrienoic acid (EETs) and 20-HETE, respectively. Finally, anandamide pathway comprises the FAAH (fatty acid amide hydrolase) to produce the endocannabinoid, anandamide [14–17, 20, 32–37].

Cell–cell interaction is important for eicosanoids synthesis since a donor cell has to transfer its unstable intermediate, e.g. PGH_2, to a recipient cell to trigger eicosanoid biosynthesis in the latter. The single donor cell should have all the enzymes necessary to produce the different eicosanoids while the recipient cell does not necessarily have all the required enzymes for AA release. Thus, for inflammation initiation, the complete set of enzymes to initiate eicosanoids production must be present in at least two cells in the injured tissue. The AA intermediate metabolites are lipophilic and require a group of special proteins called fatty acid binding proteins (FABP), specific for each cell type, which are responsible for the increase of AA intermediates export via their stabilization and lengthening their half-life time to allow them to exert their biological activities. The major action of AA metabolites is promotion of acute inflammatory response, characterized by the production of pro-inflammatory mediators, e.g. PGE_2 and PGI_2, followed by a second phase in which lipid mediators with pro-resolution activities may be generated [14–17, 20, 32–37] Resolution of inflammation is no more considered a passive process, but rather an active programmed response regulated by mediators with pro-resolving capacity.

7.3 Arachidonic Acid Metabolism in Platelets

The activation or inhibition of platelets can be modulated by many agents with a central role being played by eicosanoids, being TxA_2 and PGI_2 the main eicosanoids affecting platelets' function [38].

Arachidonic acid is metabolized to PGH_2 in the cytosol of platelets and for this, PLA_2 activity is necessary to release AA from platelet membrane [39, 40]. PGH_2 in platelets undergoes further transformations catalyzed by Tx synthase, PGD isomerase or PGE synthase to form TxA_2, PGD_2 or PGE_2 respectively (Fig. 7.1). In addition to the wide variety of eicosanoids formed by the COX pathway [41], 11-HETE and 15(S)-HETE- but not 5-HETE- are produced when AA is inserted at the active site of COX-1 in a different structural arrangement than the one necessary for PGH_2 synthesis [42]. Both products can be formed at similar levels to TxA_2 whose formation is associated with an increase in AA levels, *e.g.* at platelet hyperactivation where high concentrations of AA are released from platelet membranes. The participation of COX in 11-HETE and 15(S)-HETE formation was also demonstrated by the lack of decrease in 12-HETE when using only aspirin thus LOX is not the enzyme involved in these products formation [42]. Reduction of 12-HETE was only observed when both PGHS and $P2Y_{12}$ inhibitors were used confirming the former proposal [42].

Platelets mainly express PGHS-1, but traces of PGHS-2 have been detected, possibly carried over from megakaryocytes, the platelet precursor cells, or as a result of the transcription of residual mRNA into protein [22, 33, 43]. In addition to PGHS, AA can also be oxidized by the non-heme iron-containing enzymes LOXs. Different isoforms of LOXs are found depending on the carbon where the hydroperoxyl (-OOH) group is added. Platelets present 12-LOX which inserts oxygen primarily at C-12 of AA forming the 12S-hydroperoxy-5Z,8Z,10E,14Z-eicosatetraenoic acid [12(S)HpETE] [44]. The highly reducing environment present in the platelet rapidly reduce 12(S)-HpETE to its hydroxyl derivative 12-HETE [44]. The biological activity of 12-HETE in platelets *in vivo* remains under discussion; some authors suggest a pro-thrombotic activity by acting at vascular hypertension (Fig. 7.2) while others propose an anti-aggregant and anti-inflammatory action by decreasing the release of AA from the membrane and regulating integrin activation (Fig. 7.3). The pro-inflammatory leukotrienes precursor, being able to affect platelet activation during an inflammatory process, is formed by the 5-LOX present in neutrophils. One important difference between the AA-metabolizing enzymes due to their capacity to be inhibited is that while aspirin or other nonsteroidal anti-inflammatory drugs (NSAIDs) inhibit PGHS activity [45, 46], LOX remain active in their presence being able to oxidize AA to HpETEs.

7.3.1 Role of Distinct Eicosanoids in Platelets

7.3.1.1 Thromboxane A_2 (TxA_2)

The most directly important prostanoid for platelet function is the PGHS-1-generated TxA_2, which is synthesized by activated platelets acting in an autocrine and paracrine manner to induce thrombosis. On platelets, TxA_2 binds to the thromboxane prostanoid (TP) receptor and initiates an amplification loop leading to further platelet activation, aggregation and TxA_2 formation. In the vasculature, TxA_2 induces vasoconstriction and the proliferation of vascular smooth muscle cells (Fig. 7.2) [21, 38, 47, 48].

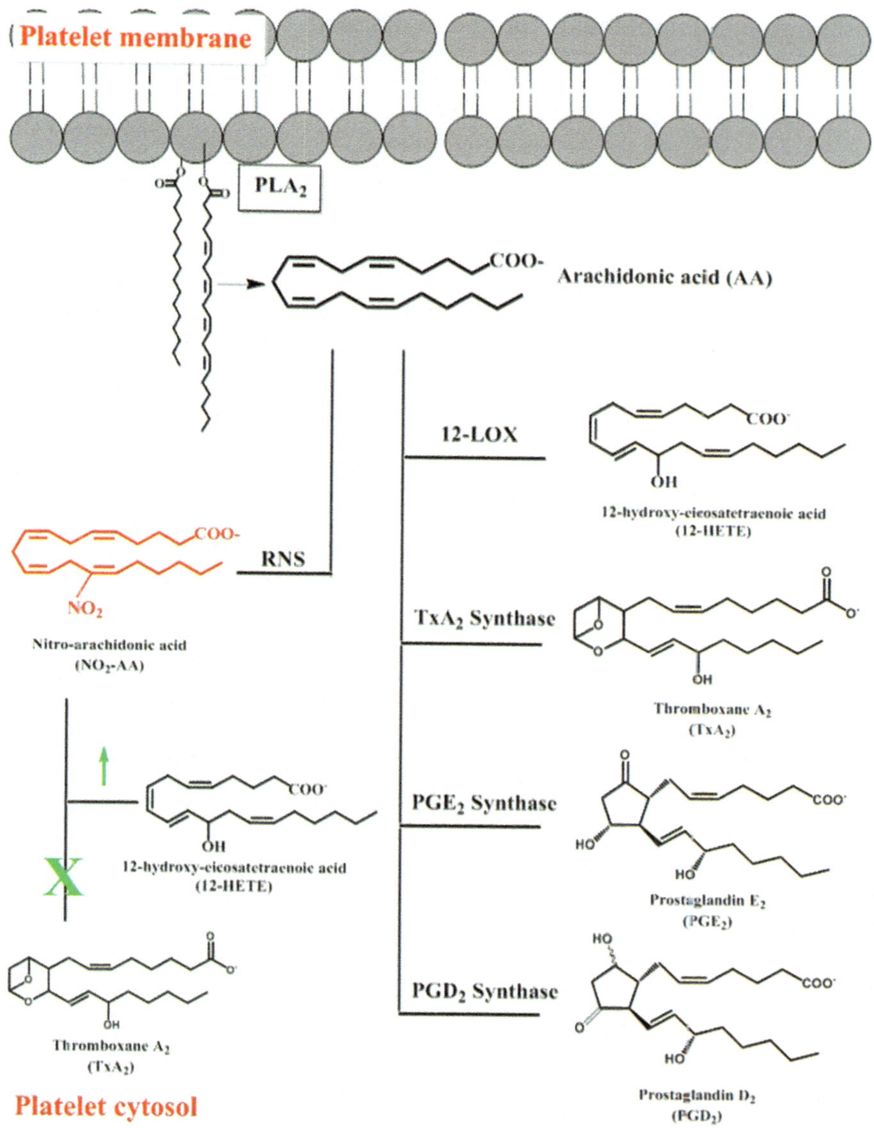

Fig. 7.1 Arachidonic acid metabolism in platelets. Arachidonic acid is esterified at the platelet membrane and released to the cytosol due to the activity of PLA_2. Then, AA is the substrate for PGHS-1 forming PGH_2 which then is the substrate for the subsequent formation of TxA_2, PGE_2 or PGD_2. If AA is oxidized by LOX, then 12-HETE is formed. When an increase of vasculature RNS formation or the appearance of an inflammatory process occur, AA is nitrated to NO_2-AA which inhibit PGHS-1 thus decreasing the pro-aggregant TxA_2 formation [40]. The increase of non-oxidized AA in the cytosol led to a subsequent oxidation of the fatty acid by the 12-LOX thus increasing the levels of 12-HETE [47]

7.3.1.2 12-Hydroxy-Eicosatetraenoic Acid (12-HETE)

12-HETE is the major 12-LOX-catalysed metabolite and the most abundant eicosanoid produced by platelets upon stimulation [41, 42], although its effects on platelet function are not completely understood. There are studies reporting that inhibition of 12-LOX led to decreased platelet aggregation correlated with a significant reduction of 12-HETE levels in response to collagen [49].

Pro-aggregant biomolecules formed in platelets		References
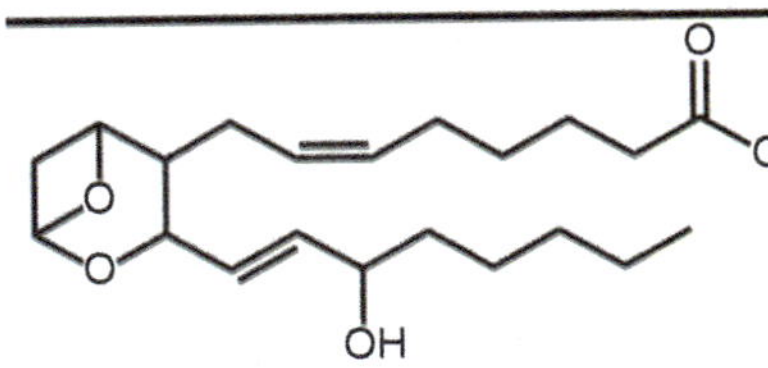 *Thromboxane A_2* *(TxA_2)*	- Binds to the thromboxane prostanoid (TP) receptor - Induce thrombosis - Amplification loop leading to further platelet activation	Holinstat *et al.* [21] Jennings [38] Rouzer and Marnett [48]
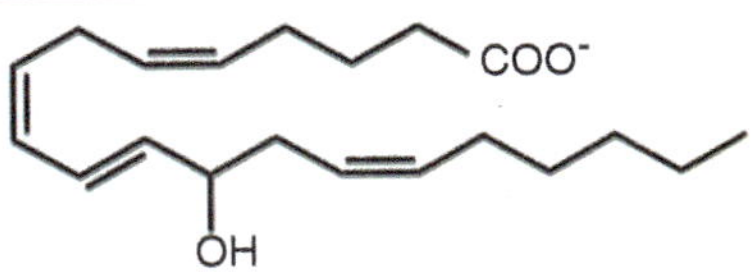 *12-hydroxy-eicosatetraenoic acid* *(12-HETE)*	-Pro-aggregant effects depend on: i) concentration, ii) estereospecificity; iii) co-incubation with different agonists	Porro *et al.* [50]
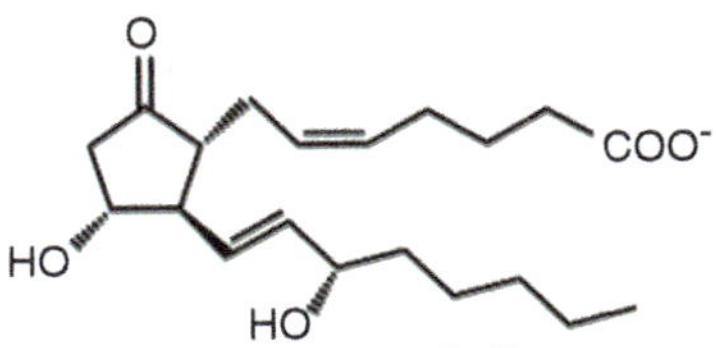 *Prostaglandin E_2* *(PGE_2)*	-At 0.1–10 μM, binds to EP3 Gi-coupled receptors to stimulate platelet aggregation by decreasing cAMP - After plaque rupture, PGE_2 enhances thrombus formation	Petrucci *et al.* [56] Glenn *et al.* [57] Friedman *et al.* [2015]

Fig. 7.2 Structure and biological pro-aggregant activities of bioactive lipid derivatives of AA formed inside platelets

A recent review concluded that 12-HETE can exert both pro- and anti-aggregant effects (Figs. 7.2 and 7.3) on platelets, depending on 12-HETE concentration, stereospecificity and co-incubation with different agonists [50].

7.3.1.3 Prostacyclin (PGI_2)

Endothelium-produced PGI_2 binds to the Gs-coupled PGI_2 receptor (IP) on platelets and generally reduces platelet reactivity, which can be critical to minimizing the risk for atherothrombotic events [51, 52]. Binding of PGI_2 to the IP receptor, results in the activation of adenylate cyclase and a subsequent rise in cAMP levels in platelets [51, 52]. This stimulates phosphorylation of PKA, which suppresses various signaling pathways involved in platelet function such as adhesion, aggregation and granule secretion.

7.3.1.4 Prostaglandin D_2 (PGD_2)

PGD_2 is well established as a macrophage product but, in lesser amounts, is also synthesized by platelets. By interaction with platelet DP1 receptors, PGD_2 increases adenylyl cyclase activity and so, like PGI_2, inhibits platelet activation [53, 54] (Fig. 7.3).

7.3.1.5 Prostaglandin E_2 (PGE_2)

PGE_2 is released by endothelial cells (ECs) and, to some extent, by activated platelets. It acts on a range of prostanoid receptors, EP1 – EP4, that differently modulate second messengers, such as cAMP and free Ca^{2+}, within platelets and exert contrasting effects on platelet function [55]. The effects on platelets of PGE_2 acting through EP receptors are concentration dependent. At low concentrations (0.1–10 μM), PGE_2 binds to Gi-coupled receptors (EP3) to enhance

Anti-aggregant biomolecules formed in platelets		References
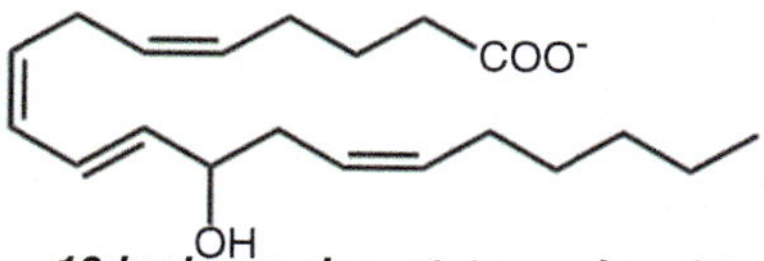 12-hydroxy-eicosatetraenoic acid (12-HETE)	-In response to collagen, 12-HETE decreased platelet aggregation	Maskrey *et al.* [49]
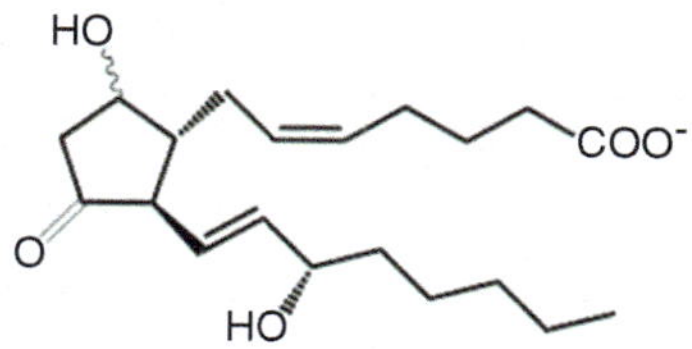 Prostaglandin D_2 (PGD_2)	-Activating DPI receptors, increases adenylyl cyclase activity and inhibits platelet activation	Gimenez-Bastida *et al.* [53] Schuligoi *et al.* [53]
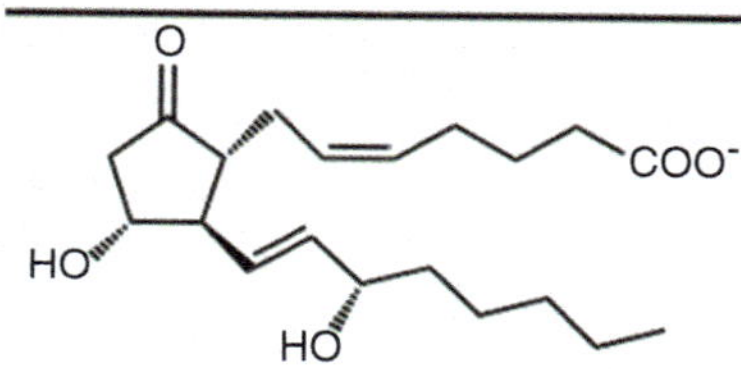 Prostaglandin E_2 (PGE_2)	-At concentrations of 100 μM, PGE_2 activates EP2 and EP4 Gs-coupled receptors that inhibit platelet aggregation	Petrucci *et al.* [56] Glenn *et al.* [57] Friedman *et al.* [58]

Fig. 7.3 Structure and biological anti-aggregant activities of bioactive lipid derivatives of AA formed inside platelets

aggregation (Fig. 7.2), whereas at higher concentrations (100 μM), it activates Gs-coupled receptors (EP2, EP4) to inhibit aggregation [56–58] (Fig. 7.3). Stimulation of EP3 receptors by PGE_2 decreases cAMP levels, thus favoring platelet aggregation, but the full effect is only seen in the presence of another platelet agonist [58]. On the other hand, the increased cAMP levels which accompany EP4 receptor activation correlate with suppressed platelet aggregation [57]. In addition to PGE_2, $PGF_2\alpha$ and PGD_2 can also bind to EP3 and EP4 receptors but with lower affinity and reversible effects [57, 58]. As well as the well-characterized effects of PGE_2 mediated through EP3 and EP4 receptors, EP1 receptors are also expressed on platelets [56, 59]. Although the signal transduction pathway is not clear, studies in several cell lines expressing EP1 receptors suggest that its activation increases Ca^{2+} influx and might thereby stimulate platelet aggregation [60]. While PGE_2 seems to both inhibit and potentiate platelet aggregation *in vitro*, a study by Gross *et al.* has elegantly shown that, *in vivo*, PGE_2 is produced by the vessel wall or after the rupture of a plaque. Under these conditions, PGE_2 activates the EP3 receptors on platelets and clearly enhances, rather than reduces, thrombus formation in the arterial vessel wall [61].

7.4 Diversion of Normal AA Metabolism in Platelets: Formation of NO_2-AA

Regarding AA metabolism in platelets and its oxidation by PGHS and/or LOXs, the participation of nitric oxide (•NO) and reactive nitrogen species (RNS) during platelet activation has been widely demonstrated. PGHS is a target for •NO and peroxynitrite, modulating prostaglandins as well as TxA_2 synthesis [62, 63]. Inflammatory processes and PUFAs are linked by eicosanoids, which represent mediators and regulators of inflammatory processes. Many of the reported protective, anti-aggregant and signaling actions of •NO are due to its interaction with iron-containing enzymes. In fact, the heme-containing enzyme PGHS is a target for •NO, modulating prostaglandin synthesis [62]. Nitric oxide can act as a reducing co-substrate of the peroxidase (POX) activity of the enzyme favoring the catalytic cycle of POX [39] or as a precursor of peroxynitrite which may also be a POX substrate or an inhibitor of the enzyme by tyrosine nitration at the enzyme active site. Indeed, •NO enhanced COX inactivation by peroxynitrite [62]. During COX catalysis, AA is oxidized being formed arachidonyl radicals which are potential targets for •NO or RNS reactivity thus influencing enzyme activity. Our reports suggest that during COX catalysis, those AA-derived radicals can be "sequestered" by RNS to form nitrogen-containing AA products decreasing enzyme substrates thus diminishing enzyme activity [62].

We have extensively described the synthesis of the nitrated derivative of AA, which we propose that besides being formed by acidic gastric nitration, can be a byproduct of COX activity in the presence of RNS. Nitroarachidonic acid (NO_2-AA) can be formed in biological membranes from AA, as explained above the most abundant fatty acid present in the 2-carbon position of phospholipids, to exert biological effects upon PLA_2 cleavage. One of the currently accepted mechanisms for RNS (e.g. $•NO_2$)-mediated oxidation and nitration for AA involve different routes including hydrogen atom abstraction and addition reactions [64]. Reaction of $•NO_2$ with PUFAs leads to the generation of isomerized, oxidized and/or nitro-allylic, nitroalkene, dinitro, or nitro-hydroxy lipid derivatives. An arachidonyl carbon-centered lipid radical and nitrous acid (HONO) are formed, and under anaerobic conditions or when O_2 tension is low, a second molecule of $•NO_2$ reacts with the carbon-centered radical generating a nitroalkane (the $-NO_2$ moiety is bound to a double bond) [64, 65].

As AA is the substrate for PGHS activity, we evaluated if nitration of the carbon chain of AA can exert any effects on PGHS normal activity of oxidizing AA. When analyzed *in vitro*, NO_2-AA inhibited both POX and COX activities of PGHS-1 while only affected POX in PGHS-2 [40], while nitration of other fatty acids was unable to modify PGHS activity. The mechanism of inhibition involves the release of heme as a result of NO_2-AA reaction with the protein [40]. Not only in vitro the effects of NO_2-AA on PGHS were analyzed. The capacity of NO_2-AA to modulate PGHS-1 activity was also evaluated in human platelets [40, 47] being a potent inhibitor of thrombin-mediated platelet aggregation with an inhibitory concentration 50 (IC_{50}) of 1.3 μM (refs. [40, 47]). Inhibiton of PGHS in platelets lead to a decrease in TxA_2 synthesis with a concomitant divertion of AA to the LOX pathway, thus AA is mainly oxidized by LOX forming 12-HETE [40, 47]. Importantly, platelet aggregation and activation was modulated by NO_2-AA in response to several membrane receptors activation, *i.e.* protease activated receptors -PARs- for thrombin, P2Y receptors for ADP and TxA_2 receptors for AA, by either TxA_2 positive retroalimentation after its synthesis or through direct stimulation of protein kinase C (PKC), indicating that NO_2-AA acts downstream membrane receptors to exert its antiplatelet effects [47].

The PKC family is centrally involved in platelet activation and aggregation. As explained previously, stimulation of PLC led to the generation of IP3 and DAG where the latter activates at the platelet membrane the cytosolic PKC isoforms (cPKC), which act in a synergistic manner with Ca^{2+}. While NO_2-AA did not mobilize Ca^{2+}, it was able to inhibit α-granule secretion [47]. The effect of the nitroalkene on PKCα was analyzed in non-activated and thrombin activated platelets. PKCα was diffusively distributed in the cytosol of untreated platelets and migrated to the plasma membrane after thrombin-stimulation in a process that was abolished by NO_2-AA in addition to a prevention of platelet shape change and cytoskeletal reorganization, characteristics of platelet activation [47]. It has been reported that the bioactive signaling activities of nitroalkenes are due to their capacity to perform Michael-addition reactions with nucleophilic Cys of His residues. cPKC isoforms contain Cys rich motifs that are duplicated as a tandem domain which are critical for its interaction with membrane phospholipids, suggesting that PKC inhibition could be mediated by electrophilic modifications of the enzyme. In addition, NO_2-AA modulation of PKC route has been shown by its capacity to inhibit platelet responses downstream to PKC activation (α-granule secretion, Erk2 phosphorylation, PKC translocation to the membrane) while not affecting upstream responses (*e.g.* Ca^{2+} mobilization). Moreover, when platelet activation is directly activated by the PKC activator PMA, NO_2-AA inhibited platelet aggregation. These observations provide a possible novel mechanism for platelet regulation under conditions where AA acts as a mild agonist for hemostasis, but adopts potent anti-platelet properties at inflammatory environments associated with increased •NO and RNS production when transformed into NO_2-AA [47].

7.5 How Lipoproteins Modulate Platelet Function

7.5.1 Native and Oxidized LDL in Cardiovascular Disease

High levels of LDL are more prone to be altered by reactive oxygen species (ROS), which are normally present in physiological processes acting as second messengers, but also induce oxidative stress in conditions such as ageing, smoking, diabetes mellitus, metabolic syndrome and dyslipidemia. ROS act on PUFAs (free or ester-bound) on the LDL particle, leading to hydroperoxides formation, which in turn, breaks down to form reactive particles as malondialdehyde (MDA) and 4-hydroxy-nonenal (4-HNE). MDA and 4-HNE react with lysine residues at the Apo lipoprotein B in LDL, together with amine groups in some phospholipids such as phosphatidylethanolamine and phosphatidylserine, forming Schiff-bases, generating neoepitopes. These processes produce oxidized LDL (OxLDL) particles, which are highly immunogenic. In fact, IgG autoantibodies against OxLDL have been found, although their role in atherogenesis and atherotrombosis remains to be elucidated [66].

Oxidized LDL has been associated to atheromatous plaque formation through four mechanisms that complement each other: endothelial dysfunction, foam cell formation, smooth muscle cell migration and proliferation, and induction of platelet adhesion and aggregation. Endothelial dysfunction occurs when its anticoagulant and vascular regulatory properties are compromised and a pro-inflammatory phenotype is produced. OxLDL can be formed and retained in subendothelial space, where it can induce adhesion molecules as ICAM-1 and VCAM-1 and the secretion of chemotactic protein MCP-1 and mCSF, favoring monocyte recruitment. Also, OxLDL is associated with a decrease in •NO production, leading to endothelial dysfunction, facilitating the atheromatous plaque formation [67]. One of the main mechanisms is that OxLDL up-regulates arginase I expression in the vascular wall without altering eNOS expression, which contributes to endothelial dysfunction by reducing l-arginine availability to eNOS for •NO production [68].

The oxidation of LDL particles increases their uptake by macrophages, so OxLDL is considered highly atherogenic by inducing foam cells formation in the first steps of the atherosclerotic lesion. Indeed, macrophages of the sub-endothelial space express scavenger receptors, which have a high affinity for OxLDL (in contrast to normal LDL) and they are not downregulated, leading to a massive fat accumulation. In turn, the foam cells present pro-inflammatory and pro-oxidative features that promote more LDL oxidization in the sub-endothelial space, and the recruitment of more monocyte to form macrophages and then, foam cells. Finally, the excessive intracellular accumulation of oxLDL induces the apoptosis or necrosis of foam cells, forming an important amount of cell debris that constitute the core of atherosclerotic plaque [67].

The vascular smooth muscle cells migrate from tunica to intima and proliferate in sub-endothelial space, as part of the process of atheromatous plaque formation. OxLDL can favor this, due to its capacity to induce platelet-derived growth factor (PDGF) and fibroblast growth factor (FGF) release form endothelial cells and macrophages. Also, OxLDL induces LOX-1 expression, a scavenger receptor also present in macrophages, allowing the entry of OxLDL in smooth muscle cells, forming new foam cells, this time derived from smooth muscle [69].

7.5.2 Native and Oxidized LDL in Platelet Function

Relevant to dyslipidemia, several groups have described an increase in platelet activation after platelet exposure to native (nLDL) and oxLDL [5, 70, 71]. Given the pro-oxidant environment that develops in dyslipidemia and associated metabolic diseases, the accepted hypothesis is that LDL is partly oxidized into oxLDL, which is the main molecule responsible for potentiating platelet response. It has been shown the importance of platelets in atheromatous plaque formation, and OxLDL also have a role in it. As mentioned above, OxLDL induces an endothelial dysfunction, which in turn favors platelet activation. But also, OxLDL have direct effects on platelets that lead to activation and thrombus formation in the aterothrombosis context. We will review this last aspect with more detail next.

Studies demonstrated that specific oxidized phospholipids that accumulate in plasma in dyslipidemia promote platelet activation by interacting with different scavenger receptors: CD36, lectin-like oxidized LDL receptor-1, scavenger receptor class B type I (SR-BI), Toll-like receptor 2 (TLR2), scavenger receptor that binds phosphatidylserine and oxidized lipoprotein/chemokine (C-X-C motif) ligand [71–75]. Although it was known for some time that platelets possessed CD36, a class B scavenger receptor that binds to several unrelated molecules, such as thrombospondin-1 and -2, [76, 77] as well as oxLDL [78], it was only on the last decade that it was shown that CD36 is the main receptor for oxLDL in platelets. Upon binding to CD36, oxLDL enhances oxidative stress and pro-thrombotic phenotype. In fact, Podrez et al. [7] were the firsts to describe the importance of CD36 in dyslipidemia, showing that genetic deletion of CD36 in murine models of dyslipidemia restored arteriole and venule occlusion times back to control levels.

Less than a year later, the first signalling pathway for oxLDL-CD36 was described. Chen et al. [79] demonstrated for the first time that mitogen-activated protein kinase kinase 4 (MKK4) and downstream c-Jun N-terminal kinase (JNK)2 were phosphorylated in oxLDL-exposed platelets. Interestingly, src family kinases Fyn and Lyn, which are also important for collagen receptor GPVI signalling, were recruited to the cytoplasmic tail of CD36 after exposure to oxLDL. Also, immobilized oxLDL increased protein tyrosine kinase Syk activation and intracellular Ca^{2+} while enhancing thrombus formation under flow in a collagen-coated surface – an effect highly dependent on CD36 signalling [80]. Platelet-bound oxLDL was increased in patients with acute coronary syndromes, while incubation of oxLDL with platelets from healthy individuals was able to enhance platelet adhesion to collagen, as well as activate endothelial cells [81]. Overall, it seems CD36 and collagen signalling pathways are intrinsically connected in the context of dyslipidemia and other metabolic diseases, prompting the discovery of novel treatments targeting CD36 and its pathway components.

In addition to the strong body of evidence showing how oxLDL induce platelet activation, platelets may induce LDL oxidation per si, prompting a positive feedback loop. For instance, supernatant collected from collagen-activated platelets was able to enhance copper-induced LDL oxidation [82]. More direct evidence showed that AA-stimulated but not resting platelets were able to oxidize LDL and HDL in a concentration-dependent manner. Moreover, the authors ascribed the oxidative stress induced by platelets to be due to secretory PLA_2, produced by activated platelets, that would then induce adverse modifications in lipoproteins [83]. Another group proposed that activated platelets oxidize LDL in an NADPH oxidase (Nox)2-dependent manner, by showing that oxLDL had no effect on platelet activation in Nox-2-deficient patients and that the addition of a Nox-2 inhibitor abrogated the production of oxLDL induced by activated platelets [84]. Finally, platelets contribute to the deleterious actions of oxLDL in other cells, such as monocytes, inducing oxLDL-dependant monocyte migration and thus contributing to vascular oxidative stress and atherosclerotic plaque instability [85]. Overall, it seems there is an intricate positive feedback loop in which platelets oxidize LDL that will then activate platelets in a continuum that also involves other blood cells and possibly the endothelium.

Besides increased oxLDL levels, patients with dyslipidemia and associated metabolic syndrome tend to produce more microparticles, derived from platelets (PMP), endothelial cells (EMP) as well as other blood components [86, 87]. Curiously, EMPs were found to bind to platelet surface CD36, since the inhibition of CD36 with monoclonal antibodies or the addition of competitive agonist oxLDL prevented EMP binding [88]. However, the majority of microparticles in blood is derived from platelets [89], which also induce a pro-thrombotic and pro-coagulant state due to PMP-induced platelet activation and the presence of negatively-charged phospholipids, like phosphatidylserine (PS) [86, 90, 91]. Hypothesizing that PS might be a ligand for CD36, Wang et al. [92] demonstrated that not only oxLDL leads to increased production of PMPs, but PMPs per se activate platelet through binding of CD36. Therefore, CD36 is a receptor for PMP binding and consequent pro-thrombotic outcome, while both contribute to the positive feedback installed in metabolic syndrome.

7.5.3 Native and Oxidized HDL in Platelet Function

Although LDL and its oxidized form oxLDL display pro-thrombotic effects through binding to CD36, HDL and oxidized HDL (oxHDL) have been shown to exert the opposite effect binding to a similar receptor. Population studies have shown that HDL is inversely correlated to thrombosis [93], whereas this observation was corroborated *in vitro* by showing that HDL reduces platelet aggregation in response to a variety of agonists, such as collagen and ADP [71]. Although HDL was firstly described to bind to platelet through fibrinogen receptor $\alpha_{IIb}\beta3$ [94, 95], later reports refuted these findings using more refined techniques [96], proposing scavenger receptor B1 (SR-B1) as the binding site of HDL in platelets [97]. It is curious to notice that the initial data on SR-B1 knockout mice showed increased platelet activation, with elevated P-selectin binding at resting and increased adherence to immobilized fibrinogen, due to dyslipidemia and platelet cholesterol overload [73]. However, in ApoE−/− dyslipidemic mice, bone marrow transfusion from SR-B1−/− mice resulted in resistance to platelet hyperactivity as well as protected the pro-thrombotic phenotype present in these dyslipidemic animals [72]. It is now widely accepted that SR-B1 is the HDL receptor in platelets and that there is an inverse relationship between levels of this receptor and platelet response to ADP [98, 99].

Interestingly, oxHDL has been shown to inhibit platelet function more potently than native HDL [71]. Valiyaveettil [71] showed that oxHDL was able to inhibit platelet aggregation induced by collagen, ADP and thrombin – an effect that was dependent on platelet SR-BI. Another study using HDL from diabetic patients have found that glycoxidized HDL was also able to inhibit platelet aggregation induced by collagen [100]. However, others have shown that the oxidation of HDL by hypochlorite produces a different form of oxHDL that is seemingly more oxidized and able to potently activate platelets through binding to CD36 [101, 102]. Hence, it seems that the degree of HDL oxidation determines whether these lipoproteins will exert an anti- or pro-thrombotic effect [103]. It is still elusive which form of oxHDL is predominant in dyslipidemia and whether or not oxHDL functions as a negative loop for the already established oxLDL-platelet positive feedback system.

7.6 Dyslipidemias and Platelet Activation

There is evidence that distinct dyslipidemias (See Box 2 for dyslipidemias definition, classification and diagnosis) are associated with increased platelet activity. Hyperactive platelets with increased platelet cholesterol may contribute to accelerated atherogenesis associated with coronary artery disease. Plasma cholesterol levels appear to have a critical role in modulating platelet activity because hypercholesterolemia increases platelet activation (GPIIb/IIIa, P-selectin and phosphatidylserine expression), platelet FXa generation and platelet tissue factor activity, more potently than hypertriglyceridemia. Thus, plasma membrane cholesterol accumulation in platelets could potentially alter the membrane structure and affect signaling via surface receptors [104–108]. As compared with 26 normal subjects, platelets from patients with the Type II hyperlipoproteinemia aggregated in response to 1/25 the mean concentration of epinephrine, one third the concentration of collagen, and one third the concentration of ADP. In contrast, platelets from type IV hyperlipoproteinemia showed normal sensitivity to ADP and collagen, and normal secretion. This suggest that platelet activation is associated with the thrombotic complications of type II hyperlipoproteinemia [6].

Low-density lipoprotein particles sensitize platelets via binding of apoB-100 to a receptor on the platelet membrane and via transfer of lipids to the platelet membrane. Oxidative modifications of LDL (oxLDL) generate a platelet-activating particle, and this interaction might contribute to the development of the atherosclerotic plaque. Preincubation of isolated platelets with oxLDL, but not with native LDL, resulted in enhanced platelet adhesion to collagen and activated endothelial cells. The oxLDL has been shown to promote platelet activation by interact with different scavenger receptors: class A scavenger receptor, CD36, lectin-like oxidized LDL receptor-1, scavenger receptor class B type I (SR-BI), Toll-like receptor 2 (TLR2), scavenger receptor that binds phosphatidylserine and oxidized lipoprotein/chemokine (C-X-C motif) ligand. Also, oxidized choline glycerophospholipids are markedly increased in plasma of hyperlipidemic mice and in plasma of subjects with low HDL level, and promote platelet activation and hyperreactivity [7, 81, 109, 110].

7.6.1 Dyslipidemias: Definition, Classification and Diagnosis

Dyslipidemias are a group of abnormalities in the lipid metabolism that leads to altered blood levels of triglycerides, lipoproteins or phospholipids. More frequently, dyslipidemias are associated with high levels of blood lipids, although an altered lipoprotein pattern is also important and not only the total lipids levels in blood. Dyslipidemias can be classified regarding their causes: Genetic (familial) or Acquired (secondary). Most of dyslipidemias are hyperlipidemic, and between them, most are caused by a combination of genetic polymorphism and dietary and lifestyle factors. An iconic example of this is the apolipoprotein *E2/E2* polymorphism, which is present in 1/200–1/500 frequency, but its dyslipidemia associated is present only in 1/5000, because it needs the association with certain lifestyles (sedentary, hypercaloric diets) and other disorders, such as obesity, diabetes and/or hypothyroidism. On another hand, only a 2% of hyperlipidemic disorders can be associated with identifiable mutations (as a strong genetic cause) and they are the most difficult to treat and with a high cardiovascular risk. Acquired dyslipidemias are hyperlipemic also, and they commonly associate with diabetes mellitus, hypothyroidism, nephrotic syndrome and use of drugs (thiazide diuretics, beta adrenergic blockers, glucocorticoids, etc.) [111].

Other important classification is based on the type of lipid and lipoprotein elevated in plasma, and it is applied to familial dyslipidemias (Table 7.1) [112]. A third classification (Table 7.2) can be made based on clinical phe-

Table 7.1 Frederickson classification of familial dyslipidemias

Dyslipidemia	Frederickson classification	Lipid elevated	Lipoprotein altered	Genetic cause
Familial hyperchylomicronemia	Type I	TG	Chylomicrons elevated	Autosomal recessive: *LPL, ApoC2, Apo A-V, LMF-1* or *GPIHBP1* mutated
Familial hypercholesterolemia	Type IIA	TC	LDL elevated	Autosomal codominant in heterozygous form: *LDL receptor*, *ApoB* or *PCSK9* mutated in one allele.
				Homozygous form: two alleles mutated of these genes.
Familial combined Hyperlipoproteinemia	Type IIB	TC and TG	VLDL and LDL elevated	Polygenic
Dysbetalipoproteinemia	Type III	TC and TG	IDL elevated	Homozygous for *Apo E2/E2*
				Heterozygous for mutations in *APO E*
Primary hypertriglyceridemia	Type IV	TG	VLDL elevated	No determined
Mixed hypertriglyceridemia	Type V	TC and TG	VLDL and chylomicrons elevated	Polygenic

TG Triglycerides, *TC* Total Cholesterol

Table 7.2 Clinical classification of dyslipidemias

Dyslipidemia	Genetic	Secondary or associated to	Environmental factors associated
Hypercholesterolemia (High TC)	Familial hypercholesterolemia	Hypothyroidism	Diet rich in saturated fats and cholesterol
	Familial combined hypercholesterolemia	Cholestasis	Drugs: androgens, progestogens, anabolic
	Polygenic hypercholesterolemia	Nephrotic syndrome	
Hypertriglyceridemia (High TG)	Familial combined	Obesity	Diet rich in high glycemic index carbs
	Hypercholesterolemia	Diabetes mellitus	High alcohol consume
	Lipoprotein lipase deficiency	Renal insufficiency	Drugs: estrogens, beta blockers, diuretics
	Primary hypertriglyceridemia	Metabolic syndrome	
		Nephrotic syndrome	
HDL-C deficiency	Apo A deficiency	Obesity	Diet rich in saturated or trans fatty acids
	ABCA1 mutations	Diabetes mellitus	Diet rich in high glycemic index carbs
			Sedentary lifestyle
			Smoking
			Drugs: beta blockers, diuretics
Mixed hyperlipidemia (high TG and TC)	Polygenic	Hypothyroidism	Diet rich in saturated fats and cholesterol
		Cholestasis	Drugs: androgens, progestogens, anabolic
		Nephrotic syndrome	Diet rich in high glycemic index carbs
		Obesity	High alcohol consume
		Diabetes mellitus	Drugs: estrogens, beta blockers, diuretics
		Renal insufficiency	
		Metabolic syndrome	

TG Triglycerides, *TC* Total Cholesterol

notype independently of causes, and includes the most frequently dyslipidemia manifestation: the genetic and environmental factors combination. In general, this classification takes into account not only elevations in LDL, VLDL or Chylomicrons, but also HDL decrease, and some of them can be associated to other pathologies [113].

The diagnosis is based mainly on clinical tests for a lipid profile, searching for abnormal levels of triglycerides, LDLc (LDL cholesterol), VDLc (VLDL cholesterol), or HDLc (HDL cholesterol). Usually, low levels of LDLc and/or triglycerides are not considered of clinical importance by themselves. A decrease below normal levels of LDLc and triglycerides are usually considered as secondary to other disorders as liver disease, malabsorption syndromes, hyperthyroidism, etc. On the other hand, high levels of HDLc as an abnormal condition, constitutes a very rare genetic disorder with not well elucidated cardiovascular risk. For these reasons, we will focus on hyperlipidemic disorders [114]. The only clinical manifestation that helps to think specifically in hyperlipidemia is the presence of xanthomas, but they are observed mainly in genetic dyslipidemias with low frequency. In these patients, hepatosplenomegaly, *lipemia retinalis* and family history of pancreatitis (when triglycerides are elevated) can be present. Independently of the dyslipidemic origin, of special concern in all of them is the cardiovascular risk associated, and its determination and management is one of the main goals in dyslipidemia treatment [114]. To evaluate the cardiovascular risk in a dyslipidemic patient is important to consider: clinical manifestations of atherosclerosis (coronary, cerebral or peripheral), age (>45 years in males, >55 in females), family background with atherosclerosis in young, smoking, hypertension, diabetes mellitus and HDLc <40 mg/dl. Of course, the blood levels of triglycerides and total cholesterol are important, but of special relevance for cardiovascular risk, are the LDLc levels. A large amount of evidence support the idea that high levels of LDL in blood, in concert with an elevated oxidative stress in the organism, leads to a high risk to the development of atherosclerotic and atherotrombotic events [115].

7.7 Closing Remarks and Perspectives

Platelets are central workers in the cardiovascular function and hemodynamics, acting as watchers of changes within the bloodstream. As shown in this chapter, lipids play an essential role on platelet metabolism and reactivity, modulating its response in accordance to the serum lipid profile. Although the focus has been placed on the circulating levels of distinct lipoproteins, with LDL particles augmenting and HDL particles decreasing platelet reactivity (Fig. 7.4), the immunological essence of platelet lipid metabolism increases its importance in disorders characterized by low-grade inflammation. Of importance, metabolic syndrome allies pro-thrombotic and pro-inflammatory statuses with dyslipidemia, whereas constitutes itself an independent risk factor for cardiovascular disease [116, 117]. We have previously revised the relationship between platelet hyperactivity and metabolic syndrome, suggesting the redox-sensing protein disulfide isomerase as a central actor [118]. Notwithstanding, the role of lipids in platelet biology has been recently widened with the application of lipidomic approaches [119]. A study just released by Peng, Geue [120] showed how changes in the platelet lipidome may modulate platelet function, uncovering novel therapeutic targets and agents. On this matter, the NO_2-AA might be additionally proposed as a potential target/agent in metabolic syndrome-derived platelet hyperactivity, since its formation depends on both redox and inflammatory processes. Thus, the tale of lipid role on platelet function and dysfunction seems to be an open novel with many interesting chapters to be written.

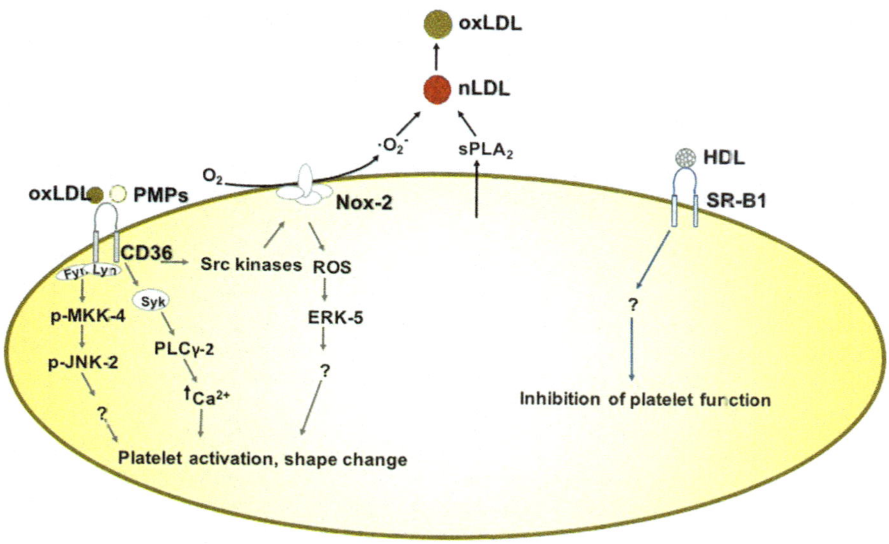

Fig. 7.4 How lipoproteins affect platelet function. Oxidized low-density lipoprotein (oxLDL) or platelet-derived microparticles (PMPs) bind to CD36 on platelets, generating three concomitant responses. On the left, active fyn and lyn will activate mitogen-activated protein kinase kinase 4 (MKK-4) and downstream c-Jun N-terminal kinase (JNK-2). Upon activation, CD36 will also stimulate platelets through a Src kinase-dependent activation of Syk and downstream phospholipase C (PLCγ-2), which will then increase intracellular Ca^{2+}. The third signaling pathway of CD36 involves activation of NADPH oxidase (Nox-2), which will generate reactive oxygen species (ROS) that will act as second messenger, inducing extracellular-signal-regulated kinase (ERK-5) phosphorylation. All of the CD36 pathways culminate in platelet activation, which produces oxidant molecules on the extracellular environment that will oxidize native LDL (nLDL) into oxLDL and perpetrate the positive feedback loop. To counter balance, high-density lipoprotein (HDL) binds to its own receptor, scavenge receptor B1 (SR-B1) to inhibit platelet function. The mechanisms of such inhibition are still poorly understood

References

1. Popkin BM, Adair LS, Ng SW (2012) Global nutrition transition and the pandemic of obesity in developing countries. Nutr Rev 70(1):3–21
2. Alberti KG, Eckel RH, Grundy SM, Zimmet PZ, Cleeman JI, Donato KA et al (2009) Harmonizing the metabolic syndrome: a joint interim statement of the International Diabetes Federation Task Force on Epidemiology and Prevention; National Heart, Lung, and Blood Institute; American Heart Association; World Heart Federation; International Atherosclerosis Society; and International Association for the Study of Obesity. Circulation 120(16):1640–1645
3. Libby P (2002) Inflammation in atherosclerosis. Nature 420(6917):868–874
4. Berger S, Raman G, Vishwanathan R, Jacques PF, Johnson EJ (2015) Dietary cholesterol and cardiovascular disease: a systematic review and meta-analysis. Am J Clin Nutr 102(2):276–294
5. Berger M, Wraith K, Woodward C, Aburima A, Raslan Z, Hindle MS et al (2018) Dyslipidemia-associated atherogenic oxidized lipids induce platelet hyperactivity through phospholipase Cgamma2-dependent reactive oxygen species generation. Platelets:1–6
6. Carvalho AC, Colman RW, Lees RS (1974) Platelet function in hyperlipoproteinemia. N Engl J Med 290(8):434–438
7. Podrez EA, Byzova TV, Febbraio M, Salomon RG, Ma Y, Valiyaveettil M et al (2007) Platelet CD36 links hyperlipidemia, oxidant stress and a prothrombotic phenotype. Nat Med 13(9):1086–1095
8. Colas R, Sassolas A, Guichardant M, Cugnet-Anceau C, Moret M, Moulin P et al (2011) LDL from obese patients with the metabolic syndrome show increased lipid peroxidation and activate platelets. Diabetologia 54(11):2931–2940
9. Massberg S, Gawaz M, Gruner S, Schulte V, Konrad I, Zohlnhofer D et al (2003) A crucial role of glycoprotein VI for platelet recruitment to the injured arterial wall in vivo. J Exp Med 197(1):41–49
10. Savage B, Saldivar E, Ruggeri ZM (1996) Initiation of platelet adhesion by arrest onto fibrinogen or translocation on von Willebrand factor. Cell 84(2):289–297

11. Mancuso ME, Santagostino E (2017) Platelets: much more than bricks in a breached wall. Br J Haematol 178(2):209–219
12. Akkerman JW (2008) From low-density lipoprotein to platelet activation. Int J Biochem Cell Biol 40(11):2374–2378
13. Coller BS (2011) Historical perspective and future directions in platelet research. J Thromb Haemost 9(Suppl 1):374–395
14. Brash AR (2001) Arachidonic acid as a bioactive molecule. J Clin Invest 107(11):1339–1345
15. Das UN (2018) Arachidonic acid in health and disease with focus on hypertension and diabetes mellitus: a review. J Adv Res 11:43–55
16. Das UN (2018) Arachidonic acid and other unsaturated fatty acids and some of their metabolites function as endogenous antimicrobial molecules: a review. J Adv Res 11:57–66
17. Hanna VS, Hafez EAA (2018) Synopsis of arachidonic acid metabolism: a review. J Adv Res 11:23–32
18. Tsai IJ, Croft KD, Puddey IB, Beilin LJ, Barden A (2011) 20-Hydroxyeicosatetraenoic acid synthesis is increased in human neutrophils and platelets by angiotensin II and endothelin-1. Am J Physiol Heart Circ Physiol 300(4):H1194–H1200
19. Neufeld EJ, Majerus PW (1983) Arachidonate release and phosphatidic acid turnover in stimulated human platelets. J Biol Chem 258(4):2461–2467
20. Davi G, Patrono C (2007) Platelet activation and Atherothrombosis. N Engl J Med 357(24):2482–2494
21. Holinstat M, Boutaud O, Apopa PL, Vesci J, Bala M, Oates JA et al (2011) Protease-activated receptor signaling in platelets activates cytosolic phospholipase A2alpha differently for cyclooxygenase-1 and 12-lipoxygenase catalysis. Arterioscler Thromb Vasc Biol 31(2):435–442
22. Maskrey BH, Bermudez-Fajardo A, Morgan AH, Stewart-Jones E, Dioszeghy V, Taylor GW et al (2007) Activated platelets and monocytes generate four hydroxyphosphatidylethanolamines via lipoxygenase. J Biol Chem 282(28):20151–20163
23. Thomas CP, Morgan LT, Maskrey BH, Murphy RC, Kuhn H, Hazen SL et al (2010) Phospholipid-esterified eicosanoids are generated in agonist-activated human platelets and enhance tissue factor-dependent thrombin generation. J Biol Chem 285(10):6891–6903
24. Dana R, Leto TL, Malech HL, Levy R (1998) Essential requirement of cytosolic phospholipase A2 for activation of the phagocyte NADPH oxidase. J Biol Chem 273(1):441–445
25. Han C, Demetris AJ, Michalopoulos G, Shelhamer JH, Wu T (2002) 85-kDa cPLA(2) plays a critical role in PPAR-mediated gene transcription in human hepatoma cells. Am J Physiol Gastrointest Liver Physiol 282(4):586–597
26. Hurt-Camejo E, Camejo G, Peilot H, Oorni K, Kovanen P (2001) Phospholipase A(2) in vascular disease. CircRes 89(4):298–304
27. Das UN (2018) Ageing: is there a role for arachidonic acid and other bioactive lipids? A review. J Adv Res 11:67–79
28. Baldassare JJ, Henderson PA, Burns D, Loomis C, Fisher GJ (1992) Translocation of protein kinase C isozymes in thrombin-stimulated human platelets. Correlation with 1,2-diacylglycerol levels. J Biol Chem 267(22):15585–15590
29. Kolesnick RN, Hemer MR (1990) Physiologic 1,2-diacylglycerol levels induce protein kinase C-independent translocation of a regulatory enzyme. J Biol Chem 265(19):10900–10904
30. Yamamoto Y, Kambayashi Y, Ito T, Watanabe K, Nakano M (1997) 1,2-diacylglycerol hydroperoxides induce the generation and release of superoxide anion from human polymorphonuclear leukocytes. FEBS Lett 412(3):461–464
31. Smith WL, DeWitt DL, Garavito RM (2000) Cyclooxygenases: structural, cellular, and molecular biology. Annu Rev Biochem 69:145–182
32. Ikei KN, Yeung J, Apopa PL, Ceja J, Vesci J, Holman TR et al (2012) Investigations of human platelet-type 12-lipoxygenase: role of lipoxygenase products in platelet activation. J Lipid Res 53(12):2546–2559
33. Marnett LJ, Rowlinson SW, Goodwin DC, Kalgutkar AS, Lanzo CA (1999) Arachidonic acid oxygenation by COX-1 and COX-2. Mechanisms of catalysis and inhibition. J Biol Chem 274(33):22903–22906
34. Mollace V, Muscoli C, Masini E, Cuzzocrea S, Salvemini D (2005) Modulation of prostaglandin biosynthesis by nitric oxide and nitric oxide donors. Pharmacol Rev 57(2):217–252
35. Murphy RC, Gijon MA (2007) Biosynthesis and metabolism of leukotrienes. Biochem J 405(3):379–395
36. Nascimento-Silva V, Arruda MA, Barja-Fidalgo C, Villela CG, Fierro IM (2005) Novel lipid mediator aspirin-triggered lipoxin A4 induces heme oxygenase-1 in endothelial cells. Am J Physiol Cell Physiol 289(3):C557–C563
37. O'Donnell VB, Maskrey B, Taylor GW (2009) Eicosanoids: generation and detection in mammalian cells. Methods Mol Biol 462:5–23
38. Jennings LK (2009) Mechanisms of platelet activation: need for new strategies to protect against platelet-mediated atherothrombosis. Thromb Haemost 102(2):248–257
39. O'Donnell VB, Coles B, Lewis MJ, Crews BC, Marnett LJ, Freeman BA (2000) Catalytic consumption of nitric oxide by prostaglandin H synthase-1 regulates platelet function. J Biol Chem 275(49):38239–38244
40. Trostchansky A, Bonilla L, Thomas CP, O'Donnell VB, Marnett LJ, Radi R et al (2011) Nitroarachidonic acid, a novel peroxidase inhibitor of prostaglandin endoperoxide H synthases 1 and 2. J Biol Chem 286(15):12891–12900
41. Kirkby NS, Reed DM, Edin ML, Rauzi F, Mataragka S, Vojnovic I et al (2015) Inherited human group IVA cytosolic phospholipase A2 deficiency abolishes

platelet, endothelial, and leucocyte eicosanoid generation. FASEB J Off Publ Fed Am Soc Exp Biol 29(11):4568–4578
42. Rauzi F, Kirkby NS, Edin ML, Whiteford J, Zeldin DC, Mitchell JA et al (2016) Aspirin inhibits the production of proangiogenic 15(S)-HETE by platelet cyclooxygenase-1. FASEB J Off Publ Fed Am Soc Exp Biol 30(12):4256–4266
43. Marnett LJ (2002) Recent developments in cyclooxygenase inhibition. Prostaglandins Other Lipid Mediat 68-69:153–164
44. Brash AR (1999) Lipoxygenases: occurrence, functions, catalysis, and acquisition of substrate. J Biol Chem 274(34):23679–23682
45. Kalgutkar AS, Crews BC, Rowlinson SW, Garner C, Seibert K, Marnett LJ (1998) Aspirin-like molecules that covalently inactivate cyclooxygenase-2. Science 280(5367):1268–1270
46. Kalgutkar AS, Kozak KR, Crews BC, Hochgesang GP Jr, Marnett LJ (1998) Covalent modification of cyclooxygenase-2 (COX-2) by 2-acetoxyphenyl alkyl sulfides, a new class of selective COX-2 inactivators. J Med Chem 41(24):4800–4818
47. Bonilla L, O'Donnell VB, Clark SR, Rubbo H, Trostchansky A (2013) Regulation of protein kinase C by nitroarachidonic acid: impact on human platelet activation. Arch Biochem Biophys 533(1–2):55–61
48. Rouzer CA, Marnett LJ (2008) Non-redundant functions of cyclooxygenases: oxygenation of endocannabinoids. J Biol Chem 283(13):8065–8069
49. Maskrey BH, Rushworth GF, Law MH, Treweeke AT, Wei J, Leslie SJ et al (2014) 12-hydroxyeicosatetraenoic acid is associated with variability in aspirin-induced platelet inhibition. J Inflamm 11(1):33
50. Porro B, Songia P, Squellerio I, Tremoli E, Cavalca V (2014) Analysis, physiological and clinical significance of 12-HETE: a neglected platelet-derived 12-lipoxygenase product. J Chromatogr B Analyt Technol Biomed Life Sci 964:26–40
51. Midgett C, Stitham J, Martin K, Hwa J (2011) Prostacyclin receptor regulation--from transcription to trafficking. Curr Mol Med 11(7):517–528
52. Stitham J, Midgett C, Martin KA, Hwa J (2011) Prostacyclin: an inflammatory paradox. Front Pharmacol 2:24
53. Gimenez-Bastida JA, Boeglin WE, Boutaud O, Malkowski MG, Schneider C (2018) Residual cyclooxygenase activity of aspirin-acetylated COX-2 forms 15 R-prostaglandins that inhibit platelet aggregation. FASEB J Off Publ Fed Am Soc Exp Biol:fj201801018R
54. Schuligoi R, Schmidt R, Geisslinger G, Kollroser M, Peskar BA, Heinemann A (2007) PGD2 metabolism in plasma: kinetics and relationship with bioactivity on DP1 and CRTH2 receptors. Biochem Pharmacol 74(1):107–117
55. Deeb RS, Upmacis RK, Lamon BD, Gross SS, Hajjar DP (2008) Maintaining equilibrium by selective targeting of cyclooxygenase pathways: promising offensives against vascular injury. Hypertension 51(1):1–7
56. Petrucci G, De Cristofaro R, Rutella S, Ranelletti FO, Pocaterra D, Lancellotti S et al (2011) Prostaglandin E2 differentially modulates human platelet function through the prostanoid EP2 and EP3 receptors. J Pharmacol Exp Ther 336(2):391–402
57. Glenn JR, White AE, Iyu D, Heptinstall S (2012) PGE(2) reverses G(s)-mediated inhibition of platelet aggregation by interaction with EP3 receptors, but adds to non-G(s)-mediated inhibition of platelet aggregation by interaction with EP4 receptors. Platelets 23(5):344–351
58. Friedman EA, Ogletree ML, Haddad EV, Boutaud O (2015) Understanding the role of prostaglandin E2 in regulating human platelet activity in health and disease. Thromb Res 136(3):493–503
59. Kauskot A, Hoylaerts MF (2012) Platelet receptors. In: Handbook of experimental pharmacology, vol 210. Springer-Verlag, Berlin Heidelberg, pp 23–57
60. Whittle BJ, Silverstein AM, Mottola DM, Clapp LH (2012) Binding and activity of the prostacyclin receptor (IP) agonists, treprostinil and iloprost, at human prostanoid receptors: treprostinil is a potent DP1 and EP2 agonist. Biochem Pharmacol 84(1):68–75
61. Gross S, Tilly P, Hentsch D, Vonesch JL, Fabre JE (2007) Vascular wall-produced prostaglandin E2 exacerbates arterial thrombosis and atherothrombosis through platelet EP3 receptors. J Exp Med 204(2):311–320
62. Trostchansky A, O'Donnell VB, Goodwin DC, Landino LM, Marnett LJ, Radi R et al (2007) Interactions between nitric oxide and peroxynitrite during prostaglandin endoperoxide H synthase-1 catalysis: a free radical mechanism of inactivation. Free Radic Biol Med 42(7):1029–1038
63. Trostchansky A, Souza JM, Ferreira A, Ferrari M, Blanco F, Trujillo M et al (2007) Synthesis, isomer characterization, and anti-inflammatory properties of nitroarachidonate. Biochemistry 46(15):4645–4653
64. O'Donnell VB, Eiserich JP, Chumley PH, Jablonsky MJ, Krishna NR, Kirk M et al (1999) Nitration of unsaturated fatty acids by nitric oxide-derived reactive nitrogen species peroxynitrite, nitrous acid, nitrogen dioxide, and nitronium ion. Chem Res Toxicol 12(1):83–92
65. Trostchansky A, Rubbo H (2008) Nitrated fatty acids: mechanisms of formation, chemical characterization, and biological properties. Free Radic Biol Med 44(11):1887–1896
66. Koenig W, Karakas M, Zierer A, Herder C, Baumert J, Meisinger C et al (2011) Oxidized LDL and the risk of coronary heart disease: results from the MONICA/KORA Augsburg study. Clin Chem 57(8):1196–1200
67. Di Pietro N, Formoso G, Pandolfi A (2016) Physiology and pathophysiology of oxLDL uptake by vascular wall cells in atherosclerosis. Vasc Pharmacol 84:1–7

68. Wang W, Hein TW, Zhang C, Zawieja DC, Liao JC, Kuo L (2011) Oxidized low-density lipoprotein inhibits nitric oxide-mediated coronary arteriolar dilation by up-regulating endothelial arginase I. Microcirculation 18(1):36–45
69. Pirillo A, Norata GD, Catapano AL (2013) LOX-1, OxLDL, and atherosclerosis. Mediat Inflamm 2013:152786
70. Hackeng CM, Franke B, Relou IA, Gorter G, Bos JL, van Rijn HJ et al (2000) Low-density lipoprotein activates the small GTPases Rap1 and Ral in human platelets. Biochem J 349(Pt 1):231–238
71. Valiyaveettil M, Kar N, Ashraf MZ, Byzova TV, Febbraio M, Podrez EA (2008) Oxidized high-density lipoprotein inhibits platelet activation and aggregation via scavenger receptor BI. Blood 111(4):1962–1971
72. Ma Y, Ashraf MZ, Podrez EA (2010) Scavenger receptor BI modulates platelet reactivity and thrombosis in dyslipidemia. Blood 116(11):1932–1941
73. Korporaal SJ, Meurs I, Hauer AD, Hildebrand RB, Hoekstra M, Cate HT et al (2011) Deletion of the high-density lipoprotein receptor scavenger receptor BI in mice modulates thrombosis susceptibility and indirectly affects platelet function by elevation of plasma free cholesterol. Arterioscler Thromb Vasc Biol 31(1):34–42
74. Biswas S, Zimman A, Gao D, Byzova TV, Podrez EA (2017) TLR2 plays a key role in platelet Hyperreactivity and accelerated thrombosis associated with hyperlipidemia. Circ Res 121(8):951–962
75. Assinger A, Koller F, Schmid W, Zellner M, Koller E, Volf I (2010) Hypochlorite-oxidized LDL induces intraplatelet ROS formation and surface exposure of CD40L--a prominent role of CD36. Atherosclerosis 213(1):129–134
76. Silverstein RL, Nachman RL (1987) Thrombospondin binds to monocytes-macrophages and mediates platelet-monocyte adhesion. J Clin Invest 79(3):867–874
77. Simantov R, Febbraio M, Silverstein RL (2005) The antiangiogenic effect of thrombospondin-2 is mediated by CD36 and modulated by histidine-rich glycoprotein. Matrix Biol 24(1):27–34
78. Podrez EA, Poliakov E, Shen Z, Zhang R, Deng Y, Sun M et al (2002) Identification of a novel family of oxidized phospholipids that serve as ligands for the macrophage scavenger receptor CD36. J Biol Chem 277(41):38503–38516
79. Chen K, Febbraio M, Li W, Silverstein RL (2008) A specific CD36-dependent signaling pathway is required for platelet activation by oxidized low-density lipoprotein. Circ Res 102(12):1512–1519
80. Nergiz-Unal R, Lamers MM, Van Kruchten R, Luiken JJ, Cosemans JM, Glatz JF et al (2011) Signaling role of CD36 in platelet activation and thrombus formation on immobilized thrombospondin or oxidized low-density lipoprotein. J Thromb Haemost 9(9):1835–1846
81. Stellos K, Sauter R, Fahrleitner M, Grimm J, Stakos D, Emschermann F et al (2012) Binding of oxidized low-density lipoprotein on circulating platelets is increased in patients with acute coronary syndromes and induces platelet adhesion to vascular wall in vivo--brief report. Arterioscler Thromb Vasc Biol 32(8):2017–2020
82. Aviram M, Dankner G, Brook JG (1990) Platelet secretory products increase low density lipoprotein oxidation, enhance its uptake by macrophages, and reduce its fluidity. Arteriosclerosis 10(4):559–563
83. Blache D, Gautier T, Tietge UJ, Lagrost L (2012) Activated platelets contribute to oxidized low-density lipoproteins and dysfunctional high-density lipoproteins through a phospholipase A2-dependent mechanism. FASEB J Off Publ Fed Am Soc Exp Biol 26(2):927–937
84. Carnevale R, Bartimoccia S, Nocella C, Di Santo S, Loffredo L, Illuminati G et al (2014) LDL oxidation by platelets propagates platelet activation via an oxidative stress-mediated mechanism. Atherosclerosis 237(1):108–116
85. Badrnya S, Schrottmaier WC, Kral JB, Yaiw KC, Volf I, Schabbauer G et al (2014) Platelets mediate oxidized low-density lipoprotein-induced monocyte extravasation and foam cell formation. Arterioscler Thromb Vasc Biol 34(3):571–580
86. Helal O, Defoort C, Robert S, Marin C, Lesavre N, Lopez-Miranda J et al (2011) Increased levels of microparticles originating from endothelial cells, platelets and erythrocytes in subjects with metabolic syndrome: relationship with oxidative stress. Nutr Metab Cardiovasc Dis 21(9):665–671
87. Zhang X, McGeoch SC, Johnstone AM, Holtrop G, Sneddon AA, MacRury SM et al (2014) Platelet-derived microparticle count and surface molecule expression differ between subjects with and without type 2 diabetes, independently of obesity status. J Thromb Thrombolysis 37(4):455–463
88. Ghosh A, Li W, Febbraio M, Espinola RG, McCrae KR, Cockrell E et al (2008) Platelet CD36 mediates interactions with endothelial cell-derived microparticles and contributes to thrombosis in mice. J Clin Invest 118(5):1934–1943
89. Hunter MP, Ismail N, Zhang X, Aguda BD, Lee EJ, Yu L et al (2008) Detection of microRNA expression in human peripheral blood microvesicles. PLoS One 3(11):e3694
90. Aleman MM, Gardiner C, Harrison P, Wolberg AS (2011) Differential contributions of monocyte- and platelet-derived microparticles towards thrombin generation and fibrin formation and stability. J Thromb Haemost 9(11):2251–2261
91. Geddings JE, Mackman N (2014) New players in haemostasis and thrombosis. Thromb Haemost 111(4):570–574
92. Wang H, Wang ZH, Kong J, Yang MY, Jiang GH, Wang XP et al (2012) Oxidized low-density lipoprotein-dependent platelet-derived microvesicles

trigger procoagulant effects and amplify oxidative stress. Mol Med 18:159–166
93. Ageno W, Becattini C, Brighton T, Selby R, Kamphuisen PW (2008) Cardiovascular risk factors and venous thromboembolism: a meta-analysis. Circulation 117(1):93–102
94. Koller E, Koller F, Binder BR (1989) Purification and identification of the lipoprotein-binding proteins from human blood platelet membrane. J Biol Chem 264(21):12412–12418
95. Nofer JR, Walter M, Kehrel B, Seedorf U, Assmann G (1995) HDL3 activates phospholipase D in normal but not in glycoprotein IIb/IIIa-deficient platelets. Biochem Biophys Res Commun 207(1):148–154
96. Pedreno J, de Castellarnau C, Masana L (2001) Platelet HDL(3) binding sites are not related to integrin alpha(IIb)beta(3) (GPIIb-IIIa). Atherosclerosis 154(1):23–29
97. Brodde MF, Korporaal SJ, Herminghaus G, Fobker M, Van Berkel TJ, Tietge UJ et al (2011) Native high-density lipoproteins inhibit platelet activation via scavenger receptor BI: role of negatively charged phospholipids. Atherosclerosis 215(2):374–382
98. Imachi H, Murao K, Cao W, Tada S, Taminato T, Wong NC et al (2003) Expression of human scavenger receptor B1 on and in human platelets. Arterioscler Thromb Vasc Biol 23(5):898–904
99. Nofer JR, van Eck M (2011) HDL scavenger receptor class B type I and platelet function. Curr Opin Lipidol 22(4):277–282
100. Le QH, El Alaoui M, Vericel E, Segrestin B, Soulere L, Guichardant M et al (2015) Glycoxidized HDL, HDL enriched with oxidized phospholipids and HDL from diabetic patients inhibit platelet function. J Clin Endocrinol Metab 100(5):2006–2014
101. Assinger A, Schmid W, Eder S, Schmid D, Koller E, Volf I (2008) Oxidation by hypochlorite converts protective HDL into a potent platelet agonist. FEBS Lett 582(5):778–784
102. Assinger A, Koller F, Schmid W, Zellner M, Babeluk R, Koller E et al (2010) Specific binding of hypochlorite-oxidized HDL to platelet CD36 triggers proinflammatory and procoagulant effects. Atherosclerosis 212(1):153–160
103. van der Stoep M, Korporaal SJ, Van Eck M (2014) High-density lipoprotein as a modulator of platelet and coagulation responses. Cardiovasc Res 103(3):362–371
104. Wang N, Tall AR (2016) Cholesterol in platelet biogenesis and activation. Blood 127(16):1949–1953
105. Ravindran R, Krishnan LK (2007) Increased platelet cholesterol and decreased percentage volume of platelets as a secondary risk factor for coronary artery disease. Pathophysiol Haemost Thromb 36(1):45–51
106. Relou IA, Hackeng CM, Akkerman JW, Malle E (2003) Low-density lipoprotein and its effect on human blood platelets. Cell Mol Life Sci 60(5):961–971
107. Kumar S, Vikram A, Kim YR (2014) J SJ, Irani K. P66Shc mediates increased platelet activation and aggregation in hypercholesterolemia. Biochem Biophys Res Commun 449(4):496–501
108. Panes O, González C, Hidalgo P, Valderas JP, Acevedo M, Contreras S et al (2017) Platelet tissue factor activity and membrane cholesterol are increased in hypercholesterolemia and normalized by rosuvastatin, but not by atorvastatin. Atherosclerosis 257:164–171
109. Tsimikas S, Brilakis ES, Miller ER, McConnell JP, Lennon RJ, Kornman KS et al (2005) Oxidized phospholipids, Lp(a) lipoprotein, and coronary artery disease. N Engl J Med 353(1):46–57
110. Chen K, Li W, Major J, Rahaman SO, Febbraio M, Silverstein RL (2011) Vav guanine nucleotide exchange factors link hyperlipidemia and a prothrombotic state. Blood 117(21):5744–5750
111. Ramasamy I (2016) Update on the molecular biology of dyslipidemias. Clin Chim Acta 454:143–185
112. Beaumont JL, Carlson LA, Cooper GR, Fejfar Z, Fredrickson DS, Strasser T (1970) Classification of hyperlipidaemias and hyperlipoproteinaemias. Bull World Health Organ 43(6):891–915
113. AACE (2017) Guidelines. American Association of Clinical Endocrinologists and American College of Endocrinology guidelines for management of dyslipidemia and prevention of cardiovascular disease. Endocr Pract 23(2):1–87
114. Tada H, Kawashiri MA, Yamagishi M (2017) Clinical perspectives of genetic analyses on dyslipidemia and coronary artery disease. J Atheroscler Thromb 24(5):452–461
115. Nelson RH (2013) Hyperlipidemia as a risk factor for cardiovascular disease. Prim Care 40(1):195–211
116. van Rooy MJ, Duim W, Ehlers R, Buys AV, Pretorius E (2015) Platelet hyperactivity and fibrin clot structure in transient ischemic attack individuals in the presence of metabolic syndrome: a microscopy and thromboelastography study. Cardiovasc Diabetol 14:86
117. van Rooy MJ, Pretorius E (2015) Metabolic syndrome, platelet activation and the development of transient ischemic attack or thromboembolic stroke. Thromb Res 135(3):434–442
118. Gaspar RS, Trostchansky A, Paes AM (2016) Potential role of protein disulfide isomerase in metabolic syndrome-derived platelet hyperactivity. Oxidative Med Cell Longev 2016:2423547
119. McFadyen JD, Peter K (2018) Platelet lipidomics and function: joining the dots. Blood 132(5):465–466
120. Peng B, Geue S, Coman C, Munzer P, Kopczynski D, Has C et al (2018) Identification of key lipids critical for platelet activation by comprehensive analysis of the platelet lipidome. Blood 132(5):e1–e12

Dyslipidemia in Ischemia/Reperfusion Injury

8

Tamara Mazo, Veronica D'Annunzio, Martin Donato, Virginia Perez, Tamara Zaobornyj, and Ricardo J. Gelpi

Abstract

Ischemic heart disease is the main cause of morbidity and mortality in the developed world. Although reperfusion therapies are currently the best treatment for this entity, the restoration of blood flow leads, under certain circumstances, to a form of myocardial damage called reperfusion injury. Several studies have shown that age, sex, smoking, diabetes and dyslipidemia are risk factors for cardiovascular diseases. Among these risk factors, dyslipidemias are present in 40% of patients with ischemic heart disease and represent the clinical factor with the greatest impact on the prognosis of patients with cardiovascular diseases. It is known that during reperfusion the increase of the oxidative stress is perhaps one of the most important mechanisms implicated in cell damage. That is why several researchers have studied protective mechanisms against reperfusion injury, such as the ischemic pre- and post- conditioning, making emphasis mainly on the reduction of oxidative stress. However, few of these efforts have been successfully translated into the clinical setting. The controversial results in regards to the relation between cardioprotective mechanisms and dyslipidemia/hypercholesterolemia are mainly due to the difference among quality, composition and the time of administration of hypercholesterolemic diets, as well as the difference in the species used in each of the studies. Therefore, in order to compare results, it is crucial that all variables that could modify the obtained results are taken into consideration.

Keywords

Dyslipidemia · Ischemia/reperfusion · Cardioprotection · Thioredoxin · Myocardial infarction

The authors Veronica D'Annunzio, Martin Donato, Tamara Zaobornyj, and Ricardo J. Gelpi are Member of the National Council of Scientific and Technological Research of Argentina (CONICET).
The authors Tamara Mazo and Virginia Perez are Fellowship of the National Council of Scientific and Technological Research of Argentina (CONICET).

T. Mazo · V. D'Annunzio · M. Donato · V. Perez
R. J. Gelpi (✉)
Institute of Cardiovascular Physiopathology and Department of Pathology and Institute of Biochemistry and Molecular Medicine (IBIMOL UBA-CONICET), Faculty of Medicine, University of Buenos Aires, Buenos Aires, Argentina
e-mail: rgelpi@fmed.uba.ar

T. Zaobornyj
Institute of Biochemistry and Molecular Medicine (IBIMOL UBA-CONICET) School of Pharmacy and Biochemistry, University of Buenos Aires, Buenos Aires, Argentina

A. Trostchansky, H. Rubbo (eds.), *Bioactive Lipids in Health and Disease*, Advances in Experimental Medicine and Biology 1127, https://doi.org/10.1007/978-3-030-11488-6_8

8.1 General Overview

Cardiovascular disease (CVD) was the most common underlying cause of death in the world in 2017, accounting for an estimated 17.3 million of 54 million total deaths, or 31.5% of all global deaths [1]. Also, the World Health Organization report stated that CVD caused 15 million deaths annually with 85% of these occurring in low- and middle-income countries [2], between the ages of 30 and 69 years. Metabolic risk factors contribute to four key metabolic changes that increase the possibility of myocardial infarction: raised blood pressure, overweight/obesity, hyperglycemia and dyslipidemia. Although epidemiological studies have been crucial to determine risk factors associated to coronary disease, the effect of cholesterol content in the diet on the development of coronary atherosclerosis was known much earlier. In this sense, Anitschkow and Chalatow described in 1913 for the first time that rabbits, fed with an cholesterol-enriched diet developed hypercholesterolemia, and that this increase in plasma cholesterol levels contributed to the onset of atherosclerosis [3]. Given the relevance and the prevalence of this risk factor, it is interesting to perform an update on the behavior of ischemia/reperfusion injury (I/R) and the myocardial protection mechanisms in the presence of comorbidities such as hypercholesterolemia/dyslipidemia.

The treatment, which has shown to be more effective for ischemic heart disease, are reperfusion therapies that decrease infarct size and mortality due to acute myocardial infarction, but this reperfusion therapies, under certain circumstances developed damage. Reactive oxygen species (ROS) are one of the key signals that participate and produce injury in myocardial I/R. Regarding oxidative stress, it has been demonstrated that thioredoxin-1 (Trx-1) plays a central role as a regulator redox status in the heart [4–6]. Among the endogenous antioxidant mechanisms, Trx-1 system is an important component that participates in cellular defense against myocardial damage and it was shown to confer protection against I/R injury [5, 7–10]. Recently we have reported that ischemic postconditioning (PostC) prevents the degradation of Trx-1. Therefore, this protein could be involved in this cardioprotective mechanism [9]. On the other hand, it is known that dyslipidemia and lipid metabolism disorders produce an increase in ROS, which positively correlates with dyslipidemia [11], causing an exacerbation of the damage produced by I/R injury [12]. Taken all this together, in this chapter, we will focus on the study of hypercholesterolemic/dyslipidemic models that were established with enriched diets in order to understand and compare the diversity of results.

8.2 Effects of Dyslipidemia/Hypercholesterolemia in Myocardial Function

In epidemiological studies, there is a well-recognized relationship between total serum cholesterol concentration and the morbi-mortality due to myocardial infarction [13]. In addition, hypercholesterolemia has been shown to exert myocardial effects independent of the development of atherosclerosis in both clinical [14, 15] and preclinical studies [16, 17]. These myocardial effects include impaired cardiac performance and contractile dysfunction [18, 19], aggravated CVD [20], and diminished adaptation to ischemic stress [21, 22]. Results in animal models and some human clinical trials suggest a direct effect of cholesterol on the heart leading to impaired diastolic function and in some cases systolic function [16, 17, 19, 23, 24]. Hypercholesterolemia could be asymptomatic or its symptoms may not be pronounced initially, thus the myocardial ventricular dysfunction may not be detected for a long time although it follows a clear progression to heart failure [25].

It was demonstrated in a cell culture model that the increase in cholesteterol content into the membranes, resulted in a decrease of cytosolic calcium levels and impaired cardiac myocyte contractility [26]. Then, this was confirmed in cholesterol-fed rabbits and rats; showing contractile dysfunction characterized by a decreased maximum rate of shortening, decreased rate of relaxation, and increased left ventricular end-diastolic pressure [16, 17, 27]. Moreover, in hearts isolated from hypercholesterolemic apoB100 transgenic mice an impairment of cardiac performance assessed by the measurement of aortic flow was demonstrated [19]. In addition, hypercholesterolemia-induced cardiac dysfunction was further confirmed by echocardiography in humans [23]. From the above mentioned, it is clear that dyslipidemia/hypercholesterolemia produce contractile and diastolic dysfunction in experimental animal models and human.

8.3 Reversible Ischemia/Reperfusion Injury: Stunned Myocardium

As far as we know, few studies assessed the effects of dyslipidemia in the myocardial stunning. In this regard, Van de Velde et al. [28] administered an emulsion of fatty acids during reperfusion and showed an improvement of the recovery of the ventricular function and the metabolic status in the stunned myocardium in isolated rabbit hearts. Additionally, Satoh et al. [29] showed in dogs subjected to a myocardial stunning protocol, that the treatment with pitavastatin reduced plasma cholesterol levels and attenuated post- ischemic contractile dysfunction. However, we demonstrated in hearts of rabbits fed with a 1% enriched cholesterol diet during 4 weeks subjected to a myocardial stunning protocol, a significant improvement in the recovery of the contractile state, as well as attenuation in myocardial stiffness compared with the normocholesterolemic group [30]. One important aspect to highlight is that when isoproterenol was administered to match the pre-ischemic inotropic state with normocholesterolemic animals, the protective effect on post-ischemic ventricular dysfunction was abolished. Our findings suggest that the low inotropic state of the hypercholesterolemic rabbit hearts could reduce oxygen consumption thus protecting the myocardium and this is represented in Fig. 8.1.

In regards to our findings, Calabresi et al. [31], who administered a HDL cholesterol infusion in isolated and perfused rat hearts, described a cardioprotective effect in a model of stunned heart as we did. In addition, Kalaivanisailaja et al. [32] showed that a cholesterol-enriched diet causes an increase in free fatty acids and confers cardioprotection against I/R injury. Therefore, this mechanism could also be part of the protection evidenced by our experimental model. Also, Adameova et al. [33] showed that a 5-day treatment with simvastatin improved post-ischemic recovery of myocardial function in the hearts of both normal and diabetic-hypercholesterolemic rats without normalizing plasma cholesterol levels.

In relation to oxidative stress, we previously demonstrated in healthy mice that the inactivation of Trx-1 produces an exacerbation of post-ischemic ventricular dysfunction in a stunned myocardium protocol. Furthermore, in these mice, a more marked myocardial stiffness and isovolumic relaxation alterations at the end of reperfusion were observed as compared to wild type mice. In that study we showed for the first time a direct effect of cytosolic Trx-1 on systolic and diastolic ventricular function, the signaling pathway involved in relaxation impairment and their relationship to mitochondrial function. Recently we demonstrated using a model of early stages of atherosclerosis, that dyslipidemia produced changes in the redox state and in mitochondrial function in hearts from mice fed with a high- fat diet (HFD). This was associated with an augmentation of Trx-1 expression, but this protein was showed to be more oxidized, which represents an inactivation of Trx-1. Therefore, although we have not studied the effects produced by dyslipidemia and its relationship with Trx-1 in a model of myocardial stunning, [6, 9], we could speculate based on our previous findings that the lack of activity of this antioxidant, could at least in part, worsen ventricular dysfunction in a myocardial stunning protocol.

In summary, the experimental evidence suggests that dyslipidemia, during the early steps of atherosclerosis, could decrease the contractile state and it is clear that hypercholesterolemia/dyslipidemia interfere in the recovery of ventricular function in myocardial stunning. All these

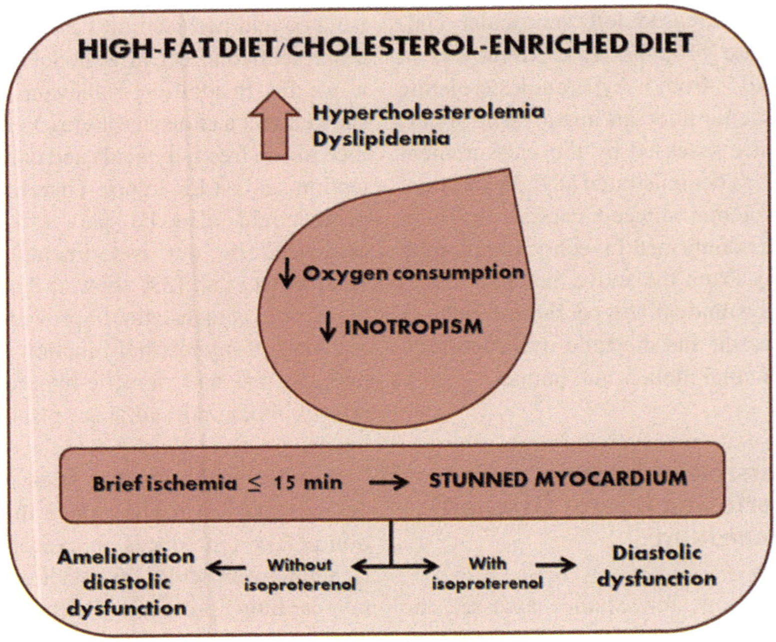

Fig. 8.1 High-fat diet/cholesterol-enriched diet effects in stunned myocardium: High-fat and/or cholesterol-enriched diets cause dyslipidemia and hypercholesterolemia. This can affect the myocardial function itself by reducing inotropism and oxygen consumption. Then the myocardium subjected to a brief period of ischemia (stunned myocardium) presents a lower baseline systolic function (LVDP: left ventricular developed pressure). Thus, in hypercholesterolemic animals, there is a decrease of the inotropism at basal conditions, and there is an attenuation of the stunned myocardium. When contractility of the normal and hypercholesterolemic animals is matched with isoproterenol addition, the beneficial effect disappears

changes depend on the type of diet administered, the time of administration and the species studied. Therefore, it is difficult to extrapolate or compare the different studies where cholesterol-enriched diets were used as a tool to develop different stages of hypercholesterolemic or atherosclerotic disease.

8.4 Irreversible Ischemia Reperfusion Injury: Myocardial Infarction

It is very well known that experimental dyslipidemia can exacerbate I/R injury and increase infarct size. In this sense, Golino et al. [34], showed that hypercholesterolemia *per se*, (3 days of 2% enriched cholesterol diet) significantly increased the infarct size. Likewise, Wang et al. [35] showed that cholesterol administration increased the infarct size and the number of apoptotic cells in the ischemic zone related to an increased of caspase- 3 activity that could be activated by intrinsic or extrinsic pathway. In addition, Szucs et al. [36] showed that a 2% enriched-cholesterol diet (9 weeks) increased the infarct size and this was probably related to an increase in metalloproteinase type-2 (MMP-2) activity during reperfusion compared to the control diet group. This result suggests an increase in oxidative stress and consequently in the peroxynitrite formation and MMP-2 activation in hypercholesterolemic animals [37]. These findings are in agreement with our previously published results, as we also showed that the infarct size was significantly larger in the hearts of rabbits fed with a 1% enriched-cholesterol diet during 4 weeks compared with normocholesterolemic animals. Although in our experimental model we did not observe atheroma plaques in the coronary arteries, we detected a clear endothelial dysfunction [38]. Also, Lv et al. demonstrated that diet-induced hypercholesterolemia enhances myocardial injury and that a selective Na^+/Ca^{++} exchanger inhibitor reduces the infarct size and apoptosis in hyperlipidemic animals through the activation of K^+ ATP channels [39], which could be related with mitochondrial transition pore modulation. In this sense, we recently demonstrated in mice fed with HFD, which developed a phenotype consistent with early-stages of atherosclerosis, the alterations in redox state are enough to increase infarct size in this dyslipidemic mice [40].

Due to the aforementioned, it is clear that in the different models where dyslipidemia is developed, in the presence or absence of atherosclerotic disease, it causes an increase in oxidative stress, and that this phenomenon exacerbates the I/R injury increasing infarct size [36, 38, 41].

8.5 Conditioning Therapies

Regarding conditioning therapies, the effects of dyslipidemia/hypercholesterolemia effects on ischemic preconditioning (PC) and PostC in different animal models are contradictory [13, 20, 41–44] (Fig. 8.2). In this section, we will discuss the findings of different kinds of diets and animal models and their effects on ischemic conditioning therapies.

8.5.1 Ischemic Preconditioning

A pioneering study that assessed ischemic PC effects in animals with hypercholesterolemia was carried out by Szilvassy et al. [43] who showed that the benefits of PC were abolished in rabbits with hypercholesterolemia and atherosclerosis (1.5% enriched-cholesterol diet during 8 weeks). When these animals were fed again with a normal diet, and blood lipid levels were returned to similar values to those at baseline, the PC protective effect was restored, even in the presence of atherosclerosis. These results indicate that hypercholesterolemia, independently to the development of atherosclerosis, interferes with the protection conferred by this powerful cardioprotective intervention.

Accordingly, Ferdinandy et al. [42], showed the lack of reduction in the infarct size in rats fed with a 2% enriched-cholesterol diet during 12 weeks, due to an alteration in nitric oxide bioavailability. These authors also showed that the PC effect was abolished due to an increase in superoxide anion levels and this phenomenon

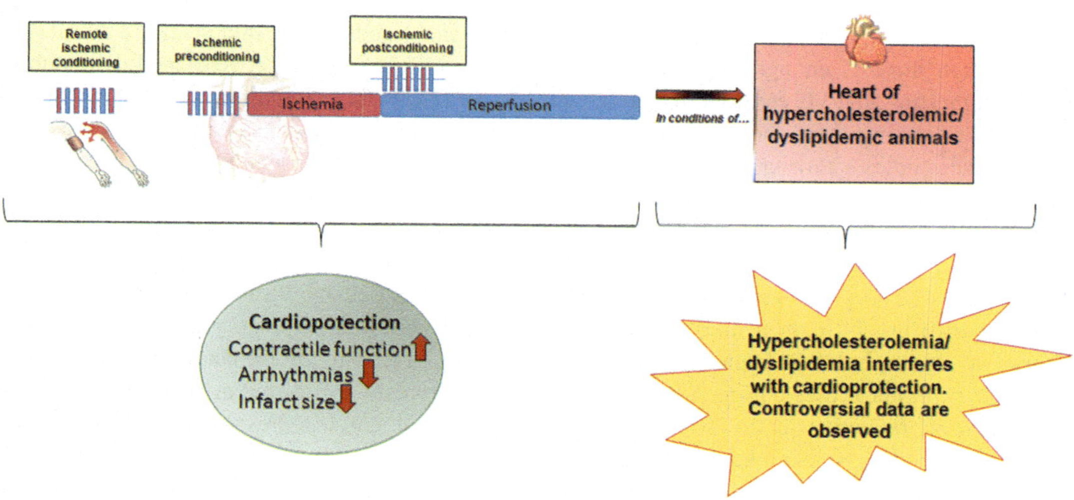

Fig. 8.2 The cardioprotective effects of conditioning therapies result in attenuation of ischemia/reperfusion injury. Major cardiovascular risk factors, such us dyslipidemia/hypercholesterolemia, have influence in the severity of ischemia/reperfusion injury and interfere with the cardioprotective effect of conditioning

occurred with a redistribution of both sarcolemmal and mitochondrial connexin 43 binding capacity in hypercholesterolemic animals [44]. Similarly, Tang et al. [45] showed that the increase in the number of PC cycles could enlarge infarct size in isolated rabbit hearts subjected to I/R after 8 weeks of cholesterol-enriched diet. The lack of success in cardioprotective mechanisms was evident also in studies carried out in patients. In this sense, Kyriakides et al. [46] described the loss of protective effect assessed through ST-segment elevation in patients that underwent a PC protocol. They also showed that the lack of cardioprotection positively correlated to LDL cholesterol plasma levels in patients. In addition, remote preconditioning failed to reduce myocardial necrosis and apoptosis as well as to increase Akt and GSK3β phosphorylation in hypercholesterolemic rat myocardium. Importantly, this study found that inhibition of GSK3β with SB216763 reduced myocardial infarct size in healthy and hypercholesterolemic hearts, but no additional cardioprotective effect was achieved when combined with remote preconditioning. Their results suggest that acute GSK3β inhibition may provide a novel therapeutic strategy for hypercholesterolemic subjects during acute myocardial infarction, whereas ischemic preconditioning is less effective due to signaling events that adversely affect GSK3β [47].

Although there is enough experimental evidence suggesting that PC cardioprotective effect is abolished during hypercholesterolemia, there are several studies showing that the beneficial effect persists even in the presence of elevated plasma cholesterol levels. In this sense, we have shown that PC significantly decreased the infarct size, without improvements in the ventricular function recovery [48], in a model of isolated heart in hypercholesterolemic rabbits (1% enriched cholesterol diet during 4 weeks). The percentage of reduction in the infarct size was higher in the hypercholesterolemic animals compared with the normal ones. In regards to our findings, Kremastinos et al. [49] used a model of rabbits fed with 2% enriched-cholesterol and 6% corn oil diet, administered during 8 weeks, and showed that PC reduces infarct size in a magnitude similar to that observed in normal animals. Iliodromitis et al. [50], who used a hypercholesterolemic diet during 6 weeks and reported that PC reduced infarct size in those rabbits, described similar findings.

8.5.2 Ischemic Postconditioning

In regards to PostC, Iliodromitis et al. [50] showed that this cardioprotective intervention did not reduce the infarct size in a model of hypercholesterolemic rabbits. In agreement, Kupai et al. [51] failed to show the protective effect of several PostC protocols (6 cycles of 10 s each) in a rat model fed with 2% cholesterol during 12 weeks.

These authors showed that PostC involves an early increase in peroxynitrite-induced nitrative stress in the triggering of cardioprotection, and in hyperlipidemia condition, the absence of this mechanism may contribute to the loss of the beneficial effects. Both groups used 12 weeks of hypercholesterolemic diet, and particularly in the study carried out by Iliodromitis et al. [50], it was shown an advanced stage in atherosclerosis, with a significant reduction in the arterial lumen. In a different direction, we showed that PostC performed with 2 cycles of reperfusion/ischemia (30 s each), reduces the infarct size in rabbits fed with a 1% cholesterol enriched-diet during 4 weeks [38]. It is important to highlight that in our model the atherosclerotic lesions in the coronary arteries were absent although a clear endothelial dysfunction could be detected.

Conversely, we recently demonstrated that in mice fed with HFD during 12 weeks, cardioprotection afforded by PostC treatment was abolished, even with an increase of Trx-1 protein content [40]. This discrepancy regarding PostC cardioprotection in different species and experimental protocols is represented in Fig. 8.3. In this sense, it had been previously reported that Trx-1 activity must be intact to induce PostC cardioprotective effects, given that in middle-aged mice, in dyslipidemic mice and in dominant negative Trx-1 transgenic mice (that overexpress inactive

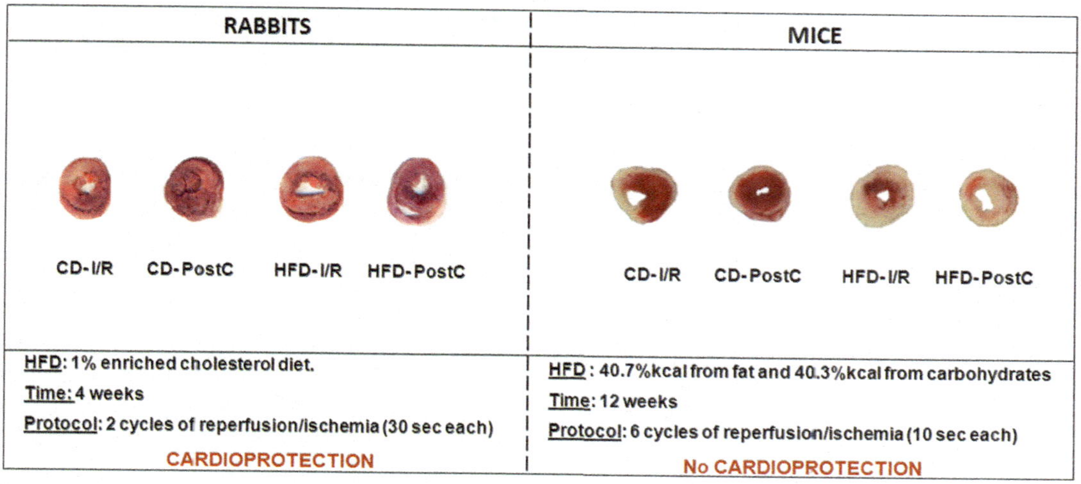

Fig. 8.3 Ischemic postconditioning effects in high fat or cholesterol enriched-diet. Comparison of experimental models (rabbits and mice) accompanied by representative cross-sections of ventricles from different groups. The cross-sections were stained with 2,3,5-triphenyl tetrazolium chloride. White area: necrotic tissue. Red area: viable tissue

Trx-1), cardioprotective effects of PostC on myocardial infarction and GSK3β phosphorylation were abolished [9, 10, 40]. Therefore, these data showed that for the well-known activation of the RISK pathway by means of a PostC protocol, an active and functional Trx-1 is necessary.

Consequently, it is clear that the results in the studies that assessed both ischemic PC and PostC are controversial. The discrepancy in the results could be attributed to the use of different experimental models that involved different animal species; different types of diets in terms of cholesterol or fat concentration and in the composition (the presence or absence of fatty acids), and the duration of the administration. Particularly, with postconditioning PostC there is another variable that should be analyzed and that is the number and duration of postconditioning cycles used. Iliodromitis et al. employed a postconditioning protocol of 6 cycles of 10 s each, and 4 cycles of 30 s each. This difference in the number of cycles as well as in the type of diet could be the cause of the discrepancy between the obtained results. In our model, we found protection using 2 cycles of reperfusion/ischemia in hypercholesterolemic rabbits [30], but in dyslipidemic mice, we used 6 cycles of R/I (10 s each) (Fig. 8.3).

Finally, it is important to mention that caution is advisable when trying to compare obtained results in models of animals with hypercholesterolemia. In this sense, it is clear that the mechanisms of cardioprotection in models with hypercholesterolemia/dyslipidemia involve a series of variables (species, type of diet, duration of the diet, presence or not of atherosclerotic disease, and the protection mechanism protocol used) which could justify the discrepancies among the obtained results.

8.6 Oxidative Stress in Experimental Dyslipidemic Models

It is widely known that dyslipidemia and lipid metabolism disorders produce an increase of oxidative stress [11], causing an exacerbation of the damage produced by I/R [40, 52]. However, the exact mechanism involved is not fully understood. Some researches propose that increased superoxide production in the hearts of hypercholesterolemic animals is a result of an increased NADPH oxidase activity, as it has been shown in cholesterol-fed Wistar rats and apoB100 transgenic mice [19]. In addition, the activity of

NADPH oxidase depends on its transcript, NOX, and hypercholesterolemia has been shown to affect the cardiac gene expression profile at the mRNA level. In this sense, Kocsis et al. [53] showed that hypercholesterolemia affects cardiac microRNA miR- 25 leading to increased NOX-4 expression and demonstrated a change in nitrosative and oxidative stress- associated genes with cholesterol-diet (Fig. 8.4).

Another mechanism through which hypercholesterolemia/dyslipidemia generates oxidative stress involves mitochondria. Indeed, together with peroxisomes, mitochondria represent the main subcellular compartments where lipid degradation occurs. Yet, the impact of dietary lipids on mitochondrial redox status and reactive oxygen species emission and their downstream mediators such as MMPs, caspase 3; and antioxidant systems such as superoxide dismutase, glutathione, and Trx-1, among others has not been completely unraveled. We previously explored the idea that mitochondrial dysfunction and redox imbalance that takes place in the cardiomyocytes after I/R are exacerbated in mice fed with HFD and developing early stages of atherosclerosis. In this sense, we showed a significant increase of the basal emission of H_2O_2 in mitochondria derived from HFD animals and a shift towards an oxidized intracellular environment using GSSG/GSH2 ratio [40]. Furthermore, in hypercholesterolemic rats, phosphorylation of eNOS was decreased compared with controls, which may result in a decrease of NO production and in the loss of the effects of cardioprotective interventions [54, 55]. Recently, it has been described both by a computational model and through experimental results that lipid oxidation and its concentration-dependent uncoupling effect, together with a partial lipid-dependent decrease in the rate of superoxide generation, modulate H_2O_2 emission as a function of oxygen consumption [56]. The authors remarked that keeping low levels of intracellular lipid is crucial for mitochondria and cells to maintain reactive oxygen species within physiological levels compatible with signaling and reliable energy supply.

It was also demonstrated in patients that oxidant species such as oxidized LDL, malondialdehyde, ROS and isoprostanes are emerging as leading mediators of the oxidative injury in subjects with familial hypercholesterolemia [57]. For this reason, targeting oxidative stress and endogenous antioxidant systems may be a promising therapeutic strategy to reduce atherogenesis in patients with dyslipidemia and myocardial infarction.

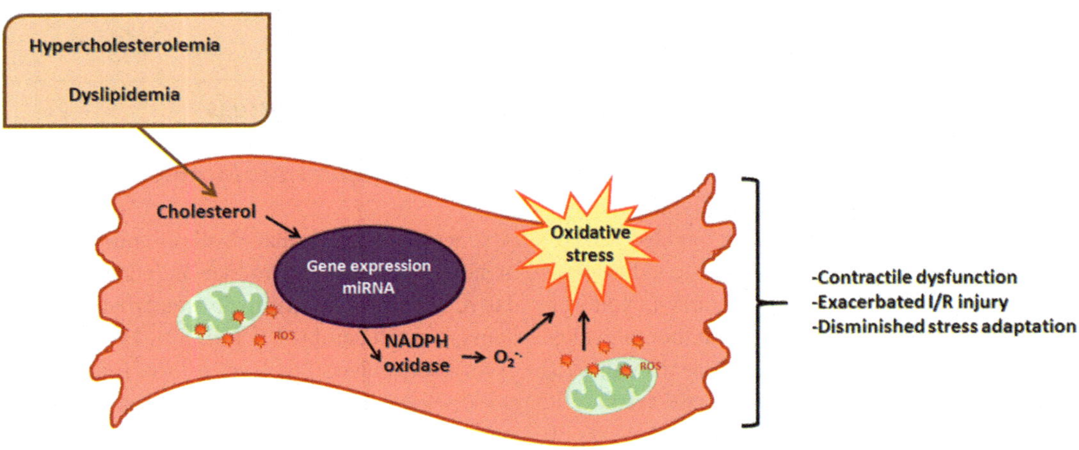

Fig. 8.4 Effects of hypercholesterolemia/dyslipidemia on ischemia/reperfusion injury and the relationship with oxidative stress. Increasing levels of cholesterol could increase oxidative stress modifying NADPH oxidase activity (NOX) by high levels of its transcript. Also, mitochondrial dysfunction cause high levels of reactive oxygen species (ROS) such as hydrogen peroxide (H_2O_2) and superoxide anion(O^-)

8.6.1 Thioredoxin System

It is known that when lipid metabolism is altered, an increase in oxidative stress coexists [11]. In relation to the latter, Trx together with glutathione (GSH) are the molecules with the highest antioxidant effect known to date (GSH) [10, 40]. Trx-1, which is present in mammalian cells, including myocytes, stands out in the Trx family; and it is located in the cytoplasm, where it fulfils its antioxidant functions. In addition to being regulated at the transcriptional level, Trx-1 activity is also regulated through post-translational modifications such as S-nitrosylation, oxidation and nitration. The first of these has beneficial effects since it enhances the protective activity of Trx-1 by reducing apoptosis and potentiating its antioxidant effect [58]. Oxidation and nitration, on the other hand, provoke the inactivation of Trx-1, inhibiting its biological function as an antioxidant in a reversible and irreversible way, respectively [8–10, 40]. In this manner, Trx-1 mediates the cellular response to the alteration of the redox state of the cellular and can act as a protector when ROS concentrations reach pathological levels.

In turn, the imbalance of intracellular redox state of the cell caused by ROS alters the function of multiple cellular mechanisms and is the underlying mechanism in the pathophysiology of many cardiovascular diseases, such as in I/R injury [30] and dyslipidemia status [11]. With respect to dyslipidemia and its relationship with Trx, Augusti et al. [59] demonstrated in a model of experimental atherosclerosis in rabbits fed with an atherogenic diet, that there was an increase in the levels of oxidized LDL, lipids and proteins. These changes were reversed with astaxanthin, a carotenoid that protects fatty acids and biological membranes from oxidative damage, through the modulation of the thioredoxin reductase-1 activity (TrxR-1). Similarly, Somacal et al. [60] who also worked in atherosclerotic rabbits, observed that the administration of the carotenoid BIX, protected from deleterious actions of ROS and RNS, as evidenced by the reduction of the formation of atherogenic lesions in the aorta. This happened at the expense of an increase of antioxidant enzymes activity, mainly of TrxR-1. Furthermore, it was demonstrated in obese and diabetic rats that hypolipidemic drug effects involve an increase in antioxidant levels in plasma, including Trx [61]. Augusti et al. [59] reported that patients with hypercholesterolemia show an imbalance in the activity of TrxR-1, probably due to an increase in oxidative stress, since it is positively associated with the oxidation of LDL molecules. Other authors, who showed that the presence of different comorbidities such as hypercholesterolemia produces an increase in oxidative stress and consequently an increase in plasma levels of Trx-1, confirmed these findings [61]. In accordance, they demonstrated the S-acetylation of Trx-1 and TrxR-1 using livers of mice fed with HFD for 13 weeks, due to an increase in oxidative stress, which produced an inactivation of these proteins [58]. With regard to what happens at the cardiac level, Liu et al. [62] showed exacerbation of the damage caused by an increase in infarct size in diabetic animals at least in part due to Trx-1 inactivation.

Similarly, we have recently used a model compatible with early-stages of atherosclerosis and detected an augmentation of oxidative stress represented by increased H_2O_2 production (Fig. 8.5a), a shift towards an oxidized intracellular environment using GSSG/GSH2 ratio (Fig. 8.5b), and changes in mitochondrial bioenergetics with decreased in oxygen consumption rates both in state 4 and state 3 [40]. This mice model is characterized by high cholesterol levels with normal triglycerides, hepatic enzymes activity, but with an endothelial dysfunction without histological lesions. In our incipient atherosclerosis model, the effects of the high-fat diet used could be reversible since no histological damage compatible with irreversible lesions in the intramyocardial vessels and in the liver was detected; also, no changes in liver enzymes between HFD and CD were observed. All these data strongly suggest that the effect of HFD could be reversible. The changes in the redox state and in mitochondrial function observed in hearts of HFD fed mice were associated with an augmentation of Trx-1 expression (Fig. 8.5c). However, this protein was shown to be more oxidized (Fig. 8.5d).

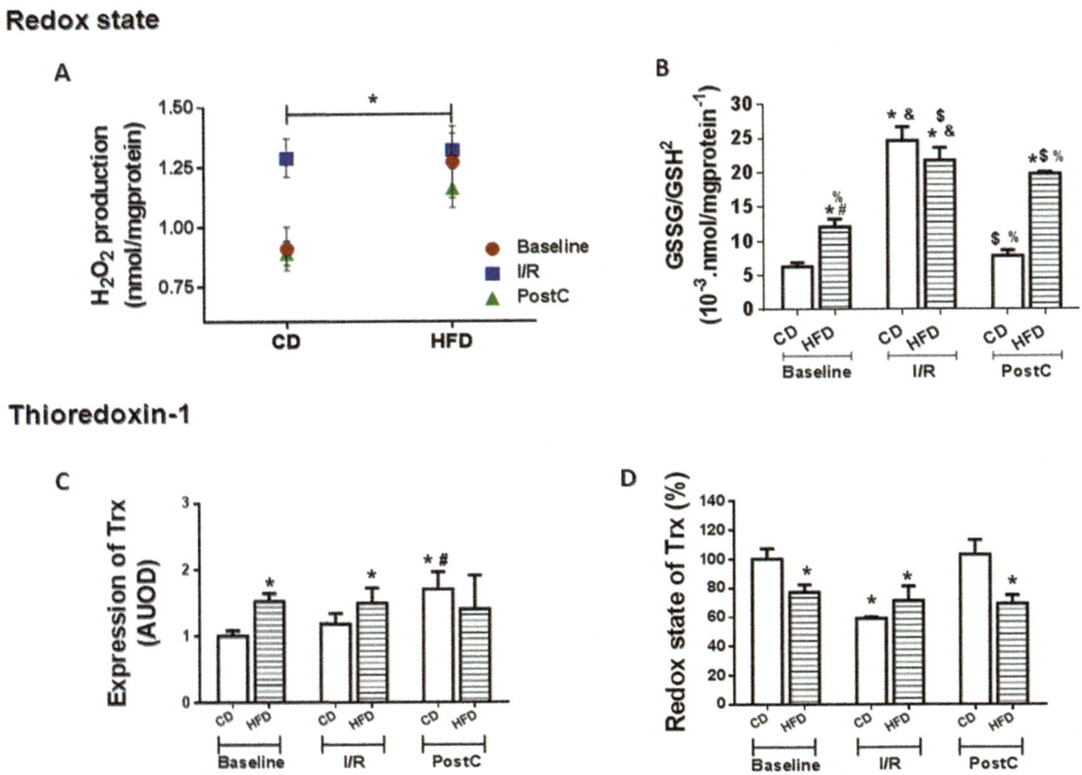

Fig. 8.5 Oxidative stress in high-fat diet fed mice. A: H_2O_2 production; B: Glutathione redox state; C: Trx-1 expression; and D: Trx-1 redox state. *P < 0.05 vs. CD-baseline; #P < 0.05 vs. CD-I/R; & P < 0.05 vs. CD-PostC; $ P < 0.05 vs. HFD-Baseline; % P < 0.05 vs. HFD-I/R. CD: control diet; HFD: High-fat diet; I/R: Ischemia/reperfusion; H_2O_2: hydrogen peroxide; PostC: ischemic postconditioning protocol. *Trx-1* Thioredoxin-1, *GSSG/GSH2* oxidized glutathione/reduced glutathione ratio. AUDO: Arbitrary units optical density (Modified from Mazo et al. [40]).

This result represents an interesting and original finding because it supports the idea that a diet that causes an initial stage of atherosclerosis is able to produce oxidative damage of such magnitude as to modify the Trx-1 activity and in this way abolish the cardioprotection conferred by this powerful antioxidant.

8.7 Conclusion

Finally, we can conclude that the hypercholesterolemia/HFD, independently of the presence or absence of atherosclerosis, modules the ventricular function under baseline conditions, as well as the post- ischemic ventricular dysfunction (stunned myocardium). It is also clear that Trx-1, in addition to its antioxidant effects, has a beneficial effect on injury due to ischemia and reperfusion (Fig. 8.6). However, results from physiological protection mechanisms (PC and PostC) are controversial as the beneficial effects depend on many variables such as diet composition, animal species studied, time of administration of the diet, and the presence or absence of atherosclerotic disease. Hence, caution is essential when we try to compare the obtained results in regards of the presence or not of beneficial effects of ischemic preconditioning and postconditioning. In this sense, only those studies that were performed using the same animal species, with the same diet composition and time of administration would be comparable. This discrepancy in the results and the variables taken

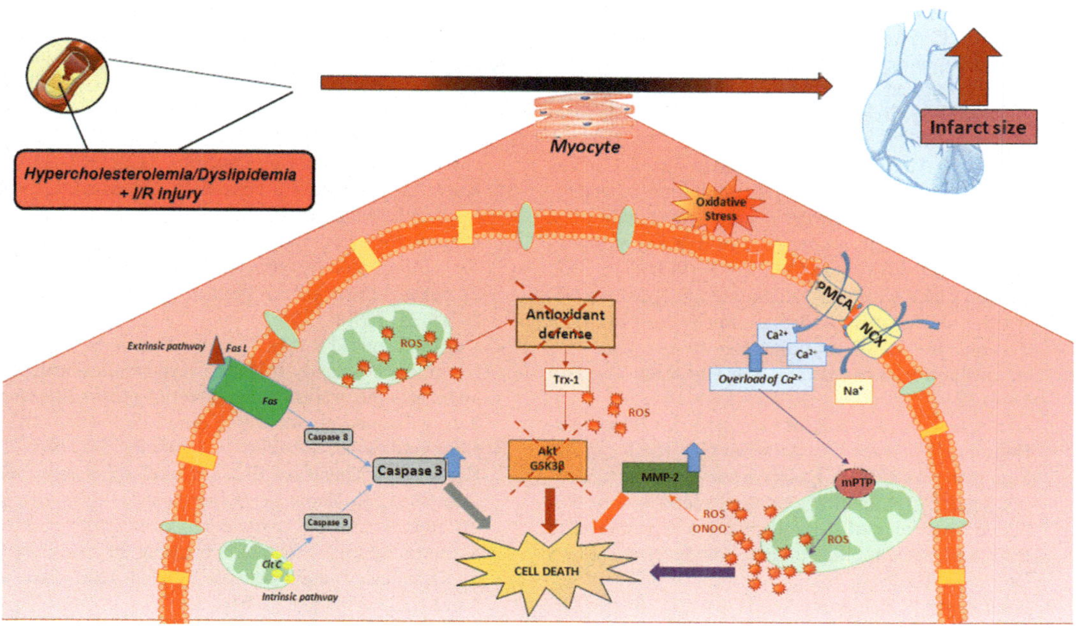

Fig. 8.6 Hypercholesterolemia/dyslipidemia effects on ischemia/reperfusion injury: Fas: apoptosis membrane protein; FasL: Ligand Fas; CitC: cytochrome C; ROS: Reactive oxygen species; *Trx-1* Thioredoxin-1, *Akt* Protein Kinase B, *GSK3β* Glycogen synthase kinase 3beta, *MMP-2* metaloproteinase type-2, *ONOO-* peroxynitrite, *mPTP* mitochondrial permeability transition pore, *PMCA* Plasma membrane Ca^{2+} ATPase, *NCX* Sodium- calcium exchanger

into consideration require a careful analysis when those findings are intended to be extrapolated into the clinical arena.

References

1. Benjamin EJ, Blaha MJ, Chiuve SE, Cushman M et al (2017) American Heart Association Statistics Committee and Stroke Statistics Subcommittee. Heart disease and stroke statistics-2017 update: a report from the American Heart Association. Circulation 135(10):e146–e603.
2. GBD 2015 Risk Factors Collaborators (2016) Global, regional, and national comparative risk assessment of 79 behavioural, environmental and occupational, and metabolic risks or clusters of risks, 1990–2015: a systematic analysis for the Global Burden of Disease Study 2015. Lancet 388(10053):1659–1724.
3. Anitschokow N, Chalatow S (1983) On experimental cholesterin steatosis and its significant in the origin of some pathological procceses. Arteriosclerosis 3:178–183
4. Shioji K, Nakamura H, Masutani H, Yodoi J (2003) Redox regulation by thioredoxin in cardiovascular diseases. Antioxid Redox Signal 5(6):795–802
5. Turoczi T, Chang VW, Engelman RM, Maulik N, Ho YS, Das DK (2003) Thioredoxin redox signaling in the ischemic heart: an insight with transgenic miceover-expressing Trx1. J Mol Cell Cardiol 35(6):695–704
6. Perez V, D'Annunzio V, Valdez LB, Zaobornyj T, Bombicino S, Mazo T, Carbajosa NL, Gironacci MM, Boveris A, Sadoshima J, Gelpi RJ (2016) Thioredoxin-1 attenuates ventricular and mitochondrial postischemic dysfunction in the stunned myocardium of transgenic mice. Antioxid Redox Signal 25(2):78–88.
7. Tao L, Gao E, Bryan NS, Qu Y, Liu HR, Hu A, Christopher TA, Lopez BL, Yodoi J, Koch WJ, Feelisch M, Ma XL (2004) Cardioprotective effects of thioredoxin in myocardial ischemia and reperfusion: role of S-nitrosation [corrected]. Proc Natl Acad Sci U S A 101(31):11471–11476. Epub 2004 July 26
8. D'Annunzio V, Perez V, Mazo T, Muñoz MC, Dominici FP, Carreras MC, Poderoso JJ, Sadoshima J, Gelpi RJ (2016) Loss of myocardial protection against myocardial infarction in middle aged transgenic mice overexpressing cardiac thioredoxin-1. Oncotarget 7(11):11889–11898.

9. Perez V, D'Annunzio V, Mazo T, Marchini T, Caceres L, Evelson P, Gelpi RJ (2016) Ischemic postconditioning confers cardioprotection and prevents reduction of Trx-1 in youngmice, but not in middle-aged and old mice. Mol Cell Biochem 415(1–2):67–76.
10. Perez V, D Annunzio V, Mazo T, Marchini T, Caceres L, Evelson P, Gelpi RJ (2016) Inhibition of endogenous thioredoxin-1 in the heart of transgenic mice does not confer cardioprotection in ischemic postconditioning. Int J Biochem Cell Biol 81(Pt B):315–322.
11. Afolabi OK, Oyewo EB, Adekunle AS, Adedosu OT, Adedeji AL (2013) Oxidative indices correlate with dyslipidemia and pro inflammatory cytokine levels in fluoride-exposed rats. Arh Hig Rada Toksikol 64(4):521–529.
12. Littlejohns B, Pasdois P, Duggan S, Bond AR, Heesom K, Jackson CL, Angelini GD, Halestrap AP, Suleiman MS (2014) Hearts from mice fed a non-obesogenic high-fat diet exhibit changes in their oxidative state, calcium and mitochondria in parallel with increased susceptibility to reperfusion injury. PLoS One 9(6):e100579.
13. Ferdinandy P, Hausenloy DJ, Heusch G, Baxter GF, Schulz R (2014) Interaction of risk factors, comorbidities, and comedications with ischemia/reperfusion injury and cardioprotection by preconditioning, postconditioning, and remote conditioning. Pharmacol Rev 66(4):1142–1174.
14. Dalen H, Thorstensen A, Romundstad PR, Aase SA, Stoylen A, Vatten LJ (2011) Cardiovascular risk factors and systolic and diastolic cardiac function:a tissue Doppler and speckle tracking echocardiographic study. J Am Soc Echocardiogr 24(3):322–32.e6.
15. Horio T, Miyazato J, Kamide K, Takiuchi S, Kawano Y (2003) Influence of low high-density lipoprotein cholesterol on left ventricular hypertrophy and diastolic function in essential hypertension. Am J Hypertens 16(11 Pt 1):938–944
16. Huang Y, Walker KE, Hanley F, Narula J, Houser SR, Tulenko TN (2004) Cardiac systolic and diastolic dysfunction after a cholesterol-rich diet. Circulation 109(1):97–102. Epub 2003 Dec 15
17. Varga ZV, Kupai K, Szűcs G, Gáspár R, Pálóczi J, Faragó N, Zvara A, Puskás LG, Rázga Z, Tiszlavicz L, Bencsik P, Görbe A, Csonka C, Ferdinandy P, Csont T (2013) MicroRNA- 25-dependent up-regulation of NADPH oxidase 4 (NOX4) mediateshypercholesterolemia- induced oxidative/nitrative stress and subsequent dysfunction in the heart. J Mol Cell Cardiol 62:111–121.
18. Csont T, Bereczki E, Bencsik P, Fodor G, Görbe A, Zvara A, Csonka C, Puskás LG, Sántha M, Ferdinandy P (2007) Hypercholesterolemia increases myocardial oxidative and nitrosative stress thereby leading to cardiac dysfunction in apoB-100 transgenic mice. Cardiovasc Res 76(1):100–109. Epub 2007 June 19
19. Osipov RM, Bianchi C, Feng J, Clements RT, Liu Y, Robich MP, Glazer HP, Sodha NR, Sellke FW (2009) Effect of hypercholesterolemia on myocardial necrosis and apoptosis in the setting of ischemia-reperfusion. Circulation 120(11 Suppl):S22–S30.
20. Ferdinandy P, Szilvássy Z, Horváth LI, Csont T, Csonka C, Nagy E, Szentgyörgyi R, Nagy I, Koltai M, Dux L (1997) Loss of pacing-induced preconditioning in rat hearts: role of nitric oxide and cholesterol-enriched diet. J Mol Cell Cardiol 29(12):3321–3333
21. Kupai K, Csonka C, Fekete V, Odendaal L, van Rooyen J, Marais de W, Csont T, Ferdinandy P (2009) Cholesterol diet induced hyperlipidemia impairs the cardioprotective effect of postconditioning: role of peroxynitrite. Am J Physiol Heart Circ Physiol 297(5):H1729–H1735.
22. Ungi I, Ungi T, Ruzsa Z, Nagy E, Zimmermann Z, Csont T, Ferdinandy P (2005) Hypercholesterolemia attenuates the anti ischemic effect of preconditioning during coronaryangioplasty. Chest 128(3):1623–1628
23. Bountioukos M, Rizzello V, Krenning BJ, Bax JJ, Kertai MD, Vourvouri EC, Schinkel AF, Biagini E, Boersma E, Roelandt JR, Poldermans D (2003) Effect of atorvastatin on myocardial contractile reserve assessed by tissue Doppler imaging in moderately hypercholesterolemic patients without heart disease. Am J Cardiol 92(5):613–616
24. Rubinstein J, Pelosi A, Vedre A, Kotaru P, Abela GS (2009) Hypercholesterolemia and myocardial function evaluated via tissue doppler imaging. Cardiovasc Ultrasound 7:56.
25. Wan S-H, Vogel MW, Chen HH (2014) Preclinical diastolic dysfunction. J Am Coll Cardiol 63(5):407–416
26. Bastiaanse EM, Atsma DE, Kuijpers MM, Van der Laarse A (1994) The effect of sarcolemmal cholesterol content on intracellular calcium ion concentration in cultured cardiomyocytes. Arch Biochem Biophys 313(1):58–63
27. Bencsik P, Sasi V, Kiss K, Kupai K, Kolossváry M, Maurovich-Horvat P, Csont T, Ungi I, Merkely B, Ferdinandy P (2015) Serum lipids and cardiac function correlate with nitrotyrosine and MMP activity in coronary arterydisease patients. Eur J Clin Invest 45(7):692–701.
28. Van de Velde M, DeWolff M, Leather HA, Wouters PF (2000) Effects of lipids on the functional and metabolic recovery from global myocardial stunning in isolated rabbit hearts. Cardiovasc Res 48(1):129–137
29. Satoh K, Takaguri A, Itagaki M, Kano S, Ichihara K (2008) Effects of rosuvastatin and pitavastatin on ischemia-induced myocardial stunning in dogs. J Pharmacol Sci 106(4):593–599. Epub 2008 Apr 9
30. D'Annunzio V, Donato M, Sabán M, Sanguinetti SM, Wikinski RL, Gelpi RJ (2005) Hypercholesterolemia attenuates postischemic ventricular dysfunction in the isolated rabbit heart. Mol Cell Biochem 273(1–2):137–143
31. Calabresi L, Gomaraschi M, Villa B, Omoboni L, Dmitrieff C, Franceschini G (2002) Elevated soluble cellular adhesion molecules in subjects with low HDL-cholesterol. Arterioscler Thromb Vasc Biol 22(4):656–661

32. Kalaivanisailaja J, Manju V, Nalini N (2003) Lipid profile in mice fed a high-fat diet after exogenous leptin administration. Pol J Pharmacol 55(5):763–769
33. Adameova A, Harcarova A, Matejikova J, Pancza D, Kuzelova M et al (2009) Simvastatin alleviates myocardial contractile dysfunction and lethal ischemic injury in rat heart independent of cholesterol-lowering effects. Physiol Res 58(3):449–454
34. Golino P, Maroko PR, Carew TE (1987) The effect of acute hypercholesterolemia on myocardial infarct size and the no- reflow phenomenon during coronary occlusion-reperfusion. Circulation 75(1):292–298
35. Wang TD, Chen WJ, Su SS, Lo SC, Lin WW, Lee YT (2002) Increased cardiomyocyte apoptosis following ischemia and reperfusion in diet-induced hypercholesterolemia: relation to Bcl-2 and Bax proteins and caspase-3 activity. Lipids 37(4):385–394
36. Szucs G, Bester DJ, Kupai K, Csont T, Csonka C, Esterhuyse AJ, Ferdinandy P, Van Rooyen J (2011) Dietary red palm oil supplementation decreases infarct size in cholesterol fed rats. Lipids Health Dis 10:103
37. Wang W, Schulze CJ, Suarez-Pinzon WL, Dyck JR, Sawicki G, Schulz R (2002) Intracellular action of matrix metalloproteinase-2 accounts for acute myocardial ischemia and reperfusion injury. Circulation 106:1543–1549
38. Donato M, D'Annunzio V, Berg G, Gonzalez G, Schreier L, Morales C, Wikinski RL, Gelpi RJ (2007) Ischemic postconditioning reduces infarct size by activation of A1 receptors and K+(ATP) channels in both normal and hypercholesterolemic rabbits. J Cardiovasc Pharmacol 49(5):287–292
39. Lv Y, Ren Y, Sun L, Wang S, Wei M, Jia D (2013) Protective effect of Na(+)/Ca (2+) exchange blocker KB-R7943 on myocardial ischemia-reperfusion injury in hypercholesterolemic rats. Cell Biochem Biophys 66(2):357–363.
40. Mazo T, D'Annunzio V, Zaobornyj T, Perez V, Gomez A, Berg G, Barchuk M, Ossani G, Martinefski M, Tripodi V, Lago N, Gelpi RJ (2018) High-fat diet abolishes the cardioprotective effects of ischemic postconditioning in murine models despite increased thioredoxin-1 levels. Mol Cell Biochem.
41. Iliodromitis EK, Andreadou I, Prokovas E, Zoga A, Farmakis D, Fotopoulou T, Ioannidis K, Paraskevaidis IA, Kremastinos DT (2010) Simvastatin in contrast to postconditioning reduces infarct size in hyperlipidemic rabbits: possible role of oxidative/nitrosative stress attenuation. Basic Res Cardiol 105(2):193–203. PMID: 20066537
42. Ferdinandy P, Szilvássy Z, Horváth LI, Csont T, Csonka C, Nagy E, Szentgyörgyi R, Nagy I, Koltai M, Dux L (1997) Loss of pacing-induced preconditioning in rat hearts: role of nitric oxide and cholesterol-enriched diet. J Mol Cell Cardiol 29:3321–3333
43. Szilvassy Z, Ferdinandy P, Nagy I, Jakab I, Koltai M (1997) The effect of continuous versus intermittent treatment with transdermal nitroglycerin on pacing-induced preconditioning in conscious rabbits. Br J Pharmacol 121:491–496
44. Görbe A, Varga ZV, Kupai K, Bencsik P, Kocsis GF, Csont T, Boengler K, Schulz R, Ferdinandy P (2011) Cholesterol diet leads to attenuation of ischemic preconditioning-induced cardiac protection: the role of connexin 43. Am J Physiol Heart Circ Physiol 300(5):H1907–H1913.
45. Tang XL, Stein AB, Shirk G, Bolli R (2004) Hypercholesterolemia blunts NO donor-induced late preconditioning against myocardial infarction in conscious rabbits. Basic Res Cardiol 99:395–403
46. Kyriakides ZS, Psychari S, Iliodromitis EK, Kolettis TM, Sbarouni E, Kremastinos DT (2002) Hyperlipidemia prevents the expected reduction of myocardial ischemia on repeated balloon inflations during angioplasty. Chest 121:1211–1215
47. Ma LL, Kong FJ, Guo JJ, Zhu JB, Shi HT, Li Y, Sun RH, Ge JB (2017) Hypercholesterolemia abrogates remote ischemic preconditioning-induced cardioprotection: role of reperfusion injury salvage kinase signals. Shock 47(3):363–369.
48. D'Annunzio V, Donato M, Buchholz B, Pérez V, Miksztowicz V, Berg G, Gelpi RJ (2012) High cholesterol diet effects on ischemia-reperfusion injury of the heart. Can J Physiol Pharmacol 90(9):1185–1196.
49. Kremastinos DT, Bofilis E, Karavolias GK, Papalois A, Kaklamanis L, Iliodromitis EK (2000) Preconditioning limits myocardial infarct size in hypercholesterolemic rabbits. Atherosclerosis 150:81–89
50. Iliodromitis EK, Zoga A, Vrettou A, Andreadou I, Paraskevaidis IA, Kaklamanis L, Kremastinos DT (2006) The effectiveness of postconditioning and preconditioning on infarct size in hypercholesterolemic and normal anesthetized rabbits. Atherosclerosis 188:356–376
51. Kupai K, Csonka C, Fekete V, Odendaal L, van Rooyen J, Marais de W, Csont T, Ferdinandy P (2009) Cholesterol diet-induced hyperlipidemia impairs the cardioprotective effect of postconditioning: role of peroxynitrite. Am J Physiol Heart Circ Physiol 297(5):H1729–H1735. PMID: 19734363
52. Afolabi OK, Oyewo E, Adekunle A, Adedosu OT, Adedeji AL (2013) Oxidative indices correlate with dyslipidemia and pro-inflammatory cytokine levels in fluoride-exposed rats. Arh Hig Rada Toksikol 64(4):521–529
53. Kocsis GF, Csont T, Varga-Orvos Z, Puskas LG, Murlasits Z, Ferdinandy P (2010) Expression of genes related to oxidative/nitrosative stress in mouse hearts: effect of preconditioning and cholesterol diet. Med Sci Monit 16(1):BR32–BR39
54. Penumathsa SV, Thirunavukkarasu M, Koneru S, Juhasz B, Zhan L, Pant R, Menon VP, Otani H, Maulik N (2007) Statin and resveratrol in combination induces cardioprotection against myocardial infarction in hypercholesterolemic rat. J Mol Cell Cardiol 42(3):508–516.

55. Penumathsa SV, Thirunavukkarasu M, Koneru S, Juhasz B, Zhan L et al (2007) Statin and resveratrol in combination induces cardioprotection against myocardial infarction in hypercholesterolemic rat. J Mol Cell Cardiol 42(3):508–516.
56. Cortassa S, Sollott SJ, Aon MA (2017) Mitochondrial respiration and ROS emission during β- oxidation in the heart: an experimental-computational study. PLoS Comput Biol 13(6):e1005588.
57. Mollazadeh H, Carbone F, Montecucco F, Pirro M, Sahebkar A (2018) Oxidative burden in familial hypercholesterolemia. J Cell Physiol 233(8):5716–5725.
58. Qin H, Zhang X, Ye F, Zhong L (2014) High-fat diet-induced changes in liver thioredoxin and thioredoxin reductase as a novel feature of insulin resistance. FEBS Open Biol 31(4):928–935
59. Augusti PR, Quatrin A, Somacal S, Conterato GM, Sobieski R, Ruviaro AR, Maurer LH, Duarte MM, Roehrs M, Emanuelli T (2012) Astaxanthin prevents changes in the activities of thioredoxin reductase and paraoxonase in hypercholesterolemic rabbits. J Clin Biochem Nutr 51(1):42–49
60. Somacal S, Figueiredo CG, Quatrin A, Ruviaro AR, Conte L, Augusti PR, Roehrs M, Denardin IT, Kasten J, da Veiga ML, Duarte MM, Emanuelli T (2015) The antiatherogenic effect of bixin in hypercholesterolemic rabbits is associated to the improvement of lipid profile and to its antioxidant and anti-inflammatory effects. Mol Cell Biochem 403(1–2):243–253
61. Miwa K, Kishimoto C, Nakamura H, Makita T, Ishii K, Okuda N, Yodoi J, Sasayama S (2005) Serum thioredoxin and alpha-tocopherol concentrations in patients with major risk factors. Circ J 69(3):291–294
62. Liu Y, Qu Y, Wang R, Ma Y, Xia C, Gao C, Liu J, Lian K, Xu A, Lu X, Sun L, Yang L, Lau WB, Gao E, Koch W, Wang H, Tao L (2012) The alternative crosstalkbetween RAGE and nitrativethioredoxin inactivation during diabetic myocardial ischemia-reperfusion injury. Am J Physiol Endocrinol Metab 303(7):E841–E852

9 Epicardial Adipose Tissue in Cardiovascular Disease

Gabriela Berg, Verónica Miksztowicz, Celina Morales, and Magalí Barchuk

Abstract

Cardiovascular disease (CVD) is the main cause of morbidity and mortality in industrialized countries, despite the evolution of treatments and revascularization strategies. Obesity, also accompanied by a chronic inflammatory process, is an independent risk factor for CVD. Abdominal adipose tissue is a complex, metabolically very active organ capable of producing different adipokines and hormones, responsible for endocrine-metabolic comorbidities. The epicardial adipose tissue (EAT) has not been as extensively studied as the abdominal or subcutaneous adipose tissue. However, recent evidence associates it with an increased cardiometabolic risk due to its apposition with the heart. EAT stores triglycerides to provide energy to the myocardium and is characterized by its greater ability to release and capture free fatty acids. EAT strategic localization allows a singular cross talk with cardiomyocytes and vascular wall cells. The fact that EAT produces pro-inflammatory adipokines as well as metalloproteinases and pro-oxidant substances, highlights its possible direct impact on plaque vulnerability and heart failure, being still necessary further studies of EAT behavior in CVD.

Keywords

Epicardial adipose tissue · Cardiovascular disease · Cardiometabolic risk · Adipokines

G. Berg (✉) · V. Miksztowicz
Facultad de Farmacia y Bioquímica. Departamento de Bioquímica Clínica, Cátedra de Bioquímica Clínica I, Laboratorio de Lípidos y Aterosclerosis, Universidad de Buenos Aires, Buenos Aires, Argentina

Facultad de Farmacia y Bioquímica, Instituto de Fisiopatología y Bioquímica Clínica (INFIBIOC), Universidad de Buenos Aires, Buenos Aires, Argentina

CONICET, Facultad de Farmacia y Bioquímica, Universidad de Buenos Aires, Buenos Aires, Argentina
e-mail: gaberg@ffyb.uba.ar

C. Morales
Facultad de Medicina, Instituto de Fisiopatología Cardiovascular, Departamento de Patología, Universidad de Buenos Aires, Buenos Aires, Argentina

M. Barchuk
Facultad de Farmacia y Bioquímica. Departamento de Bioquímica Clínica, Cátedra de Bioquímica Clínica I, Laboratorio de Lípidos y Aterosclerosis, Universidad de Buenos Aires, Buenos Aires, Argentina

Facultad de Farmacia y Bioquímica, Instituto de Fisiopatología y Bioquímica Clínica (INFIBIOC), Universidad de Buenos Aires, Buenos Aires, Argentina

A. Trostchansky, H. Rubbo (eds.), *Bioactive Lipids in Health and Disease*, Advances in Experimental Medicine and Biology 1127, https://doi.org/10.1007/978-3-030-11488-6_9

9.1 Introduction

Despite advances in research and clinical care, cardiovascular disease (CVD) remains the main cause of mortality all over the world [1]. Obesity and specifically visceral adiposity are global growing health problems closely related with CVD [2]. The knowledge that visceral adipose tissue (VAT) is involved in chronic inflammation was demonstrated in the mid-1990s, when the association between obesity and increased levels of the well-known pro-inflammatory cytokine tumor necrosis factor (TNF)-α was verified [3]. Thenceforth the study of VAT has revealed that inflammation leads to metabolic and structural changes in adipose tissue including activated lipolysis, increased release of free fatty acids, hypoxia, oxidative stress, apoptosis of adipocytes and macrophages infiltration [4]. Expanded and inflamed VAT becomes a dysfunctional organ, characterized by hypertrophic but not hyperplastic adipocytes, with altered profiles of signaling molecules, such as adipokines and pro-inflammatory factors [4, 5]. Nevertheless besides, the heterogeneity of VAT composition, including white, brown and beige/brite adipocytes, and the distribution of these fat depots can be decisive for cardiometabolic disorders development [6].

While brown adipose tissue (BAT) is mainly involved in thermoregulation, mostly attributed to mitochondrial uncoupling protein 1 (UCP1) [7], white adipose tissue (WAT) is mostly related to fat storage, energy supply and the release of several hormones and adipokines with paracrine and endocrine effects on the whole-body metabolism [8]. WAT also has mechanical and thermal functions, and it can be broadly classified as VAT or subcutaneous adipose tissue (SAT). VAT surrounds organs and recently, organ-specific adiposity has renewed interest among the scientific community, because of its association with the pathophysiology of cardiometabolic disease. Ectopic adiposity, such as fat accumulation around the different organs, is considered a risk factor in cardiac disease pathogenesis [9]. Human epicardial adipose tissue (EAT) is a visceral thoracic fat due to its apposition to the heart, in direct contact with myocardium and coronary arteries [10]. Although it has been presented as an ectopic fat, EAT and perivascular adipose tissue are typical in healthy mammals, increasing its volume in obesity. Chistiakov et al. [11] consider that these tissues do not represent "ectopic fat" deposition, but rather an enlargement of normally existing anatomical formation. Considered or not an ectopic fat, it must be differentiated from myocardial fat deposition, which represents the storage of lipid droplets within cardiomyocytes, and from perivascular fat.

Finally, EAT strategic localization allows a singular cross talk with cardiomyocytes and vascular wall cells.

9.2 Histological Characteristics of EAT

In humans, there are two anatomically distinct adipose tissues around the heart. These are referred to as epicardial and pericardial adipose tissue, which includes the perivascular adipose tissue (PVAT) [12].

EAT is localized between the myocardial surface and the visceral layer of the pericardium and it is in direct connection with the myocardium [13]. In contrast, pericardial fat consists of adipose tissue between the two (visceral and parietal) pericardial layers. Vascularization is also different between epicardial and pericardial fat; vascularization for the epicardial fat is supplied by branches of the coronary arteries, whereas pericardial fat is vascularized from non-coronary sources (branches from the internal mammary artery) [14]. These anatomically distinct adipose tissues covers nearly 80% of the heart surface and this fat compartment contributes 20% to the whole heart quantity, under physiological conditions [15].

The EAT covered heart region includes atrioventricular and interventricular sulci, surrounding the main coronary arteries, and it can also accumulate on the surface of the left atrium, on the lateral and anterior walls of the right atrium, on right ventricular free wall, and apex of the left ventricle (Fig. 9.1) [14].

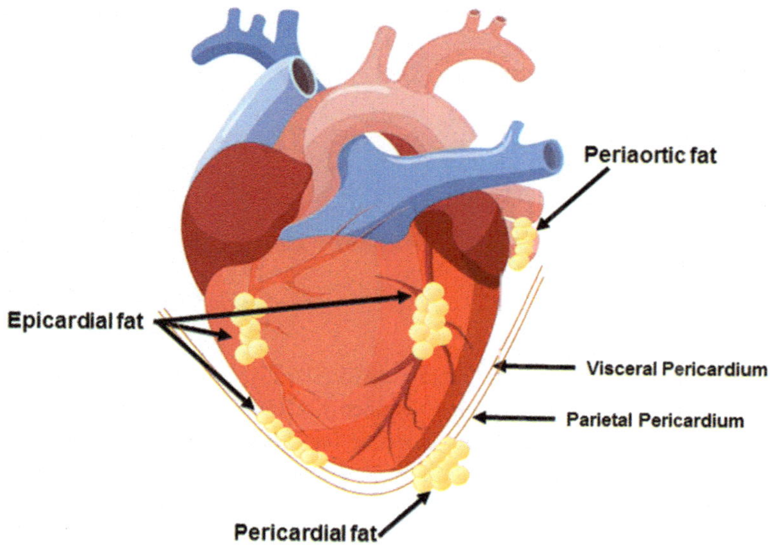

Fig. 9.1 Anatomical localization of Epicardial, Pericardial and Periaortic Adipose Tissue surrounding the heart

Gorter PM et al. showed that the distribution of EAT is mostly inhomogeneous and regional variation in the strengths of associations between different measurements of EAT and anthropometric indexes were reported as well [16]. Therefore, age, gender, body weight and ethnicity should be taken into consideration among physiological determinants of EAT.

Clinical relationships with multidetector computed tomography studies have shown that periatrial EAT volume was the most strictly associated with the incidence of atrial fibrillation, whereas EAT thickness, measured at the level of the left atrioventricular groove, was related to clustering of metabolic syndrome components and inflammatory markers [13].

The anatomical importance of EAT lies in its closeness to the myocardium without a fascia separating EAT from myocardium and coronary vessels [14]. Due to this anatomical arrangement, factors released from EAT have vasocrine and paracrine effects on the myocardium and a potential role in the pathomechanism of coronary artery disease (CAD) [17].

Furthermore, EAT and pericardial adipose tissue are embryologically and biochemically different [10]. In this regard, genetic factors can influence on the volume and quantities of these fat tissues [12, 18]. The clear distinction of epicardial fat from pericardial fat is of great clinical importance. EAT adipocytes are derived from the splanchnopleuric mesoderm and have the same embryologic origin as mesenteric and omental fat cells [19], in contrast to pericardial fat, which derives from the primitive thoracic mesenchyme [13].

Microscopically, epicardial fat is not only composed of preadipocytes and adipocytes, but also contains numerous ganglia and interconnecting nerves, resident monocytes, fibrocytes, stromovascular and immune cells as well [20]. Epicardial adipocytes are generally smaller than those in SAT and other visceral fat depot; this may be attributed to the fact that EAT contains greater number of preadipocytes than mature adipocytes [21] (Fig. 9.2).

In turn, EAT volume is not only associated with CAD [21], but also with vulnerable plaque components, which may contribute to acute coronary syndrome [22, 23]. Djaberi et al. demonstrated, by multislice computed tomographic, a relation between EAT volume and the presence of coronary artery calcium and atherosclerosis [24]. It has also been reported that high EAT volume is independently and significantly associated with the presence of coronary plaques, especially non-stenotic and non-calcified plaques using 64-slice CT [22].

Among the different cardiovascular risk factors associated with EAT volume, the positive

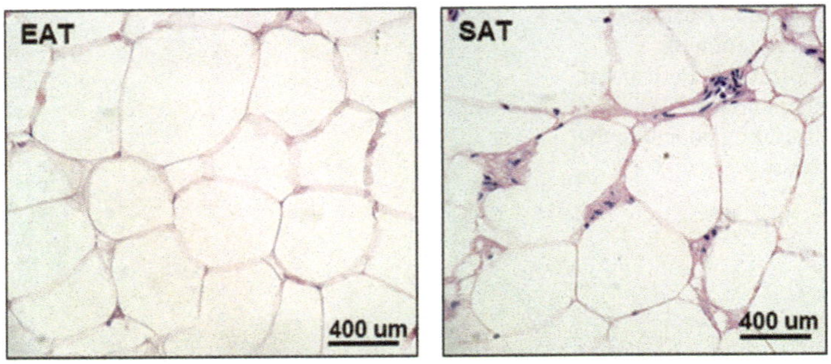

Fig. 9.2 Histological characteristics of epicardial and subcutaneous adipose tissue (EAT and SAT, respectively). Haematoxylin and eosin stain

correlations with body mass index and VAT [25], diabetes mellitus and insulin resistance [25, 26] should be highlighted. Even more, it has been reported that subjects with non calcified plaque had increased EAT volume, though plasma adipocytokine concentrations were not increased [27].

Numerous studies have confirmed the association between EAT volume and the atrial fibrillation risk, early defects in cardiac structure, volume and function [10]. Increased amount of EAT has also been associated with increased left ventricle mass and abnormal right ventricle geometry or subclinical dysfunction [10].

9.3 Metabolic Characteristics of EAT

One of the main limitations in the study of EAT becomes from the fact that the most frequent laboratory experimental animals (rodents) develop a very small amount of this tissue; only a limited amount of EAT can be found located in the atrioventricular groove of mice [28], limiting their use for investigations. Consequently, most of the studies of this tissue have been restricted to evaluate its volume through image techniques (standard dimensional transthoracic echocardiography (2D), multislice computed tomography and magnetic resonance). Therefore, the understanding of EAT metabolism becomes delayed because of the impossibility to obtain it from healthy donors. Moreover, part of the remaining controversies about the association between EAT and CAD are consequence of the variability among the different image techniques used for the measurement of EAT volume.

Nevertheless, during the last decade, huge advancements have been reported in the scientific community explaining the characteristics and behavior of EAT. One of the distinctive characteristics of this tissue in humans is that it presents brown adipose tissue features, mainly represented by the UCP1 mRNA expression (specific marker of BAT), as well as Positive Regulator Domain containing-16 (PRDM-16), peroxisome proliferator-activated receptor γ (PPAR-γ), coactivator 1-α (PGC1α) [29] and Cytochrome c Oxidase (COX-IV) [30]. UCP1 uncouples mitochondrial oxidative phosphorylation, thus producing heat instead of ATP by non-shivering thermogenesis [29]. These characteristics support the role of EAT as a mediator of thermogenesis that could protect the myocardium from developing fatal ventricular arrhythmias due to hypothermia [29]. Other physiological functions of EAT would be related to provide a mechanical protection for coronary circulation and to attenuate the torsion developed by the myocardium contraction. Similarly to other fat depots, EAT is involved in lipid trafficking and its volume has been associated with total body adiposity [31], however, the size of epicardial adipocytes is nearly half of the volume reported in other VAT depots [32], suggesting that epicardial adipocytes are not simply a benign store of excessive fat. In young adult guinea-pigs, it has been reported that the rate of free fatty acids released by EAT was twofold that observed in the perirenal fat depots, indicating an increased lipolytic activity. This fact could be probably due to the low antilipolytic effect of insulin and the increased stimulation of β-adrenergic receptors in VAT.

Nowadays EAT is considered a very active adipose tissue with multiple metabolic and endocrine functions that interacts with surrounding tissues [10], through paracrine and endocrine mechanisms. In healthy normal weight individuals, it has been proposed that EAT provides fatty acids to the underlying myocardium, acting like a buffer tissue according to the heart requirements, protecting myocardium against free fatty acids overflux. In fact, EAT presents the putative property of higher lipogenic and lipolytic rates compared with other fat depots [33]. This particularity allows EAT to accumulate lipids for storage as well as quickly release them on demand. However, as mentioned previously, EAT is not uniformly distributed over the entire surface of the heart, controverting the buffer function originally assigned to EAT for preventing myocardial steatosis [34]. Moreover, given that segments of coronary arteries that are not in contact with EAT rarely present atherosclerotic plaques [35], the participation of EAT in atherosclerotic process must not be underestimated.

9.4 EAT Expansion, Hypoxia and Inflammatory Consequences

Expansion of adipose tissue requires different modifications such as adipogenesis, angiogenesis and extracellular matrix (ECM) remodeling. In these processes, several factors are involved, being metalloproteinases (MMPs) one of the main actors. These enzymes constitute a family of more than 25 zinc dependent endopeptidases, which can be divided into five major groups: collagenases, stromelysins, matrilysins, gelatinases, and membrane-type MMPs [36], according to their substrate specificity, they are involved in ECM and basement membranes components degradation. ECM is a complex structure comprised by collagen and elastin fibers, structural glycoproteins and mucopolysaccharides, organized into a 3D network. A balance between synthesis, deposit in the extracellular environment and degradation determines the structural characteristics of ECM. MMPs, which are expressed in different cell types including fibroblasts, neutrophils, monocytes, macrophages and endothelial cells, are secreted as latent zymogens that have to be activated by proteolytic cleavage. In this regard, MMPs are regulated at different levels, including gene transcription, zymogen activation and enzyme secretion as well as by endogenous specific tissue inhibitors (TIMPs) [36].

Given the role of MMPs in degrading ECM components, an increase in MMPs activity would be necessary for EAT expansion. Studies from our Laboratory have demonstrated that MMP-2 and MMP-9 activities are increased in EAT from patients with CAD, related to higher vascular density [21]. Otherwise, more adipocytes of smaller size as well as higher vascular density have been evidenced in EAT from CAD patients compared to no-CAD [21]. Moreover, a significant association was found between vascular density and EAT MMPs activity, suggesting that the augmented vascularity would be, in part, consequence of increased in MMP-2 and -9 activity.

As mentioned above, EAT is a metabolically active organ and a major source of adipokines, which profile depends on the anti or pro-inflammatory status of the tissue, that modulate cardiac function and morphology [37]. Under low oxidative stress conditions, the normal epicardial adipocytes secrete mainly adiponectin, an anti-inflammatory adipocytokine, and in less proportion, adrenomedullin and omentin, which contribute to the regulation of vascular tone, blood pressure and proper functionality of cardiac contractility.

Adiponectin protects cardiomyocytes from hypertrophic stimuli and minimizes inflammation and fibrosis in coronary arteries and myocardium, reducing the probability of adverse clinical events [38, 39]. However, in cardiovascular patients EAT suffers great oxidative stress and exhibits associated proteomic differences. It has been reported that EAT from cardiovascular disease patients presents lower catalase levels and higher oxidative stress than SAT, and in these patients SAT and EAT differ in the post-transcriptional profiles of oxidative stress-involved proteins such as protein disulfide isomerase (PDIA1), glutathione S-transfer

(GSTP1), and phosphoglycerate mutase 1 (PGAM1) [40].

In obesity, vascularization of growing adipose tissue to provide sufficient O_2 and nutrients represents a challenge. Even though EAT presents smaller adipocytes than SAT, the idea of "hypoxia" in expanded EAT is supported [41]. Because of the hypoxic state, hypoxia-inducible factor 1α (Hif-1α) expression is induced [4]. In parallel, a modification in the expression and release of cytokines involved in the inflammatory and fibrotic alteration of EAT is reported [42]. Recently, greater expression of Hif-1α was described in EAT, which correlated with MMP-2 and MMP-9 expression in this tissue, and with collagen levels in left atrial myocardium [43]. These results, together with our observation of higher MMP-2 and MMP-9 activity in EAT [21] than SAT, reinforce the role of MMPs in EAT and atrial myocardial fibrosis.

In line with this, hypoxia promotes changes in EAT biological characteristics and shifts the adipokines profile with a decline of adiponectin release and an increase of pro-inflammatory cytokines.

The inflammatory status of EAT has a determinant effect on the cardiac and coronary vascular function [44]. Iacobellis et al. [45] showed for the first time that human EAT expresses adiponectin, and that its expression is significantly lower in epicardial fat isolated from patients with CAD, confirming the metabolically active role of this tissue. Adiponectin is an anti-inflammatory adipokine, which increases free fatty acid oxidation in the adipose tissue, improving insulin sensitivity [46] and opposite to other adipokines, it is paradoxically decreased in insulin resistant states, associated with atherosclerotic plaque vulnerability and CAD incidents [47]. This cytokine also presents anti-steatotic and anti-fibrotic effects on different organs such as liver and heart. It has been proposed that adiponectin mediates the cross-talk between adipose tissue, myocytes and vascular cells [48], developing part of its anti-inflammatory effects. In accordance with these concepts, Iwayama et al. [49] demonstrated that severe CAD patients present increased EAT volume and decreased concentration of adiponectin secreted to pericardial fluid, even in non-obese patients.

Otherwise, it is hypothesized that hypoxia promotes the development of inflammation, and it is considered as an initiating factor for ECM production, thus triggering the subsequent metabolic dysfunction of adipose tissue in obesity [4]. Therefore, the hypoxic environment promotes the infiltration of monocytes that becomes resident adipose tissue macrophages, and T lymphocytes that also contribute to the inflammatory response. These macrophages may polarize from non-inflammatory phenotype (M2) to pro-inflammatory phenotypes (M1) [50]. M1 macrophages are generally stimulated by T-helper 1 (Th1) type of cytokines, such as interferon-γ (IFN-γ), or by pathogen-associated molecular patterns (PAMPs). Therefore, M1 macrophages secrete cytokines, including IL-6, TNF-α, IL-1β [39], IL-12, and IL-23 [50]. In contrast, M2 macrophages are associated to tissue remodeling and inflammation resolution. These immune cells have immunosuppressive properties, higher phagocytic capacity, and secrete ECM components as well as angiogenic, chemotactic, anti-inflammatory and growth factors [50]. It has been demonstrated that diet-induced obesity leads to a "shift" in the activation state of adipose tissue macrophages from an M2-polarized state to an M1 pro-inflammatory state, contributing to IR [51]. Different studies have shown that infiltration of macrophages and expression of pro- and anti-inflammatory cytokines are enhanced in EAT of patients with CAD compared to non-CAD patients and that the ratio of M1/M2 macrophages in EAT correlates with the severity of CAD [52]. In accordance to this, it could be suggested that the most important factor for the inflammatory state of EAT could be not only the number of infiltrating macrophages but also the macrophage polarity [52] (Fig. 9.3).

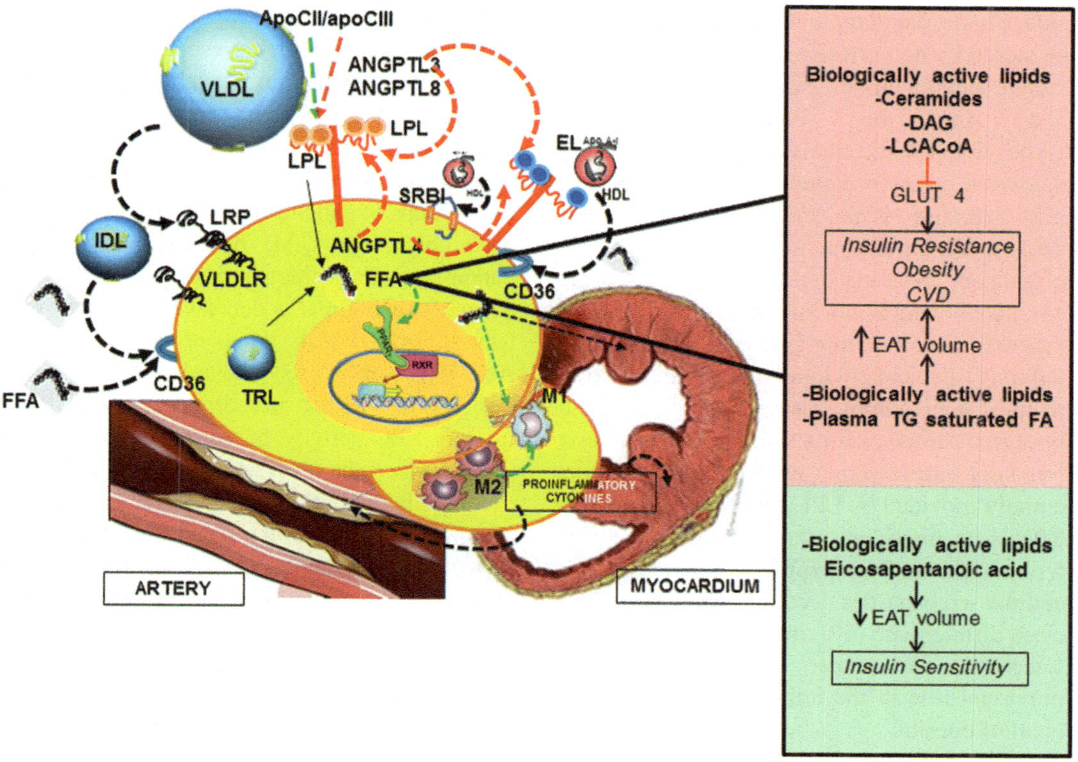

Fig. 9.3 Metabolic pathways and consequences of EAT behavior on myocardium and arteries

EAT Epicardial adipose tissue, *VLDL* Very low density lipoprotein, *HDL* High density lipoprotein, *IDL* Intermediate density lipoprotein, *TRL* Triglycerides rich lipoprotein, *LPL* Lipoprotein lipase, *EL* Endothelial lipase, *FFA* Free fatty acids, *SRBI* Scavenger receptor class B type 1, *CD36* cluster of differentiation 36, *VLDLR* VLDL-Receptor, *LRP* LDL receptor-related protein, *PPAR* Peroxisome proliferator-activated receptors, *RXR* Retinoid X receptor, *M1* M1-polarized macrophages, *M2* M2-polarized macrophages, *ANGPTLs* Angiopoietin-like, *DAG* diacylglycerol, *LCACoA* Long chain acyl-CoA, *GLUT4* Glucose transporter type 4, *CVD* Cardiovacular disease, *TG* Triglycerides

Red lines are inhibitors, Green lines are activators, Black lines indicate ligand-receptor interaction and/or transport/diffusion

FFA may activate or inhibit PPARs and M1 or M2, according to the species

9.5 Bioactive Lipids, EAT and Cardiovascular Disease

Fat deposition and mobilization from EAT depends on the activity of different lipases. Intravascular lipolysis of lipoproteins is performed mainly by lipoprotein lipase (LPL), allowing free fatty acids to get into the adipocyte. LPL belongs to extracellular triglyceride lipases family together with hepatic lipase (HL) and endothelial lipase (EL) [53]. While EL primarily hydrolyses phospholipids from High Density Lipoproteins (HDL) in several tissues [54], and HL shows an intermediate triglyceride and phospholipase activity on triglyceride-remnants lipoproteins (LRP) and HDL in the liver [53], LPL acts predominantly as triglyceride-lipase on chylomicrons and Very Low Density Lipoprotein (VLDL). LPL is synthesized by parenchymal cells and transported to the luminal side of the capillary endothelium where it is anchored by non-covalent unions to the heparan sulfate side chains of membrane proteoglycans and to the recently discovered glycosylphosphatidylinositol-anchored high density lipoprotein-binding protein 1 (GPIHBP1) [55]. Adipose tissue and muscle obtain fatty acids either via LPL mediated hydrolysis of lipoproteins or from unesterified fatty acids, which circulate bound to albumin [56]. However, more than 90% of circulating fatty acids are contained within VLDL and chylomicron and are delivered to muscle or adipose tissue by LPL-catalyzed lipolysis. Although adipose tissue can synthesize free fatty acids de novo, most of them are preferentially provided by LPL action [57]. Therefore, LPL is an important marker for adipocyte differentiation and its expression increases with cellular triglycerides accumulation as preadipocytes differentiate [57], thus this enzyme is considered a gatekeeper enzyme playing an important role in the initiation and/or development of obesity.

Otherwise, intra-adipocyte lipolysis is performed by two main lipases, hormone sensitive lipase (HSL) and the more recently described adipose triglyceride lipase (ATGL), which allow intracellular triglycerides hydrolysis and free fatty acids efflux from the tissue. For several years, HSL was believed the only lipase responsible for triglyceride hydrolysis from fat depots, mediated by the action of different hormones such as catecholamine and glucagon. Recently, Zimmermann et al. [58] reported ATGL, which transformed the dogma of HSL being the only triglyceride lipase in adipose tissue. Nowadays it is known that ATGL is the key triglyceride-hydrolase in adipose tissue as well as in other organs releasing diglycerides that are further hydrolyzed by HSL.

In the recent years, it has been demonstrated that differences in EAT volume between CAD and non-CAD patients may not be caused by a differential mRNA expression of fat mobilizing genes such as LPL, HSL and ATGL, as well as those of its modulator angiopoietin-like protein 4 (ANGPTL4) [59]. However, given that the gene expression not always represents the functional capacity of enzymes, in our laboratory we have studied EAT LPL activity from CAD and non-CAD patients. We have observed that LPL activity is increased in EAT in comparison to SAT in both groups of patients. Besides, EAT LPL activity is higher in CAD patients than no-CAD patients, whereas the expression of ANGPTL4, an inhibitor of LPL is decreased (data not published). ANGPTL4 is a novel member of ANGPTL family; its synthesis is induced by fasting via the PPAR in adipocytes. ANGPTL4 inhibits LPL activity by different mechanisms among which the most noteworthy would be the irreversibly disrupting of LPL dimerization, converting the enzyme into inactive monomers [60].

LPL is also regulated through the action of PPARs, which are involved in its gene transcription. PPARs are activated by different natural and synthetic ligands, including eicosanoids, unsaturated fatty acids, and antidiabetic agents. PPARγ is predominantly found in adipose tissue, and its expression is decreased in insulin resistant states [59].

The increased LPL activity in EAT from CAD patients would explain the expansion of adipocytes in EAT in situations of surplus of free fatty acids, such as in insulin resistant states. In our study we did not verify increased expression of PPARγ in EAT from CAD patients, although it was positively associated with LPL activity.

Besides, preliminary results from our laboratory demonstrates that the pattern of fatty acids from phospholipids and triglycerides in EAT was similar to that observed in VLDL and HDL (data not published), supporting the hypothesis that fatty acids supply to EAT derives from circulating lipoproteins (Fig. 9.3).

Adipose tissue and muscle are the main tissues involved in the development ofinsulin resistance of the whole body. Variations in the lipid content of these tissues would determine alterations in insulin stimulated glucose transporter (GLUT4) expression and glucose uptake [61], thus studies demonstrated that accumulation of intramuscular lipids is responsible for induction of insulin resistance [62]. It must be taken into account that not only lipids concentration, but also species distribution and their biochemical characteristics could be responsible of tissue insulin sensitivity or resistance. It has been suggested that abnormalities in sphingolipids in adipose tissue may contribute to the metabolic disorders associated with obesity [63]. Ceramides (central molecules in sphingolipid metabolism), diacylglycerols and long chain acyl-CoA (LCACoA) have been implicated in the pathogenesis of obesity, insulin resistance and cardiovascular disease [64]; and some evidence demonstrates that ceramide affects insulin stimulated glucose uptake in adipose tissue as well as in skeletal muscle. Studies in 3 T3-L1 adipocytes and in brown adipocytes, revealed that ceramide impairs insulin stimulated GLUT4 expression and glucose uptake and mediates the effect of TNF-α on GLUT4 mRNA content in these cells [61]. These studies proved that sphingolipids play an important role not only in regulating skeletal muscle but also in adipocyte insulin sensitivity [61]. In turn, it has been demonstrated that hepatic accumulation of diacylglycerol activates protein kinase Cξ (PKCξ), which results in the inhibition of insulin stimulated insulin receptor kinase activity, promoting insulin resistance and hepatic steatosis [65]. In addition, the study of diacylglycerol species in patients with hepatic steatosis reveled an increase in saturated and monounsaturated species in steatotic versus normal patients [66]. Besides, treatment of obese mice with palmitic acid esters of hydroxystearic acids improves glucose tolerance through the stimulation of glucagon like peptide 1 and insulin secretion, reduces inflammatory cytokine production from immune cells and ameliorate adipose inflammation in obesity [67]. These results emphasises the importance of the bioactive lipids nature in the behaviour of adipose tissue.

Even though most studies have been developed in adipose tissue, there are scarce data from EAT. Plasma lipidomic studies in healthy individuals showed a strong correlation between plasma triglycerides saturated fatty acids and EAT volume [68]. As previously mentioned, EAT thickness is directly associated with obesity, body mass index and insulin resistance, as well as with different features of the metabolic syndrome, like high fasting glucose levels, C-reactive protein, low HDL-cholesterol among others [26]. Moreover, it is also associated to myocardial lipid content and consequently may affect cardiomyocyte function and heart dysfunction [69]. Besides, it has been described that saturated fatty acids stimulate macrophages via Toll-like receptor-4 activation, and these macrophages activate pro-inflammatory NF-κB to overexpress chemotactic cytokines (i.e., MCP-1, IL-6) [70], highlighting the importance of lipoproteins metabolism on the inflammatory profile of the tissue.

Regarding bioactive lipids role in EAT volume, it has been reported that administration of eicosapentanoic acid decreases EAT and VAT volume, improving metabolic parameters such as triglycerides levels and insulin sensitivity [71]. Furthermore, Błachnio-Zabielska et al., reported that biologically active lipids are increased in EAT of obese and obese-diabetic patients and correlate with insulin resistance, which suggests that they might play some special role in the induction of whole body insulin resistance [61] (Fig. 9.3).

Taken together, these data support the deleterious and/or beneficial role of bioactive lipids concentration and distribution in EAT volume and behaviour, and their implications in insulin resistance and further metabolic disorders. However, many questions about the mechanisms by which bioactive lipids may affect insulin sensitivity still need to be answered.

9.6 EAT as Therapeutic Target

In the last years, EAT has been proposed as a new therapeutic target for CVD. It has been shown that weight loss as consequence of very low calorie diet lead to significant reduction of epicardial fat [72]. Interestingly, EAT volume decreases in a larger proportion than general adiposity, and this is usually associated with cardio-protection The introduction of bariatric surgery to the medical practice opened new expectations that have been widely overcame in obesity and diabetes treatment. Regarding EAT behavior, bariatric surgery induced an efficient reduction of EAT mass concomitant with a reduction of biological cardiovascular risk markers, such as inflammatory and insulin resistant parameters, and has also improved cardiac function [73].

Concerning lowering lipids drugs, recent evidence suggests efficacy of statin therapy, especially atorvastatin, in reducing EAT volume in patients with coronary artery disease [74]. The effects of thiazolidinediones (TZDs), insulin sensitizers that act as agonists of PPARs, have also been evaluated on epicardial fat. Type-2diabetic patients treated with pioglitazone showed a reduction of proinflammatory genes expression in epicardial fat [75]. Studies in animal models of obesity have shown that treatment with rosiglitazone, another PPARγ agonist, increases the expression of UCP-1 in EAT and also promotes mitochondrial biogenesis, inducing the tissue browning. Rosiglitazone could limit the release of fatty acids by EAT, and control the accumulation of triglycerides in myocardium [76].

Regarding new insulin sensitizing drugs, agonists for glucagon-like-protein-1 receptors (GLP-1R) improve cardiovascular and metabolic parameters, including blood pressure, lipids profile and glycaemia [77]. Specifically, liraglutide also reduces EAT mass in obese and in type-2diabetic subjects after 12 weeks of treatment [77]; this effect has not been observed with metformin. Exenatide, another GLP-1R agonist, reduces EAT and subcutaneous fat in type-2 diabetic patients, in a similar pattern than liraglutide [77].

More recently, effects of sodium–glucose co-transporter-2 (SGLT-2) inhibitors on EAT volume have been studied. In diabetic patients with CAD, dapaglifozin improved not only glycemic control but also EAT volume as well as levels of systemic inflammation markers [78].

In conclusion, the reduction of EAT trough different pharmacological treatment could be important in the management of cardiometabolic diseases, though future studies are necessary to clarify this point.

9.7 Conclusions and Future Perspectives

Obesity is a chronic health problem around the world. It is complex, heterogeneous, with epidemic behavior that shortens hope of life, generates great morbidity and increases socio-health costs. Nowadays, there are no doubts about the link between obesity and inflammation; both of them related to CVD, focusing attention on the different VAT. The volume and distribution of fat tissue condition the health of individuals. EAT represents a special kind of VAT which localization has raised concern among healthcare, and justifies its study. After years of research, EAT is still an unresolved challenge. Beyond its natural protective and nourish role, the expansion and inflammation of EAT in obesity and insulin resistant states, change its healthy purposes and negatively impact on the underlying coronary vessels and myocardium with strong deleterious effects. The certainty of the synthesis of pro-inflammatory adipokines as well as MMPs and pro-oxidant substances by EAT, highlights the possible direct impact of EAT on plaque vulnerability and heart failure through different proposed mechanism such as paracrine or vasocrine secretion. Even though huge advances in the understanding of the biology and behavior of EAT have been developed, further studies are necessary, focusing on its role as a potential target for therapeutic interventions and on the effect of different bioactive lipids as transcription regulator factors, all of which could transform the deleterious characteristics of expanded EAT.

References

1. http://www.who.int/es/news-room/fact-sheets/detail/the-top-10-causes-of-death
2. Scherer PE, Hill JA (2016) Obesity, diabetes, and cardiovascular diseases: a compendium. Circ Res 118(11):1703–1705
3. Hotamisligil GS, Arner P, Caro JF et al (1995) Increased adipose tissue expression of tumor necrosis factor-alpha in human obesity and insulin resistance. J Clin Invest 95(5):2409–2415
4. Trayhurn P (2014) Hypoxia and adipocyte physiology: implications for adipose tissue dysfunction in obesity. Annu Rev Nutr 34:207–236
5. Kim JI, Huh JY, Sohn JH et al (2015) Lipid-overloaded enlarged adipocytes provoke insulin resistance independent of inflammation. Mol Cell Biol 35:1686–1699
6. González N, Moreno-Villegas Z, González-Bris A et al (2017) Regulation of visceral and epicardial adipose tissue for preventing cardiovascular injuries associated to obesity and diabetes. Cardiovasc Diabetol 16(1):44.7
7. Oelkrug R, Polymeropoulos ET, Jastroch M (2015) Brown adipose tissue: physiological function and evolutionary significance. J Comp Physiol B 185(6):587–606
8. Rosen ED, Spiegelman BM (2006) Adipocytes as regulators of energy balance and glucose homeostasis. Nature 444:847–853
9. Lazaros G, Antonopoulos A, Antoniades C et al (2018) The role of epicardial fat in pericardial diseases. Curr Cardiol Rep 20(6):40
10. Gaborit B, Sengenes C, Ancel P et al (2017) Role of epicardial adipose tissue in health and disease: a matter of fat? Compr Physiol 7(3):1051–1082
11. Chistiakov DA, Grechko AV, Myasoedova VA et al (2017) Impact of the cardiovascular system-associated adipose tissue on atherosclerotic pathology. Atherosclerosis 263:361–368
12. Russo R, Di Iorio B, Di Lullo L et al (2018) Epicardial adipose tissue: new parameter for cardiovascular risk assessment in high risk populations. J Nephrol. doi.org/10.1007/s40620-018-0491-5
13. Guglielmi V, Sbraccia P (2017) Epicardial adipose tissue: at the heart of the obesity complications. Acta Diabetol 54(9):805–812
14. Iacobellis G, Corradi D, Sharma AM (2005) Epicardial adipose tissue: anatomic, biomolecular and clinical relationships with the heart. Nat Clin Pract Cardiovasc Med 2(10):536–543
15. Corradi D, Maestri R, Callegari S et al (2004) The ventricular epicardial fat is related to the myocardial mass in normal, ischemic and hypertrophic hearts. Cardiovasc Pathol 13:313–316
16. Gorter PM, van Lindert AS, de Vos AM et al (2008) Quantification of epicardial and peri-coronary fat using cardiac computed tomography; reproducibility and relation with obesity and metabolic syndrome in patients suspected of coronary artery disease. Atherosclerosis 197(2):896–903
17. Cherian S, Lopaschuk GD, Carvalho E (2012) Cellular cross-talk between epicardial adipose tissue and myocardium in relation to the pathogenesis of cardiovascular disease. Am J Physiol Endocrinol Metab 303:E937–E949
18. Jermendy AL, Kolossvary M, Drobni ZD et al (2018) Assessing genetic and environmental influences on epicardial and abdominal adipose tissue quantities: a classical twin study. Int J Obes 42(2):163–168
19. Talman AH, Psaltis PJ, Cameron JD et al (2014) Epicardial adipose tissue: far more than a fat depot. Cardiovasc Diagn Ther 4:416–429
20. Mazurek T, Zhang L, Zalewski A et al (2003) Human epicardial adipose tissue is a source of inflammatory mediators. Circulation 108:2460–2466
21. Miksztowicz V, Morales C, Barchuk M, López G, Póveda R, Gelpi R, Schreier L, Rubio M, Berg G (2017) Metalloproteinase 2 and 9 activity increase in epicardial adipose tissue of patients with coronary artery disease. Curr Vasc Pharmacol 15(2):135–143
22. Oka T, Yamamoto H, Ohashi N et al (2012) Association between epicardial adipose tissue volume and characteristics of non-calcified plaques assessed by coronary computed tomographic angiography. Int J Cardiol 161(1):45–49
23. Motoyama S, Sarai M, Harigaya H et al (2009) Computed tomographic angiography characteristics of atherosclerotic plaques subsequently resulting in acute coronary syndrome. J Am Coll Cardiol 54:49–57
24. Djaberi R, Schuijf JD, van Werkhoven JM et al (2008) Relation of epicardial adipose tissue to coronary atherosclerosis. Am J Cardiol 102:1602–1607
25. Wheeler GL, Shi R, Beck SR et al (2005) Pericardial and visceral adipose tissues measured volumetrically with computed tomography are highly associated in type 2 diabetic families. Investig Radiol 40:97–10
26. Iacobellis G, Leonetti F (2005) Epicardial adipose tissue and insulin resistance in obese subjects. J Clin Endocrinol Metab 90:6300–6302
27. Harada K, Amano T, Kataoka T et al (2014) Impact of abdominal and epicardial fat on the association between plasma adipocytokine levels and coronary atherosclerosis in non-obese patients. Atherosclerosis 237(2):671–676
28. Yamaguchi Y, Cavallero S, Patterson M et al (2015) Adipogenesis and epicardial adipose tissue: a novel fate of the epicardium induced by mesenchymal transformation and PPARγ activation. Proc Natl Acad Sci U S A 112(7):2070–2075

29. Sacks HS, Fain JN, Holman B et al (2009) Uncoupling protein-1 and related messenger ribonucleic acids in human epicardial and other adipose tissues: epicardial fat functioning as brown fat. J Clin Endocrinol Metab 94(9):3611–3615
30. Chechi K, Blanchard PG, Mathieu P et al (2013) Brown fat like gene expression in the epicardial fat depot correlates with circulating HDL-cholesterol and triglycerides in patients with coronary artery disease. Int J Cardiol 167(5):2264–2270
31. Iacobellis G, Assael F, Ribaudo MC et al (2003) Epicardial fat from echocardiography: a new method for visceral adipose tissue prediction. Obes Res 11:304–310
32. Marchington JM, Mattacks CA, Pond CM (1989) Adipose tissue in the mammalian heart and pericardium: structure, foetal development and biochemical properties. Comp Biochem Physiol B 94(2):225–232
33. Sacks HS, Fain JN (2011) Human epicardial fat: what is new and what is missing? Clin Exp Pharmacol Physiol 38(12):879–887
34. Rabkin SW (2007) Epicardial fat: properties, function and relationship to obesity. Obes Rev 8:253–261
35. Robicsek F, Thubrikar MJ (1994) The freedom from atherosclerosis of intramyocardial coronary arteries: reduction of mural stress–a key factor. Eur J Cardiothorac Surg 8(5):228–235
36. Berg G, Schreier L, Miksztowicz V (2014) Circulating and adipose tissue matrix metalloproteinases in cardiometabolic risk environments: pathophysiological aspects. Horm Mol Biol Clin Invest 17(2):79–87
37. Salazar J, Luzardo E, Mejías JC et al (2016) Epicardial fat: physiological, pathological, and therapeutic implications. Cardiol Res Pract 2016:1291537
38. Iacobellis G, Barbaro G (2008) The double role of epicardial adipose tissue as pro- and anti-inflammatory organ. Horm Metab Res 40:442–445
39. Antonopoulos AS, Margaritis M, Verheule S et al (2016) Mutual regulation of epicardial adipose tissue and myocardial redox State by PPAR-g/adiponectin signalling. Circ Res 118:842–855
40. Salgado-Somoza A, Teijeira-Fernández E, Fernández AL et al (2010) Proteomic analysis of epicardial and subcutaneous adipose tissue reveals differences in proteins involved in oxidative stress. Am J Physiol Heart Circ Physiol 299(1):H202–H209
41. Brahimi-Horn MC, Pouyss'egur J (2007) Oxygen, a source of life and stress. FEBS Lett 581:3582–3591
42. Akoumianakis I, Antoniades C (2017) The interplay between adipose tissue and the cardiovascular system: is fat always bad? Cardiovasc Res 113:999–1008
43. Abe I, Teshima Y, Kondo H et al (2018) Association of fibrotic remodeling and cytokines/chemokines content in epicardial adipose tissue with atrial myocardial fibrosis in patients with atrial fibrillation. Heart Rhythm 15:1717. pii: S1547-5271(18)30589-7
44. Patel VB, Basu R, Oudit GY (2016) ACE2/Ang 1-7 axis: a critical regulator of epicardial adipose tissue inflammation and cardiac dysfunction in obesity. Adipocytes 5(3):306–311
45. Iacobellis G, Pistilli D, Gucciardo M et al (2005) Adiponectin expression in human epicardial adipose tissue in vivo is lower in patients with coronary artery disease. Cytokine 29(6):251–255
46. Han YH, Buffolo M, Pires KM et al (2016) Adipocyte-specific deletion of manganese superoxide dismutase protects from diet-induced obesity through increased mitochondrial uncoupling and biogenesis. Diabetes 65(9):2639–2651
47. Barseghian A, Gawande D, Bajaj M (2011) Adiponectin and vulnerable atherosclerotic plaques. J Am Coll Cardiol 57(7):761–770
48. Hui X, Lam KS, Vanhoutte PM et al (2012) Adiponectin and cardiovascular health: an update. Br J Pharmacol 165(3):574–590
49. Iwayama T, Nitobe J, Watanabe T et al (2014) Role of epicardial adipose tissue in coronary artery disease in non-obese patients. J Cardiol 63(5):344–349
50. Castoldi A, Naffah de Souza C, Câmara NO et al (2016) The macrophage switch in obesity development. Front Immunol 6:637
51. Lumeng CN, Bodzin JL, Saltiel AR (2007) Obesity induces a phenotypic switch in adipose tissue macrophage polarization. J Clin Invest 117:175–184
52. Hirata Y, Tabata M, Kurobe H et al (2011) Coronary atherosclerosis is associated with macrophage polarization in epicardial adipose tissue. J Am Coll Cardiol 58(3):248–255
53. Miksztowicz V, Schreier L, McCoy M et al (2014) Role of SN1 lipases on plasma lipids in metabolic syndrome and obesity. Arterioscler Thromb Vasc Biol 34(3):669–675
54. McCoy MG, Sun GS, Marchadier D et al (2002) Characterization of the lipolytic activity of endothelial lipase. J Lipid Res 43:921–929
55. Young SG, Davies BS, Voss CV et al (2011) GPIHBP1, an endothelial cell transporter for lipoprotein lipase. J Lipid Res 52(11):1869–1884
56. Goldberg IJ, Eckel RH, Abumrad NA (2009) Regulation of fatty acid uptake into tissues: lipoprotein lipase- and CD36-mediated pathways. J Lipid Res 50(Suppl):S86–S90
57. Wang H, Eckel RH (2009) Lipoprotein lipase: from gene to obesity. Am J Physiol Endocrinol Metab 297:E271–E288
58. Zimmermann R, Strauss JG, Haemmerle G (2004) Fat mobilization in adipose tissue is promoted by adipose triglyceride lipase. Science 306(5700):1383–1386
59. Jaffer I, Riederer M, Shah P et al (2012) Expression of fat mobilizing genes in human epicardial adipose tissue. Atherosclerosis 220(1):122–127
60. Zhang R (2016) The ANGPTL3-4-8 model, a molecular mechanism for triglyceride trafficking. Open Biol 6(4):150272
61. Błachnio-Zabielska AU, Baranowski M, Hirnle T et al (2012) Increased bioactive lipids content in human subcutaneous and epicardial fat tissue correlates with insulin resistance. Lipids 47(12):1131–1141
62. Yu C, Chen Y, Cline GW et al (2002) Mechanism by which fatty acids inhibit insulin activation of insulin

receptorsubstrate-1 (IRS-1)-associated phosphatidylinositol 3-kinaseactivity in muscle. J Biol Chem 277:50230–50236
63. Shah C, Yang G, Lee I et al (2008) Protection from high fat diet-induced increase in ceramide in mice lacking plasminogen activator inhibitor 1. J Biol Chem 283:13538–13548
64. Holland WL, Knotts TA, Chavez JA et al (2007) Lipid mediators of insulin resistance. Nutr Rev 65:S39–S46
65. Jornayvaz F, Shulman G (2012) Diacylglycerol activation of protein kinase Ce and hepatic insulin resistance. Cell Metab 15(5):574–584
66. Gorden DL, Ivanova PT, Myers DS et al (2011) Increased diacylglycerols characterizehepatic lipid changes in progression of human nonalcoholic fatty liver disease; comparison to a murine model. PLoS One 6:e22775
67. Moraes-Vieira PM, Saghatelian A, Kahn BB (2016) GLUT4 expression in adipocytes regulates De Novo lipogenesis and levels of a novel class of lipids with antidiabetic and anti-inflammatory effects. Diabetes 65(7):1808–1815
68. Scherer M, Montoliu I, Qanadli SD et al (2015) Blood plasma lipidomic signature of epicardial fat in healthy obese women. Obesity (Silver Spring) 23(1):130–137
69. Malavazos AE, Di Leo G, Secchi F et al (2010) Relation of echocardiographic epicardial fat thickness and myocardial fat. Am J Cardiol 105(12):1831–1835
70. Kwon H, Pessin JE (2013) Adipokines mediate inflammation and insulin resistance. Front Endocrinol 4:71
71. Sato T, Kameyama T, Ohori T et al (2014) Effects of eicosapentaenoic acid treatment on epicardial and abdominal visceral adipose tissue volumes in patients with coronary artery disease. J Atheroscler Thromb 21(10):1031–1043
72. Iacobellis G, Singh N, Wharton S et al (2008) Substantial changes in epicardial fat thickness after weight loss in severely obese subjects. Obesity (Silver Spring) 16:1693–1697
73. Gaborit B, Jacquier A, Kober F et al (2012) Effects of bariatric surgery on cardiac ectopic fat: lesser decrease in epicardial fat compared to visceral fat loss and no change in myocardial triglyceride content. J Am Coll Cardiol 60:1381–1389
74. Park JH, Park YS, Kim YJ et al (2010) Effects of statins on the epicardial fat thickness in patients with coronary artery stenosis underwent percutaneous coronary intervention: comparison of atorvastatin with simvastatin/ezetimibe. J Cardiovasc Ultrasound 18(4):121–126
75. Sacks HS, Fain JN, Cheema P et al (2011) Inflammatory genes in epicardial fat contiguous with coronary atherosclerosis in the metabolic syndrome and type 2 diabetes: changes associated with pioglitazone. Diabetes Care 34:730–733
76. Distel E, Penot G, Cadoudal T et al (2012) Early induction of a brown-like phenotype by rosiglitazone in the epicardial adipose tissue of fatty Zucker rats. Biochimie 94(8):1660–1667
77. Morano S, Romagnoli E, Filardi T et al (2015) Short-term effects of glucagon-like peptide 1 (GLP-1) receptor agonists on fat distribution in patients with type 2 diabetes mellitus: an ultrasonography study. Acta Diabetol 52(4):727–732
78. Sato T, Aizawa Y, Yuasa S et al (2018) The effect of dapagliflozin treatment on epicardial adipose tissue volume. Cardiovasc Diabetol 17(1):6

Part IV

Role of Bioactive Lipids in Neurodegenerative Diseases

Bioactive Lipids and the Gut-Brain Axis: Diet as a Modulator of Bioactivity and Diversity of Lipids in the Brain

10

A. Ledo, B. S. Rocha, and J. Laranjinha

Abstract

The brain is highly rich in lipids, which accounts for roughly 50% of its dry weight. The brain lipidome, generally characterized over half a century ago, is comprised of thousands of biochemical structures expressed differentially as a function of brain region, structure, cell type and subcellular compartment. Lipids play diverse structural and functional roles in the brain, not only due to their chemical diversity but also due to the unique hydrophobic environment that they create. This lipophilic milieu promotes interactions involving reactive oxygen and nitrogen species that may not occur, at least at a similar extent, in aqueous environments.

In the present chapter, we have focused on 3 distinct types of bioactive lipids and the roles played in brain physiology and pathology: nitrated fatty acids, cholesterol and endocannabinoids. These lipids are diverse in origin and bioactivity: (1) **nitrated fatty acids** result from biochemical modification of dietary fatty acids by nutrients and are proposed to play diverse physiological roles, namely by modulating NF-kB and Nfr2-dependent signaling cascades and post-translational modification of proteins. Produced in the gastric compartment, they are absorbed into circulation and can cross the blood-brain barrier, providing a new route for the interaction between the gastrointestinal tract and the brain; (2) **cholesterol**, synthetized *de novo* in the brain, not only regulates the biophysical properties of cellular membranes, but can also physically interact with neurotransmitter receptors and other membrane proteins and enzymes such as those involved in the processing and trafficking of the amyloid precursor protein (APP) and Aß peptide; (3) **endocannabinoids**, a class of neuromodulators derived from fatty acids that are synthetized and released upon demand and incite cellular responses by binding to specific membrane receptors.

Being one of the most important and adjustable determinants of human health, our goal is to highlight the impact of diet on the bioactivity of lipids in the brain, discussing novel and provocative findings that advocate that lipids may modulate the gut-brain axis and therefore higher cortical functions such as motor function, learning and memory.

Authors A. Ledo and B. S. Rocha have equally contributed to this chapter.

A. Ledo (✉) · B. S. Rocha (✉) · J. Laranjinha (✉)
Faculty of Pharmacy, University of Coimbra, Coimbra, Portugal

Center for Neuroscience and Cell Biology, University of Coimbra, Coimbra, Portugal
e-mail: analedo@cnc.uc.pt; barbarasrocha@icloud.com; laranjin@ci.uc.pt

A. Trostchansky, H. Rubbo (eds.), *Bioactive Lipids in Health and Disease*, Advances in Experimental Medicine and Biology 1127, https://doi.org/10.1007/978-3-030-11488-6_10

Keywords
Gut-brain axis · Nitrated fatty acids · Cholesterol · Endocannabinoids

10.1 Introduction

Lipids are a class of biomolecules with a large diversity of chemical structures and biological functions, including the regulation of crucial physiological processes in the brain. Although one can identify specific structural similarities and classify the lipids in subgroups, as a whole class of biomolecules, lipids share a common physical property: they have a low solubility in water and are soluble in organic solvents. Some exhibit amphipathic nature and they are differently prone to react with oxidants [51].

Brain lipids include phospholipids, sphingolipids and cholesterol present in high proportions [96]. Lipids play a number of critical structural and signaling functions. Besides being an essential element of the myelin shaft guaranteeing insulation, they are equally essential for synaptogenesis, neuritogenesis and rapid saltatory nerve impulse conduction. The major lipid constituents of the nervous system have been known for over 60 years, and since then much research has been dedicated to understanding their functional role in the different regions, structures and cell types. Large scale –omics and animal models have more recently allowed significant advances to be made.

In the present chapter, we focus on 3 types of bioactive lipids which distinct roles in regulating brain function and dysfunction during disease, namely nitrated fatty acids, cholesterol and endocannabinoids. Diet is among the most important adjustable determinants of human health and all 3 of the bioactive lipids we have chosen to discuss here are, to different extents, impacted by diet, either as a result of precursor intake, metabolism of nutrients or due to unbalanced diet.

We will first focus on the chemical pathways that lead to unsaturated fatty acid nitration *in vivo*. Special attention will be paid to a novel driver of nitration of both endogenous and exogenous lipids, dietary nitrate. The nitrate-nitrite-nitric oxide pathway will be addressed as an additional contributor to fatty acid nitration and its physiological implications will be discussed. The signaling cascades and ensued physiological effects of nitroalkenes will be exploited both systemically and in the central nervous system (CNS). Recent translational opportunities for the use of nitrated fatty acids as therapeutic agents will close this first section. On a second note, we will discuss cholesterol metabolism in the brain and, in more detail, the crosstalk between changes in cholesterol homeostasis and Alzheimer's disease pathophysiology. The particular role of cholesterol in promoting amyloidogenic processing of amyloid precursor protein (APP) to β-amyloid peptides (Aβ) in lipid rafts will be discussed, as well as crosslines in cholesterol and APP/Aß trafficking. Finally, we will briefly discuss the fatty-acid derived neuromodulators which interact with cannabinoid receptors – the endocannabinoids. These bioactive lipids act as non-classical neurotransmitters/neuromodulators and are synthetized and released upon demand to the synaptic cleft, where they act as agonists of cannabinoid receptors, activating intracellular signaling cascades which participate in several neuronal processes.

10.2 Nitrated Fatty Acids: Oxidant-Dependent Bioactive Lipids in Health and Disease

Lipids are differently prone to react with oxidants [51]. Both free and esterified unsaturated fatty acids, as key components of cell membranes and lipoproteins, are exposed to oxidative and nitrative species due to the high diffusional spread of nitric oxide ($^{\bullet}NO$), with a reported diffusion coefficient of $2x10^5$ cm^2 s^{-1} [127], in hydrophobic environments [51, 160]. In this context, nitrogen dioxide radical ($^{\bullet}NO_2$) may be formed from $^{\bullet}NO$ auto-oxidation or from nitrous acid (HNO_2) that, at acidic pH, is produced from nitrite – up to 500 nM of nitrite are found in human plasma upon a nitrate load from dietary sources [112]. Of note, $^{\bullet}NO_2$ may induce both unsaturated fatty acid nitration and oxidation [62, 131]. The yields of the different end products are highly dependent on the partial pressure of O_2 (pO_2), as discussed below.

Regardless of the possible pathways for endogenous nitration, c.a. 500 nM of nitroalkene derivatives of fatty acids have been detected in human blood and urine under physiological conditions [6, 7, 105]. More recent studies report values of 1 nM for nitrated oleic acid (OA-NO_2) and isomers of nitrated linoleic acid (LNO_2) in the systemic circulation [188]. In addition, nitrated fatty acids have also been detected in plasma, cell membranes and tissues of both healthy and dyslipidemic patients [6, 105]. The description of the molecular mechanisms and signaling pathways modulated by nitrated fatty acids, such as the binding to Peroxisome Proliferator-Activated Receptors (PPARs) as well as the activation of Nuclear Factor (erythroid-derived 2)-like 2 (Nrf2) and inhibition of Nuclear Factor kappa-light-chain-enhancer of activated B cells (NF-κB) signaling pathways, has unveiled their citoprotective and anti-inflammatory functions [192]. Adaptive signaling responses, with possible therapeutic applications, have also recently been reported [167]. Novel interactions between nitrated fatty acids and Toll-Like Receptors (TLRs) in membrane lipid rafts of endothelial cells and macrophages, have been shown to resolve vascular inflammation and inhibit the adhesion of immune cells while inhibiting the NF-κB signaling cascade [196]. Also, nitroalkenes derivatives are now known to inhibit platelet aggregation [20], an additional mechanism that may be explored with the intent of using nitrated fatty acids to treat inflammatory and cardiovascular disorders.

10.2.1 Endogenous Synthesis: Foods as Unexpected Drivers of Endogenous and Dietary Lipid Nitration

Nitration of unsaturated fatty acids, unlike protein tyrosine nitration that is traditionally regarded as a harmful event, generates a class of pluripotent cell signaling modulators (for a recent review see [183]). Indeed, nitrated fatty acids have been shown to mediate inflammatory, cardiovascular and metabolic pathways [31, 88, 183, 185]. Chemical features, such as structural stability and the ability to nitroalkylate proteins, justify many of the biological actions of nitroalkenes. In fact, nitrated lipids are quite stale in hydrophobic environments, such as those within lipoproteins and cell membranes. Also, the nitro (NO_2) group is electronegative, withdrawing the electrons from the *β*-carbon, promoting the formation of adducts with nucleophiles such as Cys and His residues within proteins through Michael addition [10, 56]. This nitroalkylation is a reversible reaction that impacts the distribution and function of proteins involved in critical signaling pathways such as PPARs [63, 165]. Given the ability of nitrated fatty acids to modulate cell adaptive responses, it is of foremost importance to understand the pathways that may lead to endogenous fatty acid nitration. In this context, several chemical reactions may be anticipated.

Nitrogen dioxide radical mediates most of the hypothesized *in vivo* pathways of lipid nitration. Of note, $^{\bullet}NO_2$ may induce fatty acid oxidation and/or nitration both under aerobic and anaerobic conditions [186]. At high pO_2, lipid oxidation reactions, with the formation of hydroperoxides or isomerized derivatives, predominate over nitration reactions [136]. On the contrary, in the absence or at low pO_2, nitrated products are the major end products. Under these conditions, $^{\bullet}NO_2$ may react with both unsaturated fatty acids or lipid radicals yielding nitroalkenes, isomerized, nitroallylic, dinitro or nitrohydroxy derivatives [62, 131, 186]. Also, the radical-radical interaction between $^{\bullet}NO_2$ and a lipid carbon-centered radical, originated from the rearrangement of unstable oxidizing intermediates (LOONO), generates nitrated fatty acids [137]. The major recent breakthrough regarding the mechanism of lipid nitration is related to the endogenous pathways of $^{\bullet}NO_2$ formation [22, 54, 153]. In this context, $^{\bullet}NO_2$ may be produced from $^{\bullet}NO$ auto-oxidation, a reaction that occurs within cell membranes and lipoproteins, where $^{\bullet}NO$ diffuses and concentrates [127], and from the homolytic cleavage of peroxynitrous acid (ONOOH) that yields hydroxyl radical ($^{\bullet}OH$) and $^{\bullet}NO_2$, potent oxidizing, nitrosating and nitrating agents [9, 149, 159]. However, and despite the fact that incubation of fatty acids with acidified nitrite has been used for many years to produce nitroalkenes *in vitro* [103, 136, 208], it was not until the description of the nitrate-nitrite-nitric oxide pathway, a non-enzymatic mechanism of $^{\bullet}NO$ production in the gut, that the potential of gastric nitrite to induce fatty acid nitration was unveiled.

In the mid 1990s, nitrite anion, coming from the enterosalivary circulation of dietary nitrate (present in green leafy vegetables and beets), was shown to be reduced to $^{\bullet}NO$ in the human stomach [12, 116]. Up to 6000 ppb of $^{\bullet}NO$ were detected in the gastric compartment upon the ingestion of iceberg lettuce [116] and this steady state concentration has been shown to increase several fold when reductants, such as polyphenols, are also consumed [61, 141, 155]. Nitrite acidic disproportion is transiently mediated by HNO_2 and, depending on the chemical environment of the gastric milieu, several nitrogen oxides, including $^{\bullet}NO_2$, may be produced [115, 153, 156]. In accordance to what has been shown to happen with gastric proteins such as pepsin and occludin [152, 154, 157], nitrite-dependent fatty acid nitration has also been demonstrated [22, 43, 54]. Dietary oleic and conjugated-linoleic acids (OA; cLA), from olive oil, have been shown to be nitrated by nitrite under acidic digestive conditions [54]. Interestingly, both free- as well as protein-adducted nitrated fatty acids were detected, suggesting that nitroalkylation of proteins, a key pathway instigating the anti-inflammatory properties of nitrated fatty acids, is also related to nitrate and nitrite consumption and further production of $^{\bullet}NO_2$ in the acidic stomach [54]. Furthermore, Bonacci and colleagues have also elegantly shown that not only dietary cLA is nitrated by a nitrite-dependent pathway in the stomach, but also that nitrated-cLA levels are elevated in plasma, tissue and urine of rodents upon exposure to both dietary cLA and nitrite [22].

Overall, these observations support the notion hypothesized by Rocha and colleagues [153] that the nitrate-nitrite-nitric oxide pathway fuels the production of nitrating agents in the stomach that may react with exogenous fatty acids, and that nitrated fatty acids may be absorbed into the systemic circulation and trigger cell adaptive responses in virtually every organ system. Delmastro-Greenwood and colleagues have shown that dietary consumption of nitrate and nitrite as well as cLA yields electrophilic nitrated derivatives. Nitro-cLA was not only detected in plasma and urine of healthy volunteers, but was also shown to react with nucleophilic amino acids via Michael addition, suggesting that dietary concentrations of both nitrate and cLA are sufficient to generate the nitrated derivatives in the stomach, and that these species are further absorbed into the blood stream and trigger signaling cascades systemically [43]. In line with this observation, the activation of phase II enzymes, such as heme oxygenase-1 (HO-1), in the intestinal epithelium has been shown under similar circumstances [22]. It is therefore tempting to speculate that endogenous lipids may also be targeted for nitrite-dependent nitration. One example, are the lipids from the gastric mucus. The gastric mucosa, especially in post-prandial periods, secretes potentially harmful compounds into gastric lumen, such as proteolytic enzymes and HCl [171]. Thus, the gastric mucosa exhibits several degrees of protection, including pre-epithelial, epithelial and post-epithelial components [199]. The less understood of these defenses is the pre-epithelium that comprises the mucus-bicarbonate layer. The loosely bound mucus barrier is in dynamic equilibrium with the pre-formed intracellular mucus contained within the secretory granules of specialized mucous cells [142, 145, 200]. Both, the loosely and the firmly adherent mucus layers, are composed of glycoproteins embedded with water, electrolytes and glycolipids. Among these, phospholipids such as phosphatidylethanolamine, phosphatidylglycerol, diphosphatidylglycerol and phosphatidic acid can be found [132]. Since exquisite nitration conditions are achievable within the gastric lumen (due to high nitrite concentrations as well as low pH), phospholipid nitration may be expected. In fact, $^{\bullet}NO_2$ is formed from the reaction of $^{\bullet}NO$ with molecular oxygen in the gastric headspace [153]. Once formed, $^{\bullet}NO_2$

could oxidize and nitrate phospholipids from the mucus. Furthermore, this reaction may be favored by other reactive species produced from nitrite in the gastric lumen, namely $^{\cdot}OH$ and $^{\cdot}NO_2$. These radicals can be produced by ONOOH homolytic cleavage and may oxidize and nitrate mucus lipids. Although the anti-inflammatory and citoprotective effects of nitrated lipids have been shown for nitro-OA, nitro-cLA and nitrated arachidonic acid, nitration of phospholipids may also be considered. Nitration of mucus lipids can be regarded not only as a mechanism to scavenge nitrite-derived $^{\cdot}NO$ arising from the lumen, but also as a source of a new class of molecules with impact on gastric pathophysiology. Notably, at acidic pH, nitronium ion (NO_2^+) may also induce unsaturated fatty acid nitration by addition to the double bond by electrophilic substitution, an already described pathway for fatty acid nitration [136, 137].

One intriguing aspect that would deserve further attention is the pO_2 in the gastric headspace. For a pO_2 of 70 torr [76], fatty acid oxidation rather than nitration would be expected to be the dominant reaction. Hence, further studies would be required to ascertain as to whether oxidized lipid derivatives would also be formed and what biological consequences should be expected.

10.2.2 Molecular Mechanisms Underlying the Signaling Functions of Nitrated Fatty Acids

Since the beginning of 2000s, several research groups have provided solid evidence that nitration of unsaturated fatty acids generates a new class of signaling biomolecules [6, 10, 185]. Figure 10.1 summarizes the molecular mecha-

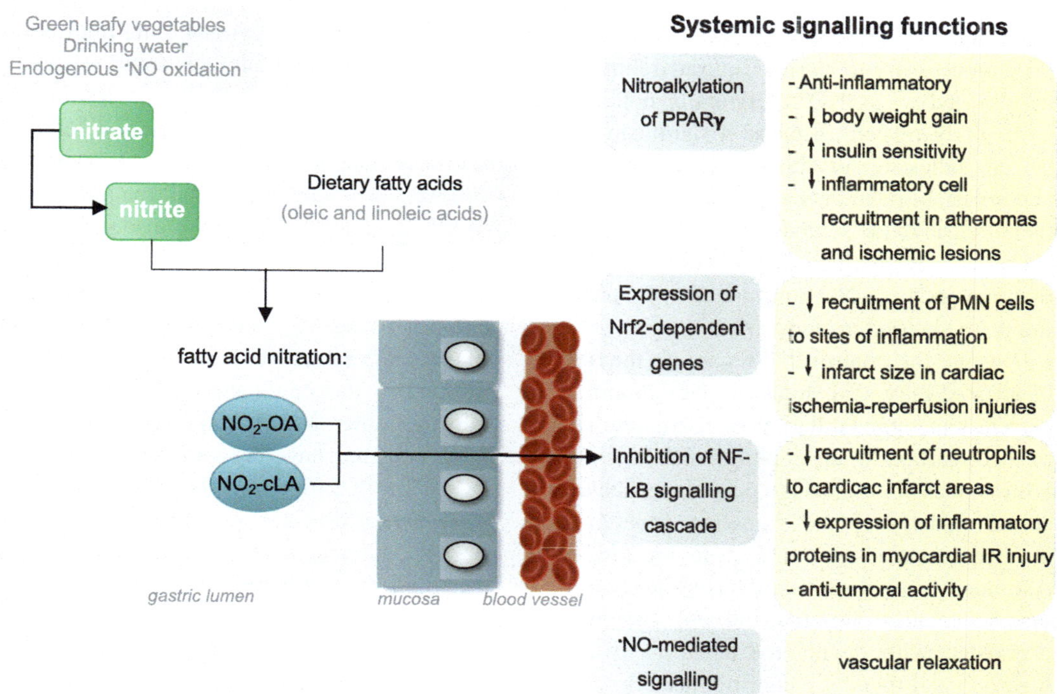

Fig. 10.1 Signaling functions of nitrated fatty acids. Nitro-OA and nitro-cLA may be formed from the reaction of OA and cLA with dietary nitrite in the stomach. These nitrated derivatives are absorbed into systemic circulation and activate different signaling cascades with consequent anti-inflammatory and citoprotective actions. [22, 43, 54, 107, 161, 165, 201]

nisms underlying the biological functions of nitrated fatty acids. The chemical stability within lipophilic environments and the ability to nitroalkylate proteins are now recognized as crucial aspects in nitroalkene bioactivity [186]. Initial studies showed that nitrated fatty acids could mediate cGMP-dependent vascular relaxation, supporting a role as ˙NO reservoirs *in vivo*, but extremely low stoichiometric concentrations of ˙NO were released [6, 66, 103, 105, 121]. Accordingly, recent reports show that the intravenous infusion of nitro-OA does not decrease neither blood pressure not heart rate [213], suggesting that ˙NO release from nitrated fatty acids is not the major mechanism of action of these signaling molecules. In this context, Coles and colleagues have demonstrated that nitrolinoleate inhibits platelet activation as well as neutrophil degranulation by reducing intracellular calcium mobilization and enhancing cAMP/adenyl cyclase signaling [32, 33], thereby demonstrating cGMP-independent signaling mechanisms of nitrated fatty acids. Further studies are necessary to clarify if nitrated fatty acids may indeed act as ˙NO reservoirs or not. This is particularly relevant because a significant number of studies support this hypothesis [104, 121] and in fact, nitrated fatty acids may activate endothelial nitric oxide synthase as well as induce the expression of phase II enzymes, such as HO-1, that may increase ˙NO bioavailability by preventing its oxidation reactions [60, 92].

Despite the controversy regarding the ability of nitrated fatty acids to act as ˙NO donors, it is currently recognized that the electrophilic attack of the ß-carbon to nucleophiles such as Cys and His residues of proteins is a pivotal mechanism of action underlying the signaling functions of nitrated fatty acids (for a recent review see [60]). As already mentioned, the NO_2 group is one of the most electronegative moieties known in chemistry, withdrawing the electrons from the ß-carbon and thereby rendering this atom highly avid for nucleophiles. This allows rapid and reversible protein nitroalkylation, a post-translational modification that may impact on protein structure and function [5, 10, 164]. Notably, many of the proteins targeted for nitroalkylation are transcription factors or modulate the activation of transcription factors [56]. One interesting example is the nitroalkylation of the subtype γ of PPAR (PPARγ). PPARs are nuclear receptors that, among other functions (out of the scope of the present chapter; for a comprehensive review see [192]), respond to both lipid metabolites and glucose and, in particular subtype γ, decreases fatty acid oxidation and transport [56, 192]. Indeed, PPARs are now known to modulate bioenergetic and inflammatory pathways *in vivo* [192]. Interestingly, nitro-OA and nitro-LA activate all PPAR subtypes although the highest activity is observed for subtype γ with a potency that depends on the position of nitration [6, 166]. In rodents, the administration of nitro-OA increases both plasma nitro-OA to 270 nM and the expression of PPARγ itself and PPARγ-dependent genes such as Signal Transducer and Activator of Transcription 1 (STAT-1) and Fatty Acid Binding Protein-2 (FABP2), suggesting that nitrated fatty acids are PPAR activators *in vivo* [23]. As mentioned previously, the relevance of PPARγ activation in biological settings is related to its anti-inflammatory activity. Several molecular mechanisms have been proposed to justify these beneficial properties, such as the expression of PPARγ-dependent genes, but the inhibition of NF-kB signaling cascade plays also an important role. Interestingly, nitrated fatty acids may also prevent NF-kB activation under inflammatory conditions [38, 55], suggesting that the anti-inflammatory effects of these compounds may be also due to the modulation of other signaling pathways rather than PPARγ activation.

Nitroalkenes have been shown to elicit salutary effects in different models of metabolic and inflammatory diseases such as type 2 diabetes, obesity, atherosclerosis, renal and cardiac ischemia-reperfusion injury and vascular disorders [107, 161, 165, 201]. Although in some of these models the activation of PPARγ has been shown to occur, the modulation of other pathways, namely activation of Nrf2 and HO-1 signaling and the inhibition of NF-kB, have also been demonstrated. Of note, the administration of low concentrations of nitro-OA (2 mg kg^{-1} day^{-1}) for extended periods of time (21 days) prevents neointimal hyperplasia in a mice model of vascular injury with increasing HO-1 expression [31]. On the other hand, in an animal model of atherosclerosis (APOE$^{-/-}$ mice), nitro-OA reduces the infiltration of inflammatory cells within the atheroma through a mechanism compatible with the activation of Nrf2 or inhibition of NF-kB signaling cascades [161].

The structural stability within membranes and the potential to form Michael adducts with proteins, including critical transcription factors, endow nitrated fatty acids with unique chemical features to control inflammatory pathways, modulate lipid and glucose metabolism and promote cell adaptation to hostile, oxidative, environments. These molecular actions justify the biological implications of these electrophiles in many disease models, including (CNS) disorders.

10.2.3 The Gut-Brain-Axis and the Impact of Nitroalkenes in the Central Nervous System

The past two decades have witnessed a change in the understanding of the Biology of dietary nitrate due to the observation that nitrate anion generates $^{\cdot}NO$ in the human stomach out of enzymatic control [12, 116]. In summary, nitrate, from green leafy vegetables, undergoes an enterosalivary circulation *in vivo*, being secreted into the oral cavity. Here, commensals from posterior tongue clefts reduce nitrate to nitrite which, once mixed and swallowed with saliva, is readily reduced to $^{\cdot}NO$ in the stomach [49, 114, 115]. This so-called *nitrate-nitrite-nitric oxide pathway* sparked a new field of research since nitrite-derived $^{\cdot}NO$ has been shown to promote not only gastrointestinal but also systemic protective mechanisms. In the gut, this non-enzymatic pathway increases mucosal vasodilation, mucus thickness and halts inflammatory and ulcerogenic pathways. Interestingly, systemic cardiovascular and metabolic effects have also been reported (for recent reviews see [28, 113]). But it was not until recently that the potential of nitrite to produce other nitrogen oxides in the acidic gastric milieu gained attention. In the beginning of 2010s, three independent groups showed that nitrite, originated from dietary nitrate, generates nitrating agents in the stomach [22, 54, 153]. Not only the nitration of endogenous gastric proteins has been observed *in vivo*, a reaction dependent on the generation of $^{\cdot}NO_2$ in the gastric lumen [152, 154, 157], but also OA and cLA have been shown to be nitrated by acidified nitrite [22, 54]. Dietary concentrations of cLA and nitrite, when administered by oral gavage to mice, led to the production of nitro-cLA which was detected in plasma, urine and tissues. Moreover, this observation was paralleled by the induction of HO-1 expression in the colon, demonstrating that dietary components generate nitrated lipid derivatives with signaling functions *in vivo* [22]. Given that nitrated fatty acids react with nucleophilic amino acids via Michael addition, changing protein structure and function, the possibility to form citoprotective biomolecules in the gut, with systemic anti-inflammatory actions, represents a groundbreaking achievement with physiological and therapeutic consequences yet to be fully understood. In this context, Delmastro-Greenwood and collaborators have elegantly shown that the consumption of $^{15}NO_3^-$ or $^{15}NO_2^-$ with cLA increases plasma levels of nitrated cLA from 0.14 nM (detected even without cLA consumption) to 2.5 nM (for $^{15}NO_2^-$) or 3.3 nM (for $^{15}NO_3^-$) [43]. Also, the plasma concentration of $^{15}NO_2$-cLA remained elevated for 6 h suggesting that the consumption of dietary doses of nitrate and cLA induces the formation of potent electrophilic derivatives, that may be absorbed into the blood stream, and alter protein function through post-translational modifications as result of Michael addition. This observation is of foremost relevance since gastric-generated nitrated fatty acids may nitroalkylate proteins, such as PPARγ, as well as modulate Nrf2 and NF-kB signaling systemically.

In line with the previous *rationale*, nitrated fatty acids may also impact on brain function, representing a novel and surprising group of biomolecules that modulate the gut-brain axis, the bidirectional neurohumoral crosstalk between the gastrointestinal tract and the brain. Accordingly, Trostchansky et al. have recently shown that the administration of nitro-OA reduces the levels of pro-inflammatory prostaglandins and oxidized products in the brain of a mice model of amyotrophic lateral sclerosis (ALS) while improving animal motor performance. It is also shown that nitro-OA crosses the blood brain barrier (BBB) since 4.33 ± 1.12 pmol mg^{-1} tissue of nitro-OA and 203.11 ± 1.09 pmol mg^{-1} tissue of its ß-oxidation product were detected in the brain upon subcutaneous injection [184]. Of note, despite the route of administration, the ability of nitroalkenes to permeate the BBB reinforces the hypothesis that upon gastric production and absorption into the systemic circulation, these compounds may reach the CNS and activate local citoprotective mechanisms. Another hypothesis is that dietary nitrite, which remains elevated for up to 6 h after a nitrate load [112], may reach the CNS and induce the production of nitrated fatty acids locally. However, nitrite-induced nitration requires low, non-physiological, pH. It is expected that under ischemic conditions, nitrite could be reduced to •NO, either by local enzymes or acidic disproportion, triggering the aforementioned mechanisms of fatty acid nitration.

In this regard, it is worth mentioning that MRI studies in a group of elderly human individuals exposed to increased intake of nitrate rich foods showed increases of cerebral blood flow (CBF) to key areas of the brain involved in executive functions, [146, 147]. More recently, an investigation with 40 healthy adults showed that dietary nitrate modulates CBF response to task performance in the frontal cortex and cognitive performance [204] and intrathecal levels of nitrate in patients with vascular dementia are inversely correlated to the degree of intellectual impairment [182].

In summary, dietary nitrate, through the generation of nitrite in the stomach, can fuel the nitration of exogenous and endogenous fatty acids. These nitrated derivatives may be absorbed into the systemic circulation and modulate Nrf2 and NF-κB signaling cascades as well as induce post-translational modifications of proteins, such as PPAR-γ, via Michael addition in every organ system. The CNS is no exception and, since nitrated fatty acids cross the BBB, anti-inflammatory actions are also expected in different brain regions, as it has been demonstrated in an animal model of amyotrophic lateral sclerosis. Finally, since nitrate consumption increases regional cerebral blood flow in human volunteers, it is expected that increased •NO production in these areas can also promote fatty acid nitration within hydrophobic environments, with the aforementioned signaling functions. Figure 10.2 summarizes the possible pathways for nitrite-dependent fatty acid nitration in the CNS.

10.2.4 Translation of the Bioactivity of Nitroalkenes into Clinical Care

The citoprotective and anti-inflammatory properties of both endogenous and exogenous nitroalkenes, make these compounds extremely attractive drug candidates. Moreover, from a chemical point of view, nitrated fatty acids have three structural characteristics that further support their use for the development of new drugs: (1) nitroalkenes are strong electrophiles thereby activating the expression of proteins encoded by Nrf2-dependent genes; (2) nitroalkenes undergo similar oxidative metabolism as non-nitrated fatty acids, yielding bioactive metabolites with putative applications in lowering pulmonary and ocular hypertension and finally, (3) nitroalkenes are incorporated into plasma lipoproteins as glycerol esters being distributed systemically. Interestingly, this hydrophobic microenvironment also allows the stabilization of the electrophilic moiety and therefore ensures their signaling features in all organ systems (for a recent review see [167, 183]).

In fact, intravenous and oral formulations of a positional isomer of nitro-OA (CXA-10, 10-nitro-OA) has shown promising results in the treatment of fibrotic glomerulopathies as well as pulmonary hypertension in Phase 1 Clinical Trials [84, 91]. The additional absence of significant adverse events and the distribution of CXA-

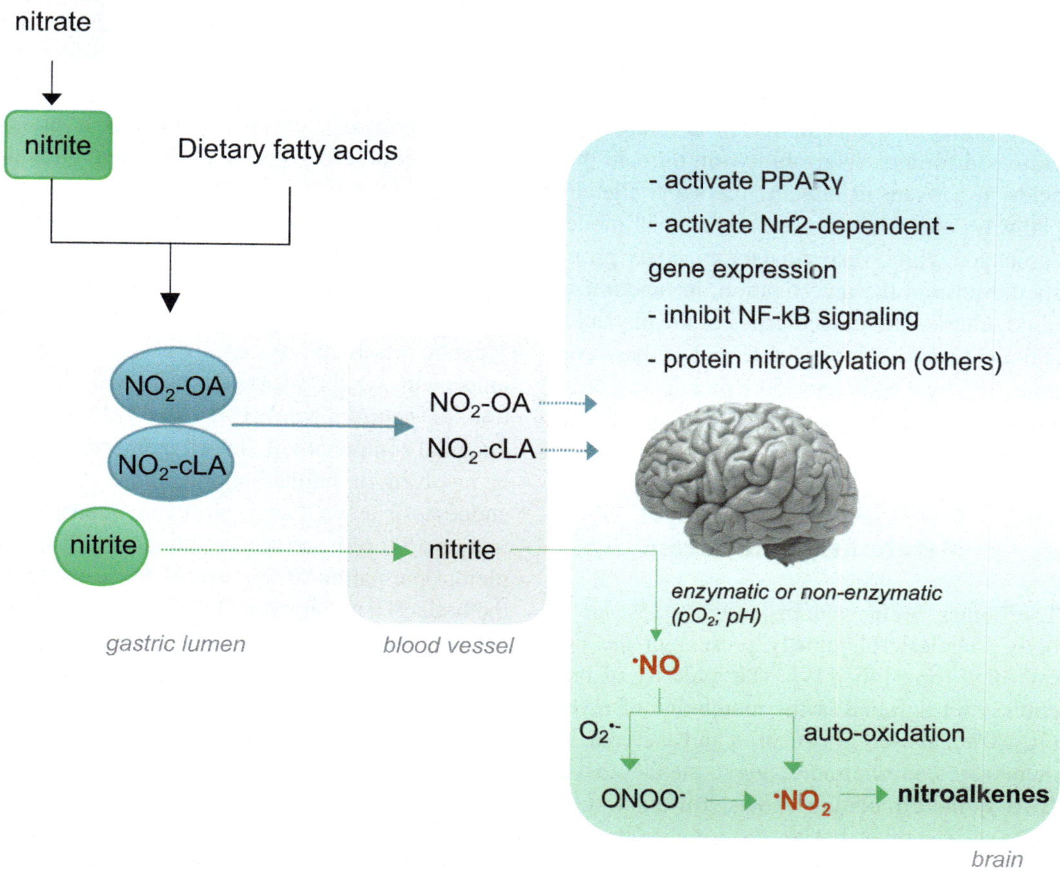

Fig. 10.2 Pathways for nitrated fatty acids formation in the CNS. Nitroalkenes may be produced from dietary lipids and nitrite in the stomach and be absorbed into the systemic circulation. Since they can cross the BBB, nitrated fatty acids may promote cell adaptive responses within the CNS (upper panel). Alternatively, dietary nitrite is likely to increase ˙NO steady state concentration in the brain through a still not fully elucidated mechanism. Nitrite may be reduced by enzymes that may acquire nitrite-reductase activity along a decreasing gradient of pO_2 (e.g., xanthine oxidase, nitric oxide synthases, haemoglobin, among others) or, alternatively, several authors hypothesise that transient and localised ischemic events may sufficiently decrease the pH to allow non-enzymatic reduction of nitrite to ˙NO. The increase of cerebral blood flow in elderly volunteers support this notion [147]. Since ˙NO diffuses and concentrates in hydrophobic environments, such as cell membranes, it can induce the formation of ˙NO_2 through oxidation reactions or the formation of $ONOO^-$ upon reaction with $O_2^{\bullet -}$, thereby promoting fatty acid nitration

10 metabolites in the liver, heart and adipose tissues, organs with high metabolic rate, increase the pharmacological interest of these compounds even further [167]. Of note, the oxidative status of these organs, which generate high steady state concentrations of oxygen and nitrogen species, would greatly benefit from the presence of nitroalkenes as these molecules would increase the expression of Nrf2-dependent genes and inhibit the production of inflammatory cytokines encoded by NF-κB-associated genes.

Hence, the almost two decades of basic and pre-clinical, academic research on the synthesis, structure, detection and biological properties of nitroalkenes, is now culminating in the development of a new class of drugs that target diseases devoid of effective pharmacological treatment. It is now tempting to speculate whether other biological functions of nitroalkenes such as the inhibition of platelet aggregation and intimal hyperplasia may improve the outcome of patients with coagulopathies and other thrombotic diseases, some of them implicated in neurodegenerative disorders. The ability of nitrated fatty acids to activate in humans the same signaling pathways that they engage in animal models, associated with a very satisfactory safety profile, should prompt the investigation, in randomized, blind, multicenter clinical trials of the physiological and pharmacological potential of these compounds to treat disorders addressed in pre-clinical studies.

10.3 Overview of Cholesterol Metabolism in the Brain

The human brain contains roughly 25% of the body's cholesterol, mostly present in the non-esterified form [46, 119]. The majority of brain cholesterol is found in the membrane of myelin (70–80%), as well as in neuron and astrocyte cell membranes. Furthermore, due to the fact that the BBB limits circulating cholesterol entry to the brain, it is synthetized *de novo* within the CNS from acetyl-CoA which is converted to 3-hydroxy-3-methyl-glutaryl-coenzyme A (HMG-CoA) and then to mevalonate by HMG-CoA reductase, the rate limiting and irreversible step in cholesterol synthesis. A series of reactions involving over 20 enzymes convert mevalonate to 3-isopenenyl pyrophosphate, farnesyl pyrophosphate, squalene, lanosterol, and so on until the final product cholesterol is obtained [14]. Cholesterol synthesis occurs primarily in the endoplasmic reticulum, from where it is then transferred to the plasma membrane. Both neurons and astrocytes are capable of cholesterol biosynthesis, although in neurons sterols are synthesized via the Kandutsch-Russel pathway, while in astrocytes, synthesis occurs via the Bloch pathway [135]. Furthermore, while during development, both neurons and astrocytes synthetize cholesterol, in the mature brain, this process occurs mainly in the astrocytic compartment [191]. The astrocytes release cholesterol in ApoE complexes, which are up taken by neurons to satisfy their needs. Interestingly, the transcription of ApoE is regulated by the cholesterol oxidation product 23-hydroxycholesterol (24-OH), which is a ligand of liver receptor X (LXR), a nuclear receptor that also regulates the expression of other genes involved in cholesterol metabolism, including ATP-binding cassette transporters A1 and G1 (ABCA1 and ABCG1) [135, 144, 172].

Functionally, cholesterol plays as critical role in neuronal development and maintenance of synaptic plasticity, synapse formation, neurite outgrowth, synaptic vesicle transport and regulation of neurotransmitter release [143]. As an essential component of the cell membrane, it also is involved in regulating ion homeostasis and endocytosis as well as synapse and dendrite formation [57, 67] and axonal guidance [39]. The pleiotropic nature of this lipids' functionality in the brain as a major regulator of neuronal activity means that cholesterol homeostasis must be tightly maintained through a dynamic balance between de novo synthesis, transport, storage and removal.

Consequent to its essential role in maintaining neuronal physiology, both during development and in the adult stage, depletion of cholesterol has been show to limit synaptic vesicle endocytosis, neurotransmission and neuronal activity in general, resulting in degeneration of spines and synapses [106]. Furthermore, changes in cholesterol metabolism/homeostasis have been implicated in the pathophysiology of central nervous system disorders and diseases such as Alzheimer's disease [45], Parkinson's disease [48], Huntington's disease [101] and Niemann-Pick C disease [118]. In the context of the present chapter, we will focus on the cross talk between cholesterol metabolism and Alzheimer's disease.

10.3.1 Alzheimer's Disease and Cholesterol Metabolism

Alzheimer's disease (AD) is the main form of dementia. This fatal neurodegenerative disease is clinically characterized by progressive memory loss and altered behavior, while histopathological hallmarks included extracellular deposition of senile plaques composed of aggregated of Aß peptides and intracellular inclusions of hyperphosphorylated tau (neurofibrillary tangles – NFT), accompanied by loss of synapses and neurons [170]. Aß peptides occur as the result of sequential cleavage of APP by ß-secretase – ß-site APP cleavage enzyme 1 (BACE-1) and by the γ-secretase complex [71, 181]. The Aß peptide that is produced in this process is released either into the lumen of intracellular organelles or into the extracellular space, interacting with cellular membranes and proteins while folding into higher-order structures with varying toxicity potentials [1, 122]. The familial form of AD can result from mutations in the genes which encode for either APP as well as presenilin 1 or presenilin 2, which are part of the catalytic domain of γ-secretase. Pathological hyperphosphorylation of tau, a protein that binds and stabilizes microtubules, leads to its detachment and formation of insoluble aggregates of paired helical filaments [8].

One third pathological hallmark of AD is the occurrence of "lipid inclusions" or "lipoid granules" in the brains of patients, suggestive of changes in lipid metabolism with either disease onset or progression. In fact, most classes of lipids have been implicated in the pathophysiology of AD, which is not surprising when one considers that they regulate the activity and trafficking of membrane bound proteins, including APP, BACE1 and presenilins. Furthermore, lipids modulate the propensity to aggregate of both Aß and tau proteins.

Accordingly, in the early 1990s, Sparks and collaborators observed that the brains of patients having suffered of advanced coronary heart disease presented senile plaques resembling those found in the brains of AD patients [175], subsequently reporting that such plaques could be induced in the brains of rabbits by feeding them a high-cholesterol diet [176]. Increased plasma/serum cholesterol was proposed to be a risk factor for development of AD [140]. This hypothesis has since been revised, linking mid-life hypercholesterolemia to an increased risk of developing AD in later life [173]. Furthermore, several studies using animal models have shown that diet-induced hypercholesterolemia can produce AD-like dementia phenotype [40, 41, 129, 151, 174].

The relationship between AD and lipid metabolism was further strengthened when the ε4-allele of the apolipoprotein E (*APOE*) was identified as the strongest genetic risk factor for spontaneous AD [16, 17, 34, 64], the form of AD for which advanced age is the major risk factor. Apolipoprotein E is responsible not only for regulating system triglyceride metabolism but, more relevantly, is crucial in the regulation of cholesterol metabolism in the brain by mediating the uptake of lipoproteins via the low-density lipoprotein (LDL) receptor related protein (LRP) and the very low-density family lipoprotein receptor.

One conundrum concerning the hypothesis that hypercholesterolemia is linked to AD pathophysiology relates to disparate observations when comparing systemic cholesterol levels between AD and healthy controls, ranging from reports of higher to lower cholesterol levels in AD patients, with others reporting no differences between AD and healthy individuals. Longitudinal studies aimed at determining the potential risk for AD in later life associated with hypercholesterolemia in mid-life, when analyzed in detail, reveal that, although individuals later (late-60s) diagnosed with AD or vascular dementia did indeed suffer of increased cholesterol levels in mid-age (early 40s), the same was true of individuals with no signs of dementia at later life (reviewed in [207]). A more cautious statement may, then, be that, midlife hypercholesterolemia is one amongst several other lifestyle factors which increased the likelihood of later-life AD.

The question then is, how does cholesterol metabolism cross paths with AD pathophysiology? The answer is, not surprisingly, complex. The most obvious answer is related to the fact that key proteins such as APP, BACE1 and components of the γ-secretase complex are transmembrane, and consequently their trafficking and proteolytic activities are regulated by lipid bilayer composition, organization and biophysical properties [73]. In fact, cholesterol directly modulates secretase activity and Aß generation. Removal of cholesterol from the cell membrane has been shown to decrease the activity of both BACE1 and γ-secretase and consequently decrease in Aß production [73, 169, 195]. Of note, cholesterol- and sphingolipid-enriched microdomains within the cell membrane are denominated "lipid rafts". Although APP, BACE1 and presenilins can be found both in raft and non-raft regions of the membrane, amyloidogenic APP processing appears to occur preferentially within raft regions, while non-amyloidogenic processing occurs in non-raft regions of the membrane [73, 195]. Indeed, palmitoylation of the transmembrane and cytosolic domains of BACE1 contributes to its localization within lipid raft domains [13, 194], while removal of cholesterol from membranes contributes to non-raft localization of the APP processing protein [50]. Presenilins are also associated with lipid raft domains [189, 195, 197].

In sum, changes in the biophysical properties of cell membranes, in particular of raft domains, by increasing cholesterol content, may shift APP processing from non-amyloidogenic to amyloidogenic, with increased Aß production.

On the other hand, increased cholesterol levels in the membranes may favor the binding of Aß to membranes [85] and accelerate recruitment and oligomerization and formation of toxic aggregates [53, 210]. This hypothesis has, however, been rebutted by others [2, 148].

Cholesterol metabolism and trafficking in the brain can also impact APP processing [26, 81, 98]. Excess cellular cholesterol is converted to cholesterol esters (CE) by acyl-CoA:cholestrol acyltransferase 1 (ACAT1) and either accumulated as intracellular lipid droplets or excreted to the extracellular milieu [29]. Several lines of evidence have shown that increasing CE levels promotes Aß production while decreased ACAT1 activity or enzyme level can reduce Aß levels and pathology and even cognitive impairment in a rodent model of AD [24, 130]. Loss of ACAT1 activity can lead to increase in the oxysterol 24(S)-hydroxycholesterol (24-OH) [24, 82]. It is, however, unclear whether this oxysterol participates directly in reduction of amyloidogenisis or whether, in the absence of ACAT1-dependent processing of cholesterol to CE, oxidation to the BBB-permeable oxysterol allows efflux of excess cholesterol from the brain.

ABCA1 stimulates cholesterol efflux from the intracellular compartment to extracellular particles such as APOE. Interestingly, ABCA1 can also control Aß levels as increasing ABCA1 reduces Aß and the inverse is also found to be true [95, 179, 198].

Noteworthy, the interactions between cholesterol metabolism/trafficking and APP processing in the brain are bidirectional and APP processing has also been shown to impact on lipid homeostasis, including cholesterol, sphingolipids and gangliosides [68, 69]. Either directly or via an intermediate, Aß can inhibit the synthesis of cholesterol, suggestive of a feedback loop in which APP acts as a molecular sensor for free cholesterol. This may, in fact, reconcile the contradictory notions that, while midlife increase in cholesterol appears to be a risk factor for later development of AD, cholesterol levels in AD patients are neither significantly different than those of healthy patients (and have even been reported to be lower) nor has statin treatment shown any beneficial effect AD pathology or cognitive function in randomized controlled trials [77, 163, 168]. A recent study in mice overexpressing APP and/or ApoB-100 (spontaneous hypercholesterolemic phenotype) supports the hypothesis that increased amyloidogenic APP processing is an early response to increase in free cholesterol and results in the lowering of the later [109]. This may corroborate the fact that population-based studies have shown a protective effect of statin treatment against the risk of AD, that is dependent of the age at statin treatment; statins may be protective at midlife by lowering free cholesterol, thus blunting increase in Aß production [36, 102, 206]. It is, however, important to keep in mind that statins have a number of pleiotropic effects, and thus the molecular mechanisms accounting for its beneficial effects may be multiple [58].

Additionally, oligomeric Aß appears to promote the release of cellular cholesterol and lipids from the cell in the form of Aß-lipid complexes

[212] while the intracellular domain of APP resulting from γ-secretase activity is reported to act as a transcriptional suppressor of LRP1, thus downregulating cholesterol uptake and causing synaptic failure and tau hyperphosphorylation [52, 97, 108].

The conversion of cholesterol to oxysterols is the main mechanism by which the brain eliminates excess cholesterol across the BBB to systemic circulation. The metabolite 24-hydroxicholesterol (24-OH) is produced by the activity of 24-hydroxylase present in neurons and can cross the BBB, unlike cholesterol itself. A less expressive pathway includes the conversion to 27-hydroxycholesterol (27-OH) by 27-hydoxylase, then to 7α-hydroxy-3-oxo-4-cholestenoic acid, which can also cross the BBB to systemic circulation for elimination in the liver [18, 19, 125]. In reality, most of 27-OH flows from systemic circulation into the brain, linking extra and intracerebral pools of cholesterol. Oxysterols resulting from non-enzymatic oxidation of cholesterol such as 7ß-, 7α- and 4ß-hydroxycholesetol, 5α-, 6α-, 5ß- and 6ß-epoxicholesterol as well as 7-ketocholesterol are also capable of crossing the BBB and have been identified in the brain of AD patients [193]. As such, some authors have suggested that oxysterols may be the putative link between hypercholesterolemia and AD pathophysiology. Of note, the oxysterols that, to date, have been shown to have the most relevance in AD have been 24-OH and 27-OH, both produced enzymatically.

On a final note, the link between lipid dysregulation and tau pathology is still poorly understood in the context of AD, although there appears to be an association between the occurrence of neurofibrillary tangles and changes of lipid trafficking. For example, cholesterol has been shown to regulate Aß-induced tau proteolysis by calpain [134], which appears to be an early step in tau pathology. Also, hyperphosphorylated tau can be found in lipid rafts alongside APP metabolites, BACE1, γ-secretase and APOE [79, 90]. Future studies into the correlation between tau pathology and cholesterol homeostasis may unveil further correlations.

10.3.2 Apolipoprotein E – Genetic Risk Factor for Alzheimer's Disease

As mentioned above, the presence of the ε4-allel of APOE (*APOE4* isoform) is the major genetic risk factor for spontaneous or late-onset AD [34], while the ε2-allel (*APOE2* isoform) has since been identified as protective against the disease [30]. The APOE receptors are a group of transmembrane proteins belonging to the LDL receptor family which is responsible for the endocytosis of ligands and recycling them to the cell membrane [80]. It is expressed throughout the body, with highest levels observed in the liver followed by the brain. Here, astrocytes are the main source of APOE, although it is also expressed in neurons and microglia [209]. In the CNS, APOE is the major component of lipoproteins particles that traffic both cholesterol and lipids between neurons and astrocytes, suggesting that the effects of APOE4 in AD pathophysiology may be mediated via brain lipid metabolism [75]. Indeed, studies in cultured neurons have shown that APOE4 is less efficient in transporting brain cholesterol as compared to isoforms APOE2 and APOE3 [150].

Several studies have shown that APOE modulates APP trafficking and Aß production and clearance by acting as a chaperone for the peptide [25]. The dosage of ε4-allel has been associated with increase in intracellular Aß, extracellular Aß oligomers as well as plaque accumulation [74, 94, 187]. The Aß clearance efficiency of the different APOE isoforms appears to be dependent on the lipoprotein lipidation status, which is optimal for the APOE2 isoform as opposed to APOE4 [3, 83]. On a different note, APOE-containing lipoproteins are responsible for transport of cholesterol and other lipids between astrocytes and neurons, and the APOE4 isoform is less efficient in the transport as compared to the APOE2 isoform [75, 150]. In fact, the cholesterol efflux efficiency of APOE is isoform dependent: APOE2 > APOE3 > APOE4 [72]. The APOE induced intracellular degradation of Aß is mediated by cholesterol efflux: lowering of intracellular cholesterol is thus coupled to the intracellular trafficking of Aß to lysosomes for degradation [99].

10.4 The Endocannabinoid System

The third important group of bioactive lipids we address here with relevance to brain pathophysiology is the *N*-acetylethanolamine family, which includes *N*-arachidonoylethanolamine (anandamine or AEA), the first described endogenous agonist for cannabinoid receptors (CBR), or endocannabinoid (eCB) [44]. A second class of eCB are fatty acid glycerol esters, namely 2-arachidonoylglycerol (2-AG) [126]. Both AEA and 2-AG, the most widely studied eCBs, are derived from arachidonic acid containing phospholipids, an important polyunsaturated fatty acid that also serves as a precursor for other regulatory lipids. Consequently, it is likely that factors capable of altering the fatty acid composition of cellular phospholipids, such as dietary fatty acid intake, can potentially effect eCB synthesis.

The eCBs act as intercellular messengers, meaning that the cellular response is a function of rate of synthesis/release, availability of the precursor arachidonic acid, binding affinity to the receptor, rate of removal from the extracellular space and, finally, rate of catabolism. Interestingly, contrary to classical neurotransmitters and neuromodulators such as glutamate or dopamine, both AEA and 2-AG are not stored in vesicles, but rather synthesized upon demand and released, although via a yet poorly understood mechanism.

On major difference between AEA and 2-AG is the route of synthesis. Anandamine can be synthesized in two pathways: (1) directly from arachidonic acid via *N*-acylation of ethanolamine and (2) transacylase phophodiesterase-mediated synthesis, where existing *N*-arachidonyl phosphatidyl ethanolamine (NAPE) can be converted to AEA by phospholipase D [117, 178]. The rate limiting step to AEA synthesis is the *N*-acyl transferase-catalyzed formation of NAPE, which is regulated by intracellular Ca^{2+} and cAMP levels [27, 138]. On the other hand, diacylglycerols with arachidonic acid on the 2-position are the precursors of the more abundant endocannabinoid, 2-AG. The diacylglycerol precursors result from phospholipase C hydrolysis of phosphatidylinositol or hydrolysis of phosphatidic acid [42, 117].

Uptake from the extracellular space may occur via protein transporters or diffusion across the cell membrane [59, 124] and intracellular degradation results from the enzymatic hydrolysis of the amide and ester bonds by fatty acid amide hydrolase or monoacylglycerol lipase and other lipases [37, 47]. The activity of 3-AG and AEA as well as that of CBRs is tightly regulated not only as a result of the cellular localization of the receptors, but also by subcellular segregation and distribution pattern of the enzymes involved in eCB metabolism (fatty acid amide hydrolase; diacylglycerol lipase α/β; monoacylglycerol lipase – MAGL) [89, 111, 123, 180, 211].

The bioactivity of AEA and 2-AG results from their binding to cannabinoid receptors type 1 and 2 (CB1R and CB2R), respectively. The receptors are differentially expressed throughout the mammalian body, with CB1R being widely expressed in the brain and to a lesser extent in peripheral tissues, while CB2R is found at higher levels in peripheral circulation, namely immune-related organs and cells. Both are G-protein coupled receptors of the class A superfamily, comprising 7 transmembrane domains. The relative abundancy of CB1R is similar to that of other main neurotransmitter receptors (glutamate, GABA) and can be found in the hippocampus, as well as cerebellum, cerebral cortex, basal ganglia, amygdala and sensory motor regions of the striatum [65, 78, 190]. These receptors are mostly found presynaptically in GABA-ergic neurons, in particular in hippocampal GABA-ergic interneurons [89, 177].

The bioactivity of eCBs is, however, not restricted to cannabinoid receptors, but extends to PPAR types α and γ, ligand-gated ion channels such as 5-HT3, the orphan receptor G-protein coupled receptor 55 (GPR55), type-1 transient receptor potential vanilloid (TRPV1) ion channels as well as the modulation of calcium and potassium channels [120, 128, 158, 162, 205].

In the brain, eCBs act as retrograde messengers. Both neurons and astrocytes express CB1R, although the effect of receptor activation is disparate. Activation of pre-synaptically localized CB1R receptors induced decreased neurotransmitter release [177] via either depolarization-induced suppression of inhibition (decreased release of GABA) or excitation (decreased release of glutamate). Because the later requires longer depolarization period, it is approximately 30-fold less predominant in the hippocampus. On the other hand, different eCBs may produce opposing effects: AEA has been shown to inhibit glutamate release while 2-AG inhibits the release of GABA, suggesting either the CB1R receptors respond differentially to the agonists, that CB1R differ between GABA-ergic and glutamatergic neurons or that CB1R form different complexes with other receptors in each type of synapse [87].

In astrocytes, activation of CB1R leads to increased intracellular [Ca^{2+}], stimulating glutamate release which activates metabotropic receptors (mGluR1) leading to potentiation of neurotransmitter release [35, 133]. Curiously, concurrent release of glutamate by astrocytes and •NO can induce long term potentiation (LTP) due to simultaneous activation of mGluR1 and protein kinase C (PKC) [35].

Changes in endocannabinoid system signaling have been linked to brain dysfunction. Decreased levels of 2-AG are reported to occur during brain trauma, convulsions and stroke [139, 203] while decrease in CB1R expression levels are observed in Huntington's disease [21] as well as Alzheimer's disease [86, 100]. Conversely, increased CB1R levels in the hippocampus can have a protective effect in epileptic seizures [70, 110].

A relevant point to emphasize in the context of this chapter is that the levels of eCBs can be modulated by diet, with the first evidence provided by Kirkham and colleagues showing that fasting, feeding and satiation produced changes in the levels in different regions of the brain [93]. These changes seem to be correlated with the role of eCB in promoting an appetite for palatable food, linked to overconsumption of total energy [11]. On the other hand, the fact that both AEA and 2-AG derive from arachidonic acid suggest that changes in tissue concentration of this fatty acid should impact eCB concentration and function. Accordingly, data from animal studies indicate that eCB levels in the tissue can be modulated by dietary fatty acids [4, 15, 202].

10.5 Concluding Remarks

Much has been unraveled over the past decades regarding the biological activity of lipids in the CNS. As the major constituents of the brain, and granted their unique physical-chemical properties, beyond their classic role as isolators, cell membrane components or energy reservoirs, bioactive lipids have been revealed to be regulators and modulators of neurochemical mechanisms underlying brain functions.

The regulation of signaling cascades by lipids modified by other dietary nutrients, such as nitrated fatty acids, the modulatory role played by fatty acid derived endocannabinoids or the complex crosstalk between cholesterol metabolism/trafficking/homeostasis and the pathophysiological pathways implicated in Alzheimer's disease are but a few of the myriad of functional aspects which have become evident for bioactive lipids in the brain. Relevant to the bioactivity of lipids is the fact that the composition, concentration and chemical variability of the brain lipidome can be modulated by dietary intervention, highlighting a new mechanism supporting the importance of the gut-brain axis in brain functions and as a key determinant of human health.

Acknowledgements This work was financed by the European Regional Development Fund (FEDER) funds through the Operational Program for Competitiveness and Internationalization – COMPETE and national funds by FCT – Foundation for Science and Technology under the projects POCI-01-0145-FEDER-007440, and POCI-01-0145-FEDER-029099 and through the Centro 2020 Regional Operational Program, under the project CENTRO-01-0145-FEDER-000012-HealthyAging2020.

References

1. Ariga T, McDonald MP, Yu RK (2008) Thematic review series: sphingolipids. Role of ganglioside metabolism in the pathogenesis of Alzheimer's disease—a review. J Lipid Res 49:1157–1175
2. Arispe N, Doh M (2002) Plasma membrane cholesterol controls the cytotoxicity of Alzheimer's disease Aβ(1–40) and (1–42) peptides. FASEB J
3. Arold S, Sullivan P, Bilousova T et al (2012) Apolipoprotein e level and cholesterol are associated with reduced synaptic amyloid beta in

Alzheimer's disease and apoE TR mouse cortex. Acta Neuropathol 123:39–52
4. Artmann A, Petersen G, Hellgren LI et al (2008) Influence of dietary fatty acids on endocannabinoid and N-acylethanolamine levels in rat brain, liver and small intestine. Biochim Biophys Acta Mol Cell Biol Lipids 1781:200–212
5. Baker LM, Baker PR, Golin-Bisello F et al (2007) Nitro-fatty acid reaction with glutathione and cysteine. Kinetic analysis of thiol alkylation by a Michael addition reaction. J Biol Chem 282:31085–31093
6. Baker PR, Lin Y, Schopfer FJ et al (2005) Fatty acid transduction of nitric oxide signaling: multiple nitrated unsaturated fatty acid derivatives exist in human blood and urine and serve as endogenous peroxisome proliferator-activated receptor ligands. J Biol Chem 280:42464–42475
7. Baker PR, Schopfer FJ, Sweeney S, Freeman BA (2004) Red cell membrane and plasma linoleic acid nitration products: synthesis, clinical identification, and quantitation. Proc Natl Acad Sci U S A 101:11577–11582
8. Ballatore C, Lee VM-Y, Trojanowski JQ (2007) Tau-mediated neurodegeneration in Alzheimer's disease and related disorders. Nat Rev Neurosci 8:663–672
9. Bartesaghi S, Radi R (2018) Fundamentals on the biochemistry of peroxynitrite and protein tyrosine nitration. Redox Biol 14:618–625
10. Batthyany C, Schopfer FJ, Baker PR et al (2006) Reversible post-translational modification of proteins by nitrated fatty acids in vivo. J Biol Chem 281:20450–20463
11. Bello NT, Coughlin JW, Redgrave GW et al (2012) Dietary conditions and highly palatable food access alter rat cannabinoid receptor expression and binding density. Physiol Behav 105:720–726
12. Benjamin N, O'Driscoll F, Dougall H et al (1994) Stomach NO synthesis. Nature 368:502
13. Benjannet S, Elagoz A, Wickham L et al (2001) Post-translational processing of β-secretase (β-amyloid-converting enzyme) and its ectodomain shedding: the pro- and transmembrane/cytosolic domains affect its cellular activity and amyloid-β production. J Biol Chem 276:10879–10887
14. Berg JM, Tymoczko JL, Stryer L (2006) The complex regulation of cholesterol biosynthesis takes place at several levels. In: Biochemistry. Macmillan Learning WH Freeman, New York
15. Berger A, Crozier G, Bisogno T et al (2001) Anandamide and diet: inclusion of dietary arachidonate and docosahexaenoate leads to increased brain levels of the corresponding N-acylethanolamines in piglets. Proc Natl Acad Sci U S A 98:6402–6406
16. Bertram L, Tanzi RE (2008) Thirty years of Alzheimer's disease genetics: the implications of systematic meta-analyses. Nat Rev Neurosci 9:768–778
17. Bettens K, Sleegers K, Van Broeckhoven C (2013) Genetic insights in Alzheimer's disease. Lancet Neurol 12:92–104
18. Björkhem I (2006) Crossing the barrier: oxysterols as cholesterol transporters and metabolic modulators in the brain. J Intern Med 260:493–508
19. Björkhem I, Cedazo-Minguez A, Leoni V, Meaney S (2009) Oxysterols and neurodegenerative diseases. Mol Asp Med 30:171–179
20. Blanco F, Ferreira AM, Lopez GV et al (2011) 6-Methylnitroarachidonate: a novel esterified nitroalkene that potently inhibits platelet aggregation and exerts cGMP-mediated vascular relaxation. Free Radic Biol Med 50:411–418
21. Blázquez C, Chiarlone A, Sagredo O et al (2011) Loss of striatal type 1 cannabinoid receptors is a key pathogenic factor in Huntington's disease. Brain 134:119–136
22. Bonacci G, Baker PR, Salvatore SR et al (2012) Conjugated linoleic acid is a preferential substrate for fatty acid nitration. J Biol Chem 287:44071–44082
23. Borniquel S, Jansson EA, Cole MP et al (2010) Nitrated oleic acid up-regulates PPARgamma and attenuates experimental inflammatory bowel disease. Free Radic Biol Med 48:499–505
24. Bryleva EY, Rogers MA, Chang CCY et al (2010) ACAT1 gene ablation increases 24(S)-hydroxycholesterol content in the brain and ameliorates amyloid pathology in mice with AD. Proc Natl Acad Sci 107:3081–3086
25. Bu G (2009) Apolipoprotein E and its receptors in Alzheimer's disease: pathways, pathogenesis and therapy. Nat Rev Neurosci 10:333–344
26. Burns M, Gaynor K, Olm V et al (2003) Presenilin redistribution associated with aberrant cholesterol transport enhances beta-amyloid production in vivo. J Neurosci. 23:5645–5649
27. Cadas H, di Tomaso E, Piomelli D (1997) Occurrence and biosynthesis of endogenous cannabinoid precursor, N-arachidonoyl phosphatidylethanolamine, in rat brain. J Neurosci 17:1226–1242
28. Carlstrom M, Lundberg JO, Weitzberg E (2018) Mechanisms underlying blood pressure reduction by dietary inorganic nitrate. Acta Physiol 224:e13080
29. Chang T-Y, Chang CCY, Ohgami N, Yamauchi Y (2006) Cholesterol sensing, trafficking, and esterification. Annu Rev Cell Dev Biol 22:129–157
30. Chartier-Hariln M-C, Parfitt M, Legrain S et al (1994) Apolipoprotein E, ε4 allele as a major risk factor for sporadic early and late-onset forms of Alzheimer's disease: analysis of the 19q13.2 chromosomal region. Hum Mol Genet 3:569–574
31. Cole MP, Rudolph TK, Khoo NK et al (2009) Nitro-fatty acid inhibition of neointima formation after endoluminal vessel injury. Circ Res 105:965–972
32. Coles B, Bloodsworth A, Clark SR et al (2002a) Nitrolinoleate inhibits superoxide generation, degranulation, and integrin expression by human neutrophils: novel antiinflammatory properties of nitric oxide-derived reactive species in vascular cells. Circ Res 91:375–381
33. Coles B, Bloodsworth A, Eiserich JP et al (2002b) Nitrolinoleate inhibits platelet activation by

attenuating calcium mobilization and inducing phosphorylation of vasodilator-stimulated phosphoprotein through elevation of cAMP. J Biol Chem 277:5832–5840
34. Corder EH, Saunders AM, Strittmatter WJ et al (1993) Gene dose of apolipoprotein E type 4 allele and the risk of Alzheimer's disease in late onset families. Science 261:921–923
35. Covelo A, Araque A (2016) Lateral regulation of synaptic transmission by astrocytes. Neuroscience 323:62–66
36. Cramer C, Haan MN, Galea S et al (2008) Use of statins and incidence of dementia and cognitive impairment without dementia in a cohort study. Neurology
37. Cravatt BF, Giang DK, Mayfield SP et al (1996) Molecular characterization of an enzyme that degrades neuromodulatory fatty-acid amides. Nature 384:83–87
38. Cui T, Schopfer FJ, Zhang J et al (2006) Nitrated fatty acids: endogenous anti-inflammatory signaling mediators. J Biol Chem 281:35686–35698
39. de Chaves EI, Rusiñol AE, Vance DE et al (1997) Role of lipoproteins in the delivery of lipids to axons during axonal regeneration. J Biol Chem 272:30766–30773
40. de Oliveira J, Hort MA, Moreira ELG et al (2011) Positive correlation between elevated plasma cholesterol levels and cognitive impairments in LDL receptor knockout mice: relevance of cortico-cerebral mitochondrial dysfunction and oxidative stress. Neuroscience 197:99–106
41. De Oliveira J, Moreira ELG, Dos Santos DB et al (2014) Increased susceptibility to amyloid-β-induced neurotoxicity in mice lacking the low-density lipoprotein receptor. J Alzheimers Dis 41:43–60
42. De Petrocellis L, Cascio MG, Di Marzo V (2004) The endocannabinoid system: a general view and latest additions. Br J Pharmacol 141:765–774
43. Delmastro-Greenwood M, Hughan KS, Vitturi DA et al (2015) Nitrite and nitrate-dependent generation of anti-inflammatory fatty acid nitroalkenes. Free Radic Biol Med 89:333–341
44. Devane WA, Hanuš L, Breuer A et al (1992) Isolation and structure of a brain constituent that binds to the cannabinoid receptor. Science 258:1946–1949
45. Di Paolo G, Kim T-W (2011) Linking lipids to Alzheimer's disease: cholesterol and beyond. Nat Rev Neurosci 12:284–296
46. Dietschy JM, Turley SD (2004) *Thematic review series: brain lipids.* Cholesterol metabolism in the central nervous system during early development and in the mature animal. J Lipid Res 46:1350
47. Dinh TP, Freund TF, Piomelli D (2002) A role for monoglyceride lipase in 2-arachidonoylglycerol inactivation. Chemistry and physics of lipids 121:149–158
48. Doria M, Maugest L, Moreau T et al (2016) Contribution of cholesterol and oxysterols to the pathophysiology of Parkinson's disease. Free Radic Biol Med 101:393–400
49. Duncan C, Li H, Dykhuizen R et al (1997) Protection against oral and gastrointestinal diseases: importance of dietary nitrate intake, oral nitrate reduction and enterosalivary nitrate circulation. Comp Biochem Physiol A Physiol 118:939–948
50. Ehehalt R, Keller P, Haass C et al (2003) Amyloidogenic processing of the Alzheimer β-amyloid precursor protein depends on lipid rafts. J Cell Biol. 160:113:123
51. Eiserich JP, Patel RP, O'Donnell VB (1998) Pathophysiology of nitric oxide and related species: free radical reactions and modification of biomolecules. Mol Asp Med 19:221–357
52. Fan QW, Yu W, Senda T et al (2001) Cholesterol-dependent modulation of tau phosphorylation in cultured neurons. J Neurochem. 80:178–190
53. Fantini J, Yahi N (2010) Molecular insights into amyloid regulation by membrane cholesterol and sphingolipids: common mechanisms in neurodegenerative diseases. Expert Rev Mol Med 12:e27
54. Fazzari M, Trostchansky A, Schopfer FJ et al (2014) Olives and olive oil are sources of electrophilic fatty acid nitroalkenes. PLoS One 9:e84884
55. Ferreira AM, Ferrari MI, Trostchansky A et al (2009) Macrophage activation induces formation of the anti-inflammatory lipid cholesteryl-nitrolinoleate. Biochem J 417:223–234
56. Ferreira AM, Minarrieta L, Lamas Bervejillo M, Rubbo H (2012) Nitro-fatty acids as novel electrophilic ligands for peroxisome proliferator-activated receptors. Free Radic Biol Med 53:1654–1663
57. Fester L, Zhou L, Bütow A et al (2009) Cholesterol-promoted synaptogenesis requires the conversion of cholesterol to estradiol in the hippocampus. Hippocampus 19:692–705
58. Fonseca ACRG, Resende R, Oliveira CR, Pereira CMF (2010) Cholesterol and statins in Alzheimer's disease: current controversies. Exp Neurol 223:282–293
59. Fowler CJ, Jacobsson SOP (2002) Cellular transport of anandamide, 2-arachidonoylglycerol and palmitoylethanolamide – targets for drug development? Prostaglandins Leukot Essent Fat Acids 66:193
60. Freeman BA, O'Donnell VB, Schopfer FJ (2018) The discovery of nitro-fatty acids as products of metabolic and inflammatory reactions and mediators of adaptive cell signaling. Nitric Oxide 77:106–111
61. Gago B, Lundberg JO, Barbosa RM, Laranjinha J (2007) Red wine-dependent reduction of nitrite to nitric oxide in the stomach. Free Radic Biol Med 43:1233–1242
62. Gallon AA, Pryor WA (1994) The reaction of low levels of nitrogen dioxide with methyl linoleate in the presence and absence of oxygen. Lipids 29:171–176
63. Geisler AC, Rudolph TK (2012) Nitroalkylation–a redox sensitive signaling pathway. Biochim Biophys Acta 1820:777–784

64. Genin E, Hannequin D, Wallon D et al (2011) APOE and Alzheimer disease: a major gene with semi-dominant inheritance. Mol Psychiatry 16:903–907
65. Glass M, Faull RL, Dragunow M (1997) Cannabinoid receptors in the human brain: a detailed anatomical and quantitative autoradiographic study in the fetal, neonatal and adult human brain. Neuroscience 77:299–318
66. Gorczynski MJ, Huang J, Lee H, King SB (2007) Evaluation of nitroalkenes as nitric oxide donors. Bioorg Med Chem Lett 17:2013–2017
67. Goritz C, Mauch DH, Pfrieger FW (2005) Multiple mechanisms mediate cholesterol-induced synaptogenesis in a CNS neuron. Mol Cell Neurosci. 29:190–201
68. Grimm MOW, Grimm HS, Hartmann T (2007) Amyloid beta as a regulator of lipid homeostasis. Trends Mol Med 13:337–344
69. Grimm MOW, Rothhaar TL, Hartmann T (2012) The role of APP proteolytic processing in lipid metabolism. Exp Brain Res 217:365–375
70. Guggenhuber S, Monory K, Lutz B, Klugmann M (2010) AAV vector-mediated overexpression of CB1 cannabinoid receptor in pyramidal neurons of the hippocampus protects against seizure-induced excitoxicity. PLoS One 5:e15707
71. Haass C, Selkoe DJ (2007) Soluble protein oligomers in neurodegeneration: lessons from the Alzheimer's amyloid β-peptide. Nat Rev Mol Cell Biol 8:101–112
72. Hara M, Matsushima T, Satoh H et al (2003) Isoform-dependent cholesterol efflux from macrophages by apolipoprotein E is modulated by cell surface proteoglycans. Arterioscler Thromb Vasc Biol 23:269–274
73. Hartmann T, Kuchenbecker J, Grimm MOW (2007) Alzheimer's disease: the lipid connection. J Neurochem 103:159
74. Hashimoto T, Serrano-Pozo A, Hori Y et al (2012) Apolipoprotein E, especially apolipoprotein E4, increases the oligomerization of amyloid peptide. J Neurosci 32:15181–15192
75. Hauser PS, Narayanaswami V, Ryan RO (2011) Apolipoprotein E: from lipid transport to neurobiology. Prog Lipid Res 50:62–74
76. He G, Shankar RA, Chzhan M et al (1999) Noninvasive measurement of anatomic structure and intraluminal oxygenation in the gastrointestinal tract of living mice with spatial and spectral EPR imaging. Proc Natl Acad Sci U S A 96:4586–4591
77. Heart Protection Study Collaborative Group (2002) MRC/BHF heart protection study of cholesterol lowering with simvastatin in 20 536 high-risk individuals: a randomised placebocontrolled trial. Lancet 360:7–22
78. Herkenham M, Lynn AB, Johnson MR et al (1991) Characterization and localization of cannabinoid receptors in rat brain: a quantitative in vitro autoradiographic study. J Neurosci 11:563–583
79. Hernandez P, Lee G, Sjoberg M, MacCioni RB (2009) Tau phosphorylation by cdk5 and Fyn in response to amyloid peptide Aβ25-35: involvement of lipid rafts. J Alzheimers Dis 16:149–156
80. Holtzman DM, Herz J, Bu G (2012) Apolipoprotein E and apolipoprotein E receptors: normal biology and roles in Alzheimer disease. Cold Spring Harb Perspect Med
81. Hughes TM, Lopez OL, Evans RW et al (2014) Markers of cholesterol transport are associated with amyloid deposition in the brain. Neurobiol Aging 35:802–807
82. Hutter-Paier B, Huttunen HJ, Puglielli L et al (2004) The ACAT inhibitor CP-113,818 markedly reduces amyloid pathology in a mouse model of Alzheimer's disease. Neuron 44:227–238
83. Jiang Q, Lee CYD, Mandrekar S et al (2008) ApoE promotes the proteolytic degradation of Aβ. Neuron 58:681–693
84. Jobbagy S, Tan RJ (2018) Nitrolipids in kidney physiology and disease. Nitric Oxide 78:121–126
85. Kakio A, Nishimoto SI, Yanagisawa K et al (2001) Cholesterol-dependent formation of GM1 ganglioside-bound amyloid beta-protein, an endogenous seed for Alzheimer amyloid. J Biol Chem 276:24985–24990
86. Kalifa S, Polston EK, Allard JS, Manaye KF (2011) Distribution patterns of cannabinoid CB1 receptors in the hippocampus of APPswe/PS1ΔE9double transgenic mice. Brain Res 1376:94–100
87. Kano M, Ohno-Shosaku T, Hashimotodani Y et al (2009) Endocannabinoid-mediated control of synaptic transmission. Physiol Rev 89:309–380
88. Kansanen E, Jyrkkanen HK, Levonen AL (2012) Activation of stress signaling pathways by electrophilic oxidized and nitrated lipids. Free Radic Biol Med 52:973–982
89. Katona I, Sperlágh B, Sík A et al (1999) Presynaptically located CB1 cannabinoid receptors regulate GABA release from axon terminals of specific hippocampal interneurons. J Neurosci 19:4544–4558
90. Kawarabayashi T (2004) Dimeric amyloid protein rapidly accumulates in lipid rafts followed by apolipoprotein E and phosphorylated tau accumulation in the Tg2576 mouse model of Alzheimer's disease. J Neurosci 24:3801–3809
91. Kelley EE, Baust J, Bonacci G et al (2014) Fatty acid nitroalkenes ameliorate glucose intolerance and pulmonary hypertension in high-fat diet-induced obesity. Cardiovasc Res 101:352–363
92. Khoo NK, Rudolph V, Cole MP et al (2010) Activation of vascular endothelial nitric oxide synthase and heme oxygenase-1 expression by electrophilic nitro-fatty acids. Free Radic Biol Med 48:230–239
93. Kirkham TC, Williams CM, Fezza F, Di Marzo V (2002) Endocannabinoid levels in rat limbic forebrain and hypothalamus in relation to fasting, feeding and

satiation: stimulation of eating by 2-arachidonoyl glycerol. Br J Pharmacol. 136:550–557
94. Koffie RM, Hashimoto T, Tai HC et al (2012) Apolipoprotein E4 effects in Alzheimer's disease are mediated by synaptotoxic oligomeric amyloid-β. Brain 135:2155–2168
95. Koldamova R, Staufenbiel M, Lefterov I (2005) Lack of ABCA1 considerably decreases brain ApoE level and increases amyloid deposition in APP23 mice. J Biol Chem. 280:43224–43235
96. Korade Z, Kenworthy AK (2008) Lipid rafts, cholesterol, and the brain. Neuropharmacology 55:1265–1273
97. Koudinov AR, Koudinova NV (2005) Cholesterol homeostasis failure as a unifying cause of synaptic degeneration. J Neurol Sci:233–240
98. Lane-Donovan C, Philips GT, Herz J (2014) More than cholesterol transporters: lipoprotein receptors in CNS function and neurodegeneration. Neuron 83:771–787
99. Lee CYD, Tse W, Smith JD, Landreth GE (2012) Apolipoprotein E promotes ß-amyloid trafficking and degradation by modulating microglial cholesterol levels. J Biol Chem 287:2032–2044
100. Lee JH, Agacinski G, Williams JH et al (2010) Intact cannabinoid CB1 receptors in the Alzheimer's disease cortex. Neurochem Int 57:985–989
101. Leoni V, Caccia C (2015) The impairment of cholesterol metabolism in Huntington disease. Biochim Biophys Acta Mol Cell Biol Lipids 1851:1095–1105
102. Li G, Larson EB, Sonnen JA et al (2007) Statin therapy is associated with reduced neuropathologic changes of Alzheimer disease. Neurology 69:878–885
103. Lim DG, Sweeney S, Bloodsworth A et al (2002) Nitrolinoleate, a nitric oxide-derived mediator of cell function: synthesis, characterization, and vasomotor activity. Proc Natl Acad Sci U S A 99:15941–15946
104. Lima ES, Bonini MG, Augusto O et al (2005) Nitrated lipids decompose to nitric oxide and lipid radicals and cause vasorelaxation. Free Radic Biol Med 39:532–539
105. Lima ES, Di Mascio P, Abdalla DS (2003) Cholesteryl nitrolinoleate, a nitrated lipid present in human blood plasma and lipoproteins. J Lipid Res 44:1660–1666
106. Linetti A, Fratangeli A, Taverna E et al (2010) Cholesterol reduction impairs exocytosis of synaptic vesicles. J Cell Sci 123:595–605
107. Liu H, Jia Z, Soodvilai S et al (2008) Nitro-oleic acid protects the mouse kidney from ischemia and reperfusion injury. Am J Physiol Ren Physiol 295:F942–F949
108. Liu Q, Zerbinatti CV, Zhang J et al (2007) Amyloid precursor protein regulates brain apolipoprotein E and cholesterol metabolism through lipoprotein receptor LRP1. Neuron 56:66–78
109. Löffler T, Schweinzer C, Flunkert S et al (2016) Brain cortical cholesterol metabolism is highly affected by human APP overexpression in mice. Mol Cell Neurosci 74:34–41
110. Ludanyi A, Eross L, Czirjak S et al (2008) Downregulation of the CB1 cannabinoid receptor and related molecular elements of the endocannabinoid system in epileptic human hippocampus. J Neurosci 28:2976–2990
111. Ludányi A, Hu SS-J, Yamazaki M et al (2011) Complementary synaptic distribution of enzymes responsible for synthesis and inactivation of the endocannabinoid 2-arachidonoylglycerol in the human hippocampus. Neuroscience 174:50–63
112. Lundberg JO, Govoni M (2004) Inorganic nitrate is a possible source for systemic generation of nitric oxide. Free Radic Biol Med 37:395–400
113. Lundberg JO, Weitzberg E (2013) Biology of nitrogen oxides in the gastrointestinal tract. Gut 62:616–629
114. Lundberg JO, Weitzberg E, Cole JA, Benjamin N (2004) Nitrate, bacteria and human health. Nat Rev Microbiol 2:593–602
115. Lundberg JO, Weitzberg E, Gladwin MT (2008) The nitrate-nitrite-nitric oxide pathway in physiology and therapeutics. Nat Rev Drug Discov 7:156–167
116. Lundberg JO, Weitzberg E, Lundberg JM, Alving K (1994) Intragastric nitric oxide production in humans: measurements in expelled air. Gut 35:1543–1546
117. Mackie K (2006) Cannabinoid receptors as therapeutic targets. Annu Rev Pharmacol Toxicol 46:101–122
118. Madra M, Sturley SL (2010) Niemann-Pick type C pathogenesis and treatment: from statins to sugars. Clin Lipidol 5:387–395
119. Martins IJ, Berger T, Sharman MJ et al (2009) Cholesterol metabolism and transport in the pathogenesis of Alzheimer's disease. J Neurochem 111:1275–1308
120. Marzo V, Petrocellis L (2010) Endocannabinoids as regulators of transient receptor potential (TRP) channels: a further opportunity to develop new endocannabinoid-based therapeutic drugs. Curr Med Chem 17:1430–1449
121. Mata-Perez C, Sanchez-Calvo B, Begara-Morales JC et al (2016) Nitro-linolenic acid is a nitric oxide donor. Nitric Oxide 57:57–63
122. Matsuzaki K, Kato K, Yanagisawa K (2010) Aβ polymerization through interaction with membrane gangliosides. Biochim Biophys Acta Mol Cell Biol Lipids 1801:868–877
123. Mátyás F, Urbán GM, Watanabe M et al (2008) Identification of the sites of 2-arachidonoylglycerol synthesis and action imply retrograde endocannabinoid signaling at both GABAergic and glutamatergic synapses in the ventral tegmental area. Neuropharmacology 54:95–107
124. McFarland MJ, Barker EL (2004) Anandamide transport. Pharmacol Ther 104:117–135

125. Meaney S, Heverin M, Panzenboeck U et al (2007) Novel route for elimination of brain oxysterols across the blood-brain barrier: conversion into 7α-hydroxy-3-oxo-4-cholestenoic acid. J Lipid Res 48:944–951
126. Mechoulam R, Ben-Shabat S, Hanus L et al (1995) Identification of an endogenous 2-monoglyceride, present in canine gut, that binds to cannabinoid receptors. Biochem Pharmacol 50:83–90
127. Moller M, Botti H, Batthyany C et al (2005) Direct measurement of nitric oxide and oxygen partitioning into liposomes and low density lipoprotein. J Biol Chem 280:8850–8854
128. Monias AT, Walke LM, Morrison RS, Meier DE (2005) The effect of age on medical decisions made by patients with chronic illness. J Palliat Med 2:311–317
129. Moreira ELG, De Oliveira J, Nunes JC et al (2012) Age-Related cognitive decline in hypercholesterolemic LDL receptor knockout mice (LDLr−/−): evidence of antioxidant imbalance and increased acetylcholinesterase activity in the prefrontal cortex. J Alzheimers Dis 32:495–511
130. Murphy SR, Chang CC, Dogbevia G et al (2013) Acat1 knockdown gene therapy decreases amyloid-β in a mouse model of Alzheimer's disease. Mol Ther 21:1497–1506
131. Napolitano A, Camera E, Picardo M, d'Ischia M (2000) Acid-promoted reactions of ethyl linoleate with nitrite ions: formation and structural characterization of isomeric nitroalkene, nitrohydroxy, and novel 3-nitro-1,5-hexadiene and 1,5-dinitro-1, 3-pentadiene products. J Org Chem 65:4853–4860
132. Nardone G, Laccetti P, Civiletti C, Budillon G (1993) Phospholipid composition of human gastric mucosa: a study of endoscopic biopsy specimens. Gut 34:456–460
133. Navarrete M, Araque A (2010) Endocannabinoids potentiate synaptic transmission through stimulation of astrocytes. Neuron 68:113–126
134. Nicholson AM, Ferreira A (2009) Increased membrane cholesterol might render mature hippocampal neurons more susceptible to -amyloid-induced calpain activation and tau toxicity. J Neurosci 29:4640–4651
135. Nieweg K, Schaller H, Pfrieger FW (2009) Marked differences in cholesterol synthesis between neurons and glial cells from postnatal rats. J Neurochem 109:125–134
136. O'Donnell VB, Eiserich JP, Chumley PH et al (1999) Nitration of unsaturated fatty acids by nitric oxide-derived reactive nitrogen species peroxynitrite, nitrous acid, nitrogen dioxide, and nitronium ion. Chem Res Toxicol 12:83–92
137. O'Donnell VB, Freeman BA (2001) Interactions between nitric oxide and lipid oxidation pathways: implications for vascular disease. Circ Res 88:12–21
138. Okamoto Y, Morishita J, Tsuboi K et al (2004) Molecular characterization of a phospholipase D generating anandamide and its congeners. J Biol Chem 279:5298–5305
139. Panikashvili D, Simeonidou C, Ben-Shabat S et al (2001) An endogenous cannabinoid (2-AG) is neuroprotective after brain injury. Nature 413:527–531
140. Pappolla MA, Bryant-Thomas TK, Herbert D et al (2003) Mild hypercholesterolemia is an early risk factor for the development of Alzheimer amyloid pathology. Neurology 61:199–205
141. Peri L, Pietraforte D, Scorza G et al (2005) Apples increase nitric oxide production by human saliva at the acidic pH of the stomach: a new biological function for polyphenols with a catechol group? Free Radic Biol Med 39:668–681
142. Petersson J, Phillipson M, Jansson EA et al (2007) Dietary nitrate increases gastric mucosal blood flow and mucosal defense. Am J Physiol Gastrointest Liver Physiol 292:G718–G724
143. Pfrieger FW (2003a) Cholesterol homeostasis and function in neurons of the central nervous system. Cell Mol Life Sci 60:1158–1171
144. Pfrieger FW (2003b) Outsourcing in the brain: do neurons depend on cholesterol delivery by astrocytes? BioEssays 25:72–78
145. Phillipson M, Henriksnäs J, Holm LAC (2002) The importance of mucus layers and bicarbonate transport in preservation of gastric juxtamucosal pH. Am J Physiol Gastrointest Liver Physiol 282:211–219
146. Piknova B, Kocharyan A, Schechter AN, Silva AC (2011) The role of nitrite in neurovascular coupling. Brain Res 1407:62–68
147. Presley TD, Morgan AR, Bechtold E et al (2011) Acute effect of a high nitrate diet on brain perfusion in older adults. Nitric Oxide 24:34–42
148. Qiu L, Buie C, Reay A et al (2011) Molecular dynamics simulations reveal the protective role of cholesterol in β-amyloid protein-induced membrane disruptions in neuronal membrane mimics. J Phys Chem B 115:9795–9812
149. Radi R (2004) Nitric oxide, oxidants, and protein tyrosine nitration. Proc Natl Acad Sci U S A 101:4003–4008
150. Rapp A, Gmeiner B, Hüttinger M (2006) Implication of apoE isoforms in cholesterol metabolism by primary rat hippocampal neurons and astrocytes. Biochimie 88:473–483
151. Refolo LM, M a P, Malester B et al (2000) Hypercholesterolemia accelerates the Alzheimer's amyloid pathology in a transgenic mouse model. Neurobiol Dis 7:321–331
152. Rocha BS, Correia MG, Fernandes RC et al (2016a) Dietary nitrite induces occludin nitration in the stomach. Free Radic Res 50:1257–1264
153. Rocha BS, Gago B, Barbosa RM et al (2012) Intragastric nitration by dietary nitrite: implications for modulation of protein and lipid signaling. Free Radic Biol Med 52:693–698
154. Rocha BS, Gago B, Barbosa RM et al (2013) Pepsin is nitrated in the rat stomach, acquiring antiulcerogenic activity: a novel interaction between dietary

nitrate and gut proteins. Free Radic Biol Med 58:26–34
155. Rocha BS, Gago B, Barbosa RM, Laranjinha J (2009) Dietary polyphenols generate nitric oxide from nitrite in the stomach and induce smooth muscle relaxation. Toxicology 265:41–48
156. Rocha BS, Gago B, Pereira C et al (2011) Dietary nitrite in nitric oxide biology: a redox interplay with implications for pathophysiology and therapeutics. Curr Drug Targets 12:1351–1363
157. Rocha BS, Lundberg JO, Radi R, Laranjinha J (2016b) Role of nitrite, urate and pepsin in the gastroprotective effects of saliva. Redox Biol 8:407–414
158. Ross RA (2003) Anandamide and vanilloid TRPV1 receptors. Br J Pharmacol 140:790–801
159. Rubbo H (1998) Nitric oxide and peroxynitrite in lipid peroxidation. Med (B Aires) 58:361–366
160. Rubbo H, Radi R, Trujillo M et al (1994) Nitric oxide regulation of superoxide and peroxynitrite-dependent lipid peroxidation. Formation of novel nitrogen-containing oxidized lipid derivatives. J Biol Chem 269:26066–26075
161. Rudolph TK, Rudolph V, Edreira MM et al (2010) Nitro-fatty acids reduce atherosclerosis in apolipoprotein E-deficient mice. Arter Thromb Vasc Biol 30:938–945
162. Ryberg E, Larsson N, Sjögren S et al (2009) The orphan receptor GPR55 is a novel cannabinoid receptor. Br J Pharmacol 152:1092–1101
163. Sano M, Bell KL, Galasko D et al (2011) A randomized, double-blind, placebo-controlled trial of simvastatin to treat Alzheimer disease. Neurology 77:556–563
164. Schopfer FJ, Batthyany C, Baker PR et al (2009) Detection and quantification of protein adduction by electrophilic fatty acids: mitochondrial generation of fatty acid nitroalkene derivatives. Free Radic Biol Med 46:1250–1259
165. Schopfer FJ, Cole MP, Groeger AL et al (2010) Covalent peroxisome proliferator-activated receptor gamma adduction by nitro-fatty acids: selective ligand activity and anti-diabetic signaling actions. J Biol Chem 285:12321–12333
166. Schopfer FJ, Lin Y, Baker PR et al (2005) Nitrolinoleic acid: an endogenous peroxisome proliferator-activated receptor gamma ligand. Proc Natl Acad Sci U S A 102:2340–2345
167. Schopfer FJ, Vitturi DA, Jorkasky DK, Freeman BA (2018) Nitro-fatty acids: new drug candidates for chronic inflammatory and fibrotic diseases. Nitric Oxide 79:31–37
168. Shepherd J, Blauw GJ, Murphy MB et al (2002) Pravastatin in elderly individuals at risk of vascular disease (PROSPER): a randomised controlled trial. Lancet 360:1623–1630
169. Simons M, Keller P, De Strooper B et al (1998) Cholesterol depletion inhibits the generation of -amyloid in hippocampal neurons. Proc Natl Acad Sci 95:6460–6464
170. Small SA, Duff K (2008) Linking Abeta and tau in late-onset Alzheimer's disease: a dual pathway hypothesis. Neuron 60:534–542
171. Smith ME, Morton DG (2010) The digestive system. Basic science and clinical conditions, 2nd edn. Elsevier, New York
172. Sodhi RK, Singh N (2013) Liver X receptors: emerging therapeutic targets for Alzheimer's disease. Pharmacol Res 72:45–51
173. Solomon A, Kivipelto M, Wolozin B et al (2009) Midlife serum cholesterol and increased risk of Alzheimer's and vascular dementia three decades later. Dement Geriatr Cogn Disord 28:75–80
174. Sparks DL (2008) The early and ongoing experience with the cholesterol-fed rabbit as a model of Alzheimer's disease: the old, the new and the pilot. J Alzheimers Dis 15:641–656
175. Sparks DL, Hunsaker JC, Scheff SW et al (1990) Cortical senile plaques in coronary artery disease, aging and Alzheimer's disease. Neurobiol Aging 11:601–607
176. Sparks DL, Scheff SW, Hunsaker JC et al (1994) Induction of Alzheimer-like β-amyloid immunoreactivity in the brains of rabbits with dietary cholesterol. Exp Neurol 126:88–94
177. Stempel AV, Stumpf A, Zhang H-Y et al (2016) Cannabinoid type 2 receptors mediate a cell type-specific plasticity in the hippocampus. Neuron 90:795–809
178. Sugiura T, Kobayashi Y, Oka S, Waku K (2002) Biosynthesis and degradation of anandamide and 2-arachidonoylglycerol and their possible physiological significance. Prostaglandins Leukot Essent Fatty Acids 66:173–192
179. Sun Y, Yao J, Kim T-W, Tall AR (2003) Expression of liver X receptor target genes decreases cellular amyloid β peptide secretion. J Biol Chem 278:27688–27694
180. Tanimura A, Yamazaki M, Hashimotodani Y et al (2010) The endocannabinoid 2-arachidonoylglycerol produced by diacylglycerol lipase α mediates retrograde suppression of synaptic transmission. Neuron 65:320–327
181. Tanzi RE, Bertram L (2005) Twenty years of the Alzheimer's disease amyloid hypothesis: a genetic perspective. Cell 120:545–555
182. Tarkowski E, Ringqvist A, Blennow K et al (2000) Intrathecal release of nitric oxide in Alzheimer's disease and vascular dementia. Dement Geriatr Cogn Disord 11:322–326
183. Trostchansky A, Bonilla L, Gonzalez-Perilli L, Rubbo H (2013) Nitro-fatty acids: formation, redox signaling, and therapeutic potential. Antioxid Redox Signal 19:1257–1265
184. Trostchansky A, Mastrogiovanni M, Miquel E et al (2018) Profile of arachidonic acid-derived inflammatory markers and its modulation by nitro-oleic acid in an inherited model of amyotrophic lateral sclerosis. Front Mol Neurosci 11:131

185. Trostchansky A, Rubbo H (2017) Anti-inflammatory signaling actions of electrophilic nitro-arachidonic acid in vascular cells and astrocytes. Arch Biochem Biophys 617:155–161
186. Trostchansky A, Rubbo H (2008) Nitrated fatty acids: mechanisms of formation, chemical characterization, and biological properties. Free Radic Biol Med 44:1887–1896
187. Trushina E, Nemutlu E, Zhang S et al (2012) Defects in mitochondrial dynamics and metabolomic signatures of evolving energetic stress in mouse models of familial Alzheimer's disease. PLoS One 7:e32737
188. Tsikas D, Zoerner A, Mitschke A et al (2009) Specific GC-MS/MS stable-isotope dilution methodology for free 9- and 10-nitro-oleic acid in human plasma challenges previous LC-MS/MS reports. J Chromatogr B Anal Technol Biomed Life Sci 877:2895–2908
189. Urano Y, Hayashi I, Isoo N et al (2005) Association of active γ-secretase complex with lipid rafts. J Lipid Res 46:904–912
190. Van Waes V, Beverley JA, Siman H et al (2012) CB1 cannabinoid receptor expression in the striatum: association with corticostriatal circuits and developmental regulation. Front Pharmacol 3:21
191. Vance JE, Hayashi H, Karten B (2005) Cholesterol homeostasis in neurons and glial cells. Semin Cell Dev Biol 16:193–212
192. Varga T, Czimmerer Z, Nagy L (2011) PPARs are a unique set of fatty acid regulated transcription factors controlling both lipid metabolism and inflammation. Biochim Biophys Acta 1812:1007–1022
193. Vaya J, Schipper HM (2007) Oxysterols, cholesterol homeostasis, and Alzheimer disease. J Neurochem 102:1727–1737
194. Vetrivel KS, Meckler X, Chen Y et al (2009) Alzheimer disease Aβ production in the absence of S -palmitoylation-dependent targeting of BACE1 to lipid rafts. J Biol Chem 284:3793–3803
195. Vetrivel KS, Thinakaran G (2010) Membrane rafts in Alzheimer's disease beta-amyloid production. Biochim Biophys Acta Mol Cell Biol Lipids 1801:860–867
196. Villacorta L, Chang L, Salvatore SR et al (2013) Electrophilic nitro-fatty acids inhibit vascular inflammation by disrupting LPS-dependent TLR4 signalling in lipid rafts. Cardiovasc Res 98:116–124
197. Wahrle S, Das P, Nyborg AC et al (2002) Cholesterol-dependent γ-secretase activity in buoyant cholesterol-rich membrane microdomains. Neurobiol Dis 9:11–23
198. Wahrle SE, Jiang H, Parsadanian M et al (2005) Deletion of Abca1 increases Aβ deposition in the PDAPP transgenic mouse model of Alzheimer disease. J Biol Chem 280:43236–43242
199. Wallace JL (2008) Prostaglandins, NSAIDs, and gastric mucosal protection: why doesn't the stomach digest itself? Physiol Rev 88:1547–1565
200. Wallace JL, Miller MJ (2000) Nitric oxide in mucosal defense: a little goes a long way. Gastroenterology 119:512–520
201. Wang H, Liu H, Jia Z et al (2010) Effects of endogenous PPAR agonist nitro-oleic acid on metabolic syndrome in obese zucker rats. PPAR Res 2010:601562
202. Watanabe S, Doshi M, Hamazaki T (2003) n-3 polyunsaturated fatty acid (PUFA) deficiency elevates and n-3 PUFA enrichment reduces brain 2-arachidonoylglycerol level in mice. Prostaglandins Leukot Essent Fatty Acids 69:51–59
203. Wettschureck N, van der Stelt M, Tsubokawa H et al (2006) Forebrain-specific inactivation of Gq/G11 family G proteins results in age-dependent epilepsy and impaired endocannabinoid formation. Mol Cell Biol 26:5888–5894
204. Wightman EL, Haskell-Ramsay CF, Thompson KG et al (2015) Dietary nitrate modulates cerebral blood flow parameters and cognitive performance in humans: a double-blind, placebo-controlled, crossover investigation. Physiol Behav 149:149–158
205. Witkamp R (2016) Fatty acids, endocannabinoids and inflammation. Eur J Pharmacol 785:96–107
206. Wolozin B, Kellman W, Ruosseau P et al (2000) Decreased prevalence of Alzheimer disease associated with 3-hydroxy-3-methyglutaryl coenzyme A reductase inhibitors. Arch Neurol. 57:1439–1443
207. Wood WG, Li L, Müller WE, Eckert GP (2014) Cholesterol as a causative factor in Alzheimer's disease: a debatable hypothesis. J Neurochem 129:559–572
208. Woodcock SR, Marwitz AJ, Bruno P, Branchaud BP (2006) Synthesis of nitrolipids. All four possible diastereomers of nitrooleic acids: (E)- and (Z)-, 9- and 10-nitro-octadec-9-enoic acids. Org Lett 8:3931–3934
209. Xu Q, Bernardo A, Walker D et al (2006) Profile and regulation of apolipoprotein E (ApoE) expression in the CNS in mice with targeting of green fluorescent protein gene to the ApoE locus. J Neurosci 26:4985–4994
210. Yanagisawa K (2005) Cholesterol and amyloid beta fibrillogenesis. Subcell Biochem 38:179–202
211. Yoshida T, Fukaya M, Uchigashima M et al (2006) Localization of diacylglycerol lipase-alpha around postsynaptic spine suggests close proximity between production site of an endocannabinoid, 2-arachidonoyl-glycerol, and presynaptic cannabinoid CB1 receptor. J Neurosci 26:4740–4751
212. Yu W, Zou K, Gong J-S et al (2005) Oligomerization of amyloid beta-protein occurs during the isolation of lipid rafts. J Neurosci Res 80:114–119
213. Zhang J, Villacorta L, Chang L et al (2010) Nitro-oleic acid inhibits angiotensin II-induced hypertension. Circ Res 107:540–548

11 Nitroalkylation of α-Synuclein by Nitro-Oleic Acid: Implications for Parkinson's Disease

Cecilia Chavarría, Andrés Trostchansky, Rosario Durán, Homero Rubbo, and José M. Souza

Abstract

α-Synuclein (α-syn) represents the main component of the amyloid aggregates present in Parkinson's disease and other neurodegenerative disorders, collectively named synucleinopathies. Although α-syn is considered a natively unfolded protein, it shows great structural flexibility which allows the protein to adopt highly rich beta-sheet structures like protofibrils, oligomers and fibrils. In addition, this protein can adopt alpha-helix rich structures when interacts with fatty acids or acidic phospholipid vesicle membranes. When analyzing the toxicity of α-syn, protein oligomers are thought to be the main neurotoxic species by mechanisms that involve modification of intracellular calcium levels, mitochondrial and lysosomal function. Extracellular fibrillar α-syn promotes intracellular protein aggregation and shows many toxic effects as well. Nitro-fatty acids (nitroalkenes) represent novel pleiotropic anti-inflammatory signaling mediators that could interact with α-syn to exert unraveling actions. Herein, we demonstrated that nitro-oleic acid (NO_2-OA) nitroalkylate α-syn, forming a covalent adduct at histidine-50. The nitroalkylated-α-syn exhibited strong affinity for phospholipid vesicles, moving the protein to the membrane compartment independent of composition of the membrane phospholipids. Moreover, NO_2-OA-modified α-syn showed a reduced capacity to induce α-syn fibrillization compared to the non-nitrated oleic acid. From this data we hypothesize that nitroalkenes, in particular NO_2-OA, may inhibit α-syn fibril formation exerting protective actions in Parkinson's disease.

Keywords

Parkinson's disease · α-Synuclein · Nitrofatty acid · Nitro-oleic acid · Nitroalkylation · Michael adducts

Abbreviations

AD	Alzheimer's disease
ALS	Amyotrophic Lateral Sclerosis
ARE	antioxidant responsive elements
FABP	fatty acid binding protein

C. Chavarría · A. Trostchansky · H. Rubbo
J. M. Souza (✉)
Departamento de Bioquímica and Centro de Investigaciones Biomédicas (CEINBIO), Facultad de Medicina, Universidad de la República, Montevideo, Uruguay
e-mail: jsouza@fmed.edu.uy

R. Durán
Institut Pasteur de Montevideo, Instituto de Investigaciones Biológicas Clemente Estable, Unidad de Bioquímica y Proteómica Analíticas, Montevideo, Uruguay

A. Trostchansky, H. Rubbo (eds.), *Bioactive Lipids in Health and Disease*, Advances in Experimental Medicine and Biology 1127, https://doi.org/10.1007/978-3-030-11488-6_11

GAPDH	glyceraldehyde 3-phosphate dehydrogenase
4-HNE	4-hydroxynonenal
MPTP	1-methyl-4-phenyl-1,2,3,6-tetrahydropyridine
MSA	multiple system atrophy
NO_2-OA	nitro-oleic acid
NSAID	nonsteroidal anti-inflammatory drugs
OA	oleic acid
PD	Parkinson's disease
PSD	post source decay
RNS	reactive nitrogen species
ROS	reactive oxygen species
SDS	sodium dodecyl sulfate
SNARE	soluble NSF attachment receptor
α-syn	alpha-synuclein

11.1 Introduction

α-Synuclein (α-syn) exerts a key role in a number of neurodegenerative diseases denominated synucleinopathies, *i.e.* Parkinson's disease (PD), multiple system atrophy (MSA), dementia with Lewy bodies and Lewy body variant of Alzheimer's disease (AD) [1]. Familial cases of PD present mutations at the α-syn gene or extra dosage of wild type α-syn gene supporting a central role of α-syn at the pathophysiological level of the disease [2]. In fact, histopathologically amyloid inclusions (named Lewy bodies and Lewy neurites) are rich on α-syn protein aggregates [3, 4].

α-Syn is a 140 amino acids protein, abundant at the presynaptic terminals. Its primary sequence shows no cysteine nor tryptophan residues, and it is divided in three regions: an N-terminal segment (residues 1–60) with several KTKEGV repeats; a central segment (residues 61–95) rich in hydrophobic residues; and a C-terminal segment (residues 96–140) rich in proline and acidic residues (Fig. 11.1a). The N-terminal sequence of α-syn has homology with lipoprotein sequences and shows amphipathic properties. The central segment of α-syn holds the amyloidogenic part of the protein, which allows starting and following the autofibrillization process leading to the formation of insoluble fibrils [5]. α-Syn is recognized as a natively unfolded monomer or random coil monomer [6], although there are reports who consider that *in vivo* α-syn is a folded tetramer rich in alpha-helix structures [7]. This issue is still under debate [8, 9].

By using nuclear magnetic and electron paramagnetic resonances in live cells, it has been shown the characteristic disordered monomeric nature of α-syn in the cytosol of neuronal and non-neuronal cells [10]. The disordered monomeric α-syn has an intrinsically inhibitory mechanism of the fibrillization process by long-range interactions of the N- and C-terminal regions with the central region. In order to promote fibril formation, some kind of disruption of these interactions should occur, exposing the central region of the protein allowing its contact with other monomers [11]. α-Syn fibrillization process is believed to be the basic mechanism of the formation of Lewy bodies and Lewy neurites, found in patients with synucleinopathies. *In vitro*, this is a nucleation-dependent process where the natively disordered α-syn adopts partially folded intermediate structures, which seeds the fibrillation and transforms the protein into highly ordered fibrils [12]. α-Syn dimers have been proposed as the seeding nucleus [13, 14], being the species formed from the former process α-syn oligomers, protofibrils, and protofilments. These intermediates are heterogeneous but they are rich in beta-sheet structure [15–18]. The oligomeric forms of α-syn are characterized by an annular or donut-like structure, but heterogeneous in terms of sizes, shapes, and in the different ways they can be obtained *in vitro*. Even though, most of the collected data favor the hypothesis that oligomers may represent the proximal toxic species of α-syn. Oligomers have been proposed to induce an increase in intracellular calcium by changes in membrane permeability, lysosomal leakage, microtubules disruption, or mitochondria damage [19–22]. As a strong evidence of oligomers toxicity it could be mentioned that the A56P or the triple mutant A30P/A56P/A76P α-syn are unable to form fibrils but displaying stabilization of oligomers, and high toxicity in *Drosophila* as well as *C. elegans* and cell cultures [23]. Also,

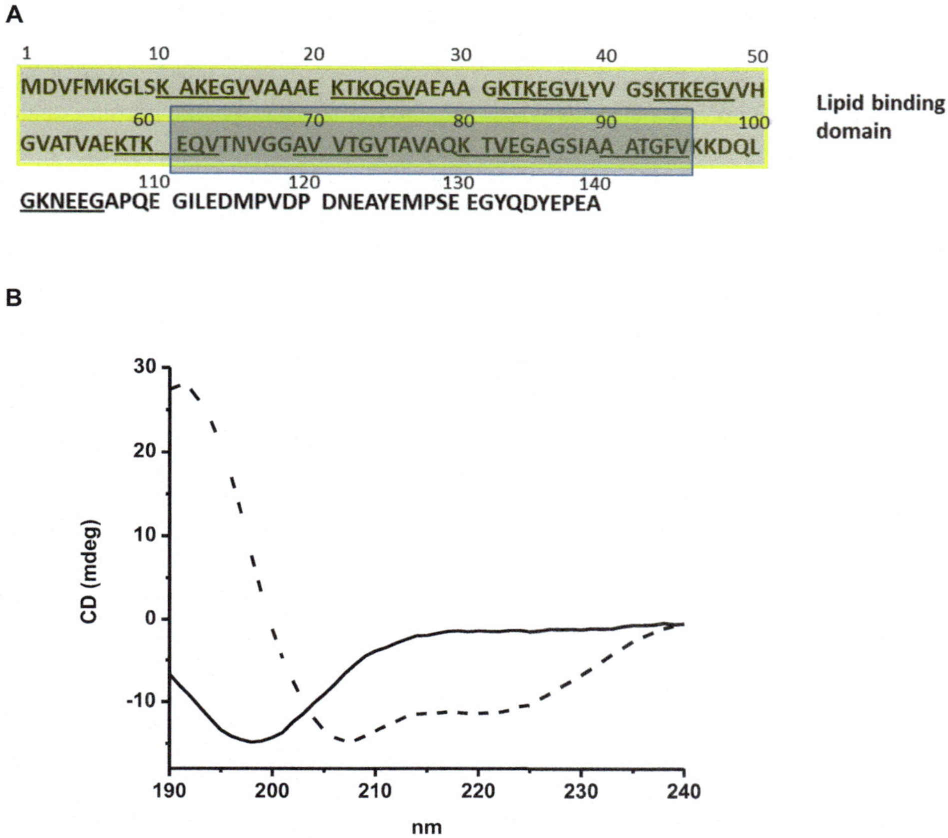

Fig. 11.1 **α-Synuclein protein sequence and structural features.** (**a**) Amino acid sequence of α-syn. The yellow box shows the lipid-binding domain and the blue box the central or amyloidogenic domain. The KTKEGV repeats are underlined. (**b**) Circular dichroism spectra of 10 μM α-synuclein in 10 mM Tris-HCl buffer pH 7.4 (solid line) or with 10 mM SDS (dashed line) are shown

dopamine interacts with α-syn inhibiting fibril formation and stabilizing the oligomers, inducing an important damage in the presynaptic terminals of dopaminergic neurons [24]. Nevertheless, evidence suggests that α-syn fibrils and the process involving the conversion of oligomers to fibrils are determinant for α-syn neurotoxicity [13]. These contradictory results are analyzed in the work of Taschenberger et al., where the authors expressed in substantia nigra of rats the same mutants (A56P and A30P/A56P/A76P) by using adenovirus-associated vectors [25]. The work showed the opposite effects; the wild type or the A30P variant-α-syn, which are capable of fibril formation, increased the loss of dopaminergic neurons in contrast to the oligomer-promoting variants [25]. Although α-syn is considered an intracellular protein, evidences support that part of α-syn oligomers and fibrils are extracellular secreted by unusual mechanisms where may interact with neurons and glial cells. Extracellular fibrillar α-syn is capable of inducing intracellular α-syn aggregates in neurons, showing a prion-like mechanism [26]. As showed by Braak et al., this behavior agrees with the staging distribution of the Lewy body pathology found in PD [27]. Both α-syn oligomers and fibrils behave as toxic species acting by different mechanisms and the output effect may differ depending on the predominant species involved [28].

Posttranslational modifications of α-syn, mediated by oxidant and enzymatic processes, have been involved in the pathogenesis of synucleinopathies [29]. These modifications include serine phosphorylation, protein truncation and oxidatively-modified amino acids such as tyrosine nitration, methionine oxidation and residue glycation [30–35]. Reactive lipid derivatives also mediate α-syn modifications an issue that will be discussed below. Reactive oxygen and nitrogen species (ROS and RNS), such as hydrogen peroxide, superoxide, hydroxyl radical, nitric oxide and peroxynitrite, have been implicated in many neurodegenerative disorders. In particular, PD shows strong correlation with an increase of the formation of ROS and RNS in neurons [36]. Dopaminergic neurons, due to neurotransmitter re-uptake systems, are susceptible to many toxins (such as rotenone, MPTP) and dopamine metabolism generating intracellular ROS. Mitochondrial dysfunction has been related with PD, where complex I inhibition has been well documented [37]. In fact, accumulation of α-syn in mitochondria induces mitochondrial dysfunction and ROS formation [38].

The physiological functions of α-syn are still unclear. α-Syn has been involved in the synaptic vesicle maintenance, neurotransmitter release and neuronal plasticity [39, 40]. Even though, α-syn knockout mice show no pathological phenotype and only present an increase of dopamine release upon electrical stimuli [41].

11.2 Alpha-Synuclein Interaction with Free Fatty Acids and Phospholipids

Early studies of α-syn determined that its primary amino acid sequence has several degenerated repeats at the N-terminal segment able to fold in amphipathic alpha-helix, each with a hydrophobic and a hydrophilic face (Fig. 11.1a). This type of helix corresponds to the A_2 class found in many apolipoproteins such as A-II, C-I, C-II and C-III [42]. These predictions have experimental correlations. α-Syn's secondary structure changes dramatically upon binding to phospholipids, from a rich random coil structure to a rich alpha-helix structure, easily visualized in the circular dichroism spectra (Fig. 11.1b). α-Syn preferentially binds to small unilamellar phospholipidic vesicles rich in negative charges such as phosphatidylserine, phosphatidylinositol and phosphatidic acid. These lipid binding properties are also observed for α-syn in the presence of micelles from detergents, such as SDS, and fatty acids (Fig. 11.1b). The lipid binding properties of α-syn are dependent of the first 102 aminoacidic residues, including the N-terminal and the central region of the protein [43]; the C-terminal region is kept in an disordered conformation capable of interacting with other proteins [44]. α-Syn binds to membranes forming an extended alpha-helix arranged parallel to the surface of the membrane [45]. The wild-type monomeric α-syn and the genetic mutants A53T and E46K are able to form ion channels in anionic membrane bilayers, although the A30P genetic variant presents lower membrane affinity and does not form ion channels [46]. The annular α-syn oligomer species, with 45 nm diameter and 2–5 nm height, disrupt ion homeostasis presumably by a pore-forming mechanism, a mechanism specific for this oligomeric species and not for the globular oligomers with a height larger than 5 nm [19].

The presynaptic vesicles have been associated with α-syn, mainly with the SNARE complexes and the depletion of α-syn induces a decrease in vesicle traffic in neurons. This is consistent with a role for α-syn in attenuating the mobility of presynaptic pool of recycling vesicles [47]. Even though, when α-syn starts its aggregation processes, a redistribution of SNARE complexes in the presynaptic button is observed in addition to a decrease of exocytosis and dopamine release [48].

α-Syn binds fatty acids but it does not behave as a typical fatty acid binding protein (FABP) [49]. The C-terminal segment of α-syn shows homology with the FABP, and the apparent K_d for oleic acid (OA) binding was estimated at 12.5 μM [50]. In α-syn knockout mice, it was shown a slight deficiency in palmitic acid brain uptake [51]. *In vitro*, prolonged exposure to polyunsaturated fatty acids (docosahexaenoic acid or arachidonic acid) induced the formation of amyloid-like fibrils and high molecular weight species as

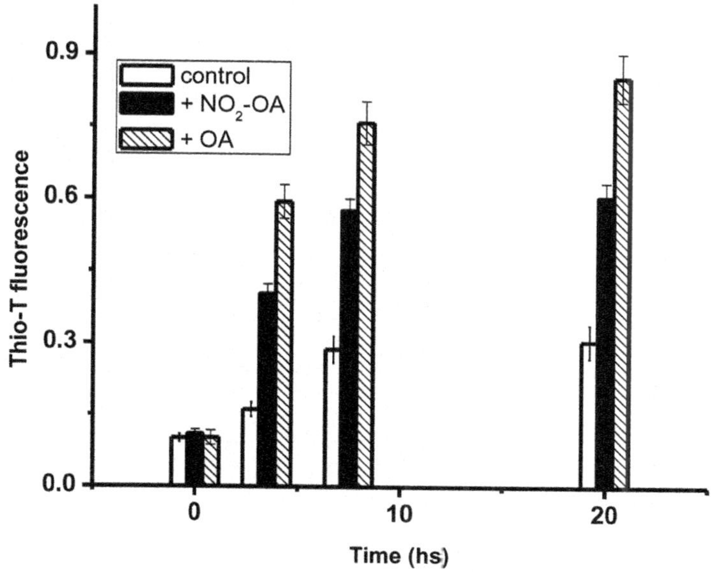

Fig. 11.2 Assessment of α-syn fibrils formation using Thioflavin T. α-Syn (0.35 mM, open bar) was pre-incubated in 50 mM potassium phosphate buffer pH 7.0 with 0.35 mM OA (diagonal stripes bar) or 0.35 mM NO_2-OA (black bar) for 18 h, and then incubated under continuous shaking at 37 °C. For fibrillation assay, aliquots were taken at different times and mixed with Thio-T at pH 8.2; fluorescence was determined using λ_{exc} = 440 nm and λ_{em} = 482 nm

shown by native-polyacrylamide electrophoresis gels [52]. Using the Thioflavin T assay to follow α-syn fibril formation, it can be observed that OA accelerates the fibrillization process (Fig. 11.2a) while similar behavior is shown with small acidic vesicles [53].

11.3 Alpha-Synuclein Adduction with Nitro-Fatty Acids

α-Syn is highly susceptible to peroxynitrite-mediated oxidation and nitration which induce tyrosine nitration and di-tyrosine formation [31]. Phospholipid vesicles interacting with α-syn exposed to peroxynitrite, showed a decrease on tyrosine nitration. In addition, the protein was modified by lipid peroxidation end-products, *e.g.* 4-hydroxynonenal (4-HNE) generating a Michael addition adduct at histidine-50 [54]. Nitroalkenes which are nitrated fatty acids whose nitro group is located at the double bond of the carbon chain, may react with nucleophilic residues by Michael addition reactions [55, 56]. Unsaturated fatty acids can be modified by either ROS or RNS species, being fatty acid nitration expected to occur in hydrophobic compartments such as the lipid bilayer of liposomes, cellular membranes or the lipophilic core of lipoproteins. RNS yields an array of hydroxyl, hydroperoxy, nitro and nitro-hydroxy lipid derivatives [57]. It has been demonstrated that the electrophilic attack of nitroalkenes to proteins occurs, being the first report GAPDH nitroalkylation by NO_2-OA altering the enzyme localization by translocating it from the cytosol to the membrane [55]. Thus, if the same happens with α-syn, the consequences on the protein folding and function may have biological relevance. As a byproduct of lipid oxidation, it has been reported that NO_2-OA can be formed in red blood cells and mitochondrial membranes [56, 58].

We assessed NO_2-OA reaction with α-syn and analyze the covalent modification of the protein, as well as changes in its localization when incubated with vesicles containing non-acidic phospholipids. As expected, NO_2-OA incubation with α-syn determined the formation of a covalent adduct that was detected by mass spectrometry (Fig. 11.3). Under our experimental conditions, a product with an increase of 327 Da appeared (14784.8 Da *vs* 14457.5 Da) which corresponded to the adduction of one molecule of NO_2-OA to α-syn. In order to determine which residue was modified by NO_2-OA, α-syn and α-syn incubated with NO_2-OA were digested with trypsin and the peptides analyzed by MALDI-TOF. A peptide holding the histi-

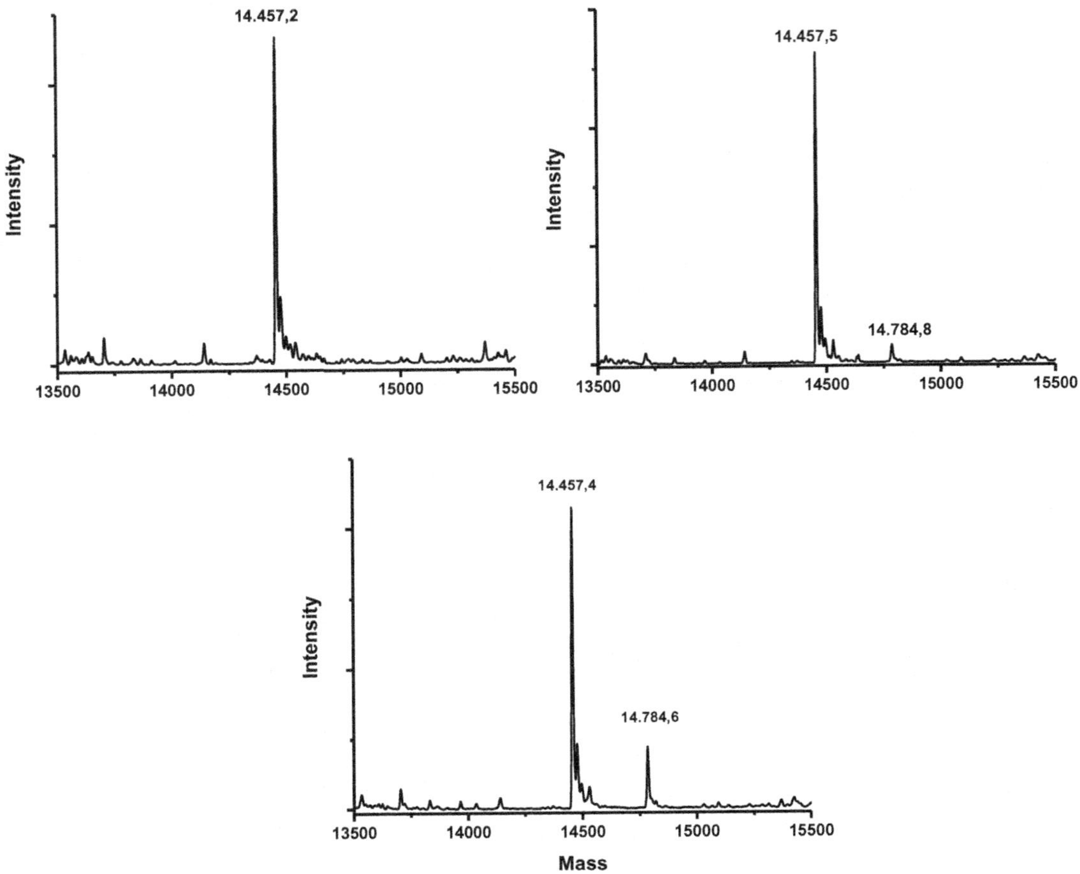

Fig. 11.3 Mass spectrometric analysis of α-syn nitroalkylation. α-Syn (100 μM, 14457,2 Da) was incubated at 25 °C in 50 mM phosphate buffer pH 7.4 with NO_2-OA (500 μM). Samples were taken at 0, 1 or 24 h and formation of lipid/protein adducts (14784,8 Da) were analyzed by mass spectrometry

dine-50 adducted with NO_2-OA was detected (Fig. 11.4). The modified peptide corresponded to E46GVVHGVATVAEK58 (*m/z* = 1295.7) plus NO_2-OA which matched with the observed mass of 1622.8 Da (Fig. 11.4). The modified peptide at histidine-50 was confirmed by post source decay (PSD) sequence analysis showing the formation of the immonium ion (*m/z* = 437), characteristic of nitroalkylated histidine (Fig. 11.4) [55]. The nitroalkylation of a single histidine residue, and not a cysteine residue, is in accordance to the lack of cysteine residues in α-syn protein and the presence of only one histidine in the primary amino acidic sequence of the protein (Fig. 11.1a) [55]. The reaction of NO_2-OA with α-syn is a slow process and only modifies a small portion of the protein (Fig. 11.3). This can be in part due to the decay of NO_2-OA in the buffer aqueous solution [58]. However, α-syn modification by NO_2-OA affects the α-syn fibrillization process with a decrease in the fibril formation rate when compared to the condition of the non-nitrated OA (Fig. 11.2). As explained above, α-syn is unable to interact with phosphatidylcholine vesicles [54], and nitroalkylated proteins could be translocated to the membrane moiety. Under our experimental conditions, the adducted α-syn moved to the phospholipid face even when using phosphatylcholine vesicles (Fig. 11.5).

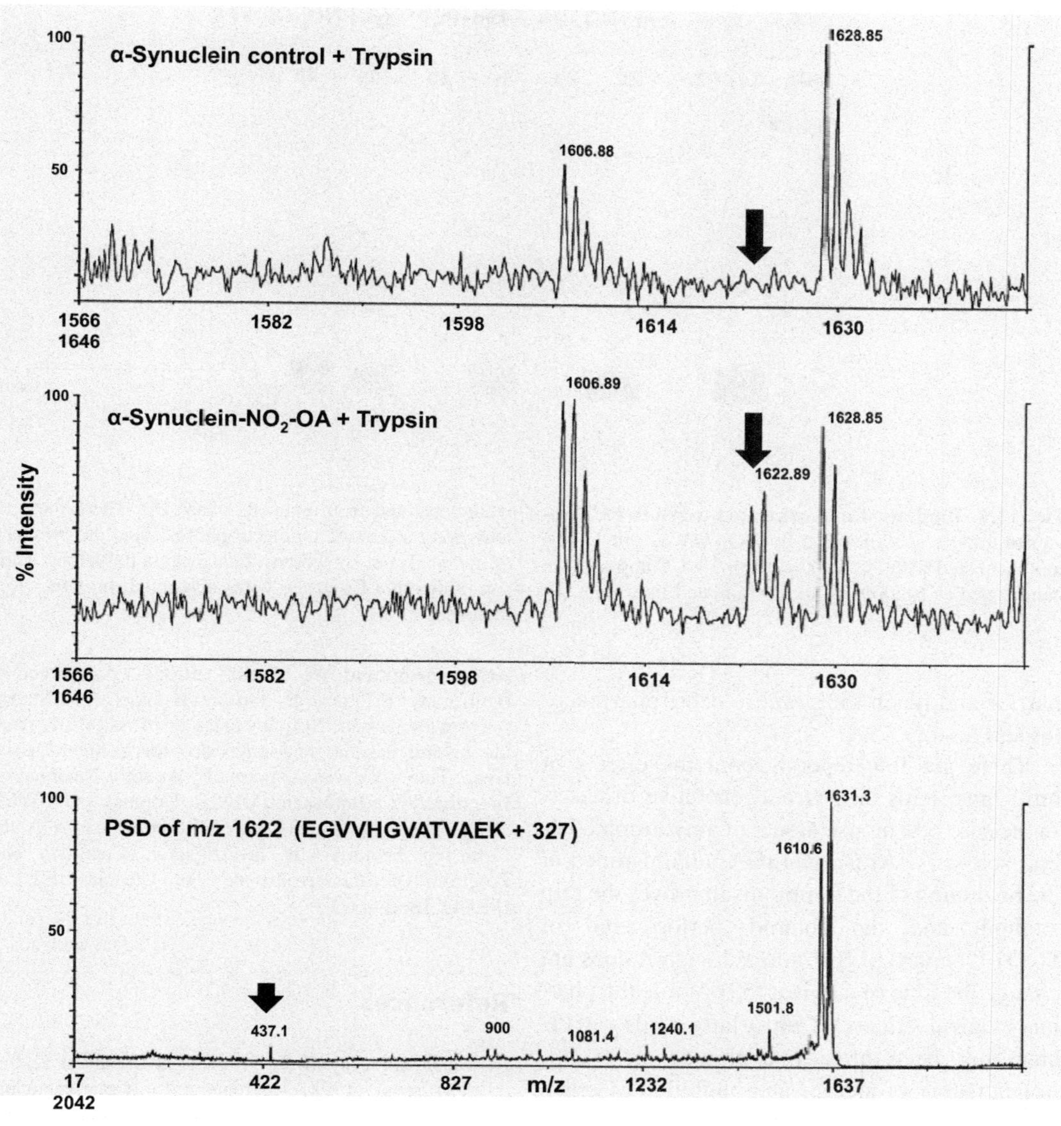

Fig. 11.4 Nitro-oleic acid modified α-syn at His-50. α-Syn was incubated with NO_2-OA for 24 h as in Fig. 11.3; then, the protein was digested with trypsin and analyzed by MALDI-TOF MS. The nitroalkylated peptide E46-K58 (m/z 1622) was identified due to the formation of the typical immonium ion of m/z 437, characteristic of nitroalkylated His

11.4 Conclusions and Perspectives

Nitro-fatty acids show pleiotropic anti-inflammatory actions by mechanisms including down regulation of nitric oxide synthase-2 and inhibition of pro-inflammatory cytokines, the increased expression of heme oxygenase-1 and other antioxidant responsive elements (ARE) as well as the increasing of glutathione [59, 60]. NO_2-OA and other nitroalkenes have been proposed as alternative nonsteroidal anti-inflammatory drugs (NSAID), with probably limited side effects. Regarding neurological diseases, we have recently demonstrated that nitro-fatty acids are able to cross blood brain

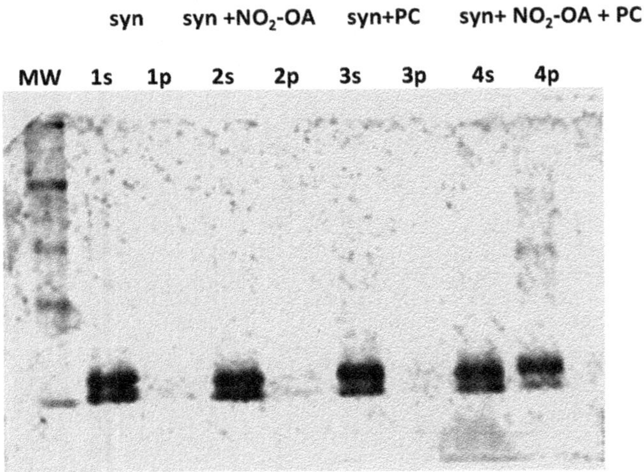

Fig. 11.5 Binding of nitroalkylated-α-syn to PC liposomes. α-Syn was modified by NO_2–OA as previously, and incubated with PC liposomes for 18 h. Samples were centrifuged at 100000 × *g* for 30 min, and liposomes (P) were separated from aqueous phase (S). Then, the proteins were separated by electrophoresis and the presence of syn analyzed by western blot using a polyclonal anti-syn antibody. Controls with unreacted protein were included

barrier and reach the brain to exert pharmacological actions [59].

There are few reports about the effects of nitro-fatty acids on neurodegenerative diseases. In the case of a mouse model of amyotrophic lateral sclerosis (ALS), NO_2-OA administrated at the beginning of the symptoms improves the grip strength and the rotarod performance of $SOD1^{G93A}$ mice ALS strain, although it does not change the time of survival [59]. While the pharmacological effects of nitro-fatty acids on PD have not been already explored, their anti-inflammatory properties and inhibition of astrocytes activation should be of potential benefit. The nitroalkylated-α-syn showed high affinity for phosphatidylcholine vesicles, when α-syn itself does not interact with neutral or positive charge phospholipids. So, the modified α-syn will be transferred to the membrane phase, independent of the affinity for acidic phospholipids, generating an additional hydrophobic anchor. NO_2-OA adducted to α-syn decreased the capacity to promote α-syn fibril formation compared to OA suggesting an scenario where nitro-fatty acids may prevent synucleinophaties by decreasing α-syn fibrillization.

Acknowledgment We thank Bruce A. Freeman (University of Pittsburgh, Pittsburgh, PA, United States) for provide us with NO_2-OA and Carlos Batthyány from the Institut Pasteur de Montevideo for technical assistance. This work was supported by Agencia Nacional de Investigación e Innovación (ANII), Uruguay (ANII-FCE 2011_1_6260 to J.M.S.) and fellowships from ANII, Comisión Sectorial de Investigación Científica and Programa de Desarrollo de las Ciencias Básicas (PEDECIBA) to C.C.

References

1. Baba M, Nakajo S, Tu PH, Tomita T, Nakaya K, Lee VM et al (1998) Aggregation of alpha-synuclein in Lewy bodies of sporadic Parkinson's disease and dementia with Lewy bodies. Am J Pathol 152(4):879–884
2. Si X, Pu J, Zhang B (2017) Structure, distribution, and genetic profile of α-synuclein and their potential clinical application in Parkinson's disease. J Mov Disord 10(2):69–79
3. Spillantini MG, Schmidt ML, Lee VMY, Trojanowski JQ, Jakes R, Goedert M (1997) Alpha-synuclein in Lewy bodies. Nature 388:839–840
4. Goedert M (2001) Alpha-synuclein and neurodegenerative diseases. Nat Rev Neurosci 2:492–501
5. Giasson BI, Murray IVJ, Trojanowski JQ, Lee VMY (2001) A hydrophobic stretch of 12 amino acid residues in the middle of α-synuclein is essential for filament assembly. J Biol Chem 276(4):2380–2386

6. Weinreb PH, Zhen W, Poon AW, Conway KA, Lansbury PT (1996) NACP, a protein implicated in Alzheimer's disease and learning, is natively unfolded. Biochemistry 35(43):13709–13715
7. Bartels T, Choi JG, Selkoe DJ (2011) α-Synuclein occurs physiologically as a helically folded tetramer that resists aggregation. Nature 477(7362):107–110
8. Gould N, Mor DE, Lightfoot R, Malkus K, Giasson B, Ischiropoulos H (2014) Evidence of native α-synuclein conformers in the human brain. J Biol Chem 289(11):7929–7934
9. Fauvet B, Mbefo MK, Fares MB, Desobry C, Michael S, Ardah MT et al (2012) α-Synuclein in central nervous system and from erythrocytes, mammalian cells, and Escherichia coli exists predominantly as disordered monomer. J Biol Chem 287(19):15345–15364
10. Theillet F, Binolfi A, Bekei B, Martorana A, Rose HM, Stuiver M (2016) Structural disorder of monomeric α-synuclein persists in mammalian cells. Nature 530 (7588):45–50
11. Bertoncini CW, Jung Y, Fernandez CO, Hoyer W, Griesinger C, Jovin TM et al (2005) Release of long-range tertiary interactions potentiates aggregation of natively unstructured alpha-synuclein. Proc Natl Acad Sci U S A 102(5):1430–1435
12. Wood SJ, Wypych J, Steavenson S, Louis J, Citron M, Biere AL (1999) Alpha-synuclein fibrillogenesis is nucleation-dependent. Biochemistry 274:19509–19512
13. Lashuel HA, Overk CR, Oueslati A, Masliah E (2013) The many faces of α-synuclein: from structure and toxicity to therapeutic target. Nat Rev Neurosci 14(1):38–48
14. Hodara R, Norris EH, Giasson BI, Mishizen-Eberz AJ, Lynch DR, Lee VMY et al (2004) Functional consequences of α-synuclein tyrosine nitration: diminished binding to lipid vesicles and increased fibril formation. J Biol Chem 279(46):47746–47753
15. Horvath I, Weise CF, Andersson EK, Chorell E, Sellstedt M, Bengtsson C et al (2012) Mechanisms of protein oligomerization: inhibitor of functional amyloids templates α-synuclein fibrillation. J Am Chem Soc 134(7):3439–3444
16. Pieri L, Madiona K, Melki R. (2016) Structural and functional properties of prefibrillar α-synuclein oligomers. Sci Rep 6:24526
17. Fauvet B, Fares MB, Samuel F, Dikiy I, Tandon A, Eliezer D et al (2012) Characterization of semisynthetic and naturally N α- acetylated α-synuclein in vitro and in intact cells: implications for aggregation and cellular properties of α-synuclein. J Biol Chem 287(34):28243–28262
18. Lashuel HA, Petre BM, Wall J, Simon M, Nowak RJ, Walz T et al (2002) Alpha-Synuclein, especially the Parkinson'S disease-associated mutants, forms pore-like annular and tubular Protofibrils. J Mol Biol 322(5):1089–1102
19. Danzer KM, Haasen D, Karow AR, Moussaud S, Habeck M, Giese A et al (2007) Different species of alpha-synuclein oligomers induce calcium influx and seeding. J Neurosci 27(34):9220–9232
20. Decressac M, Mattsson B, Weikop P, Lundblad M, Jakobsson J, Björklund A (2013) TFEB-mediated autophagy rescues midbrain dopamine neurons from α-synuclein toxicity. Proc Natl Acad Sci U S A 110:1–10
21. Alim MA, Ma Q-L, Takeda K, Aizawa T, Matsubara M, Nakamura V et al (2004) Demonstration of a role for α-synuclein as a functional microtubule-associated protein. J Alzheimer's Dis 6(4):435–42
22. Hsu LJ, Sagara Y, Arroyo A, Rockenstein E, Sisk A, Mallory M et al (2000) Alpha-synuclein promotes mitochondrial deficit and oxidative stress. Am J Pathol 157(2):401–410
23. Karpinar DP, Balija MBG, Kügler S, Opazo F, Rezaei-Ghaleh N, Wender N et al (2009) Pre-fibrillar α-synuclein variants with impaired α-structure increase neurotoxicity in Parkinson's disease models. EMBO J 28(20):3256–3268
24. Mor DE, Tsika E, Mazzulli JR, Gould NS, Kim H, Daniels MJ et al (2017) Dopamine induces soluble alpha-synuclein oligomers and nigrostriatal degeneration. Nat Neuroscie 20(11):1560–1568
25. Taschenberger G, Garrido M, Tereshchenko Y, Bähr M, Zweckstetter M, Kügler S (2012) Aggregation of α-synuclein promotes progressive in vivo neurotoxicity in adult rat dopaminergic neurons. Acta Neuropathol 123(5):671–683
26. Luk KC, Kehm V, Carroll J, Zhang B, O'Brien P, Trojanowski JQ et al (2012) Pathological alpha-synuclein transmission initiates Parkinson-like neurodegeneration in nontransgenic mice. Science (80) 338:949–953
27. Braak H, Del Tredici K (2017) Neuropathological staging of brain pathology in sporadic Parkinson's disease: separating the wheat from the chaff. J Park Dis 7(s1):S73–S87
28. Chavarría C, Rodriguez-Bottero S, Quijano C, Cassina P, Souza JM (2018) Impact of monomeric, oligomeric and fibrillar alpha-synuclein on astrocyte reactivity and toxicity to neurons. Biochem J 475(19):3153-3169
29. Chavarría C, Souza JM (2013) Oxidation and nitration of α-synuclein and their implications in neurodegenerative diseases. Arch Biochem Biophys 533(1–2):25–32
30. Oueslati A (2016) Implication of alpha-synuclein phosphorylation at S129 in synucleinopathies: what have we learned in the last decade? J Park Dis 6(1):39–51
31. Souza JM, Giasson BI, Chen Q, Lee VMY, Ischiropoulos H (2000) Dityrosine cross-linking promotes formation of stable α-synuclein polymers: implication of nitrative and oxidative stress in the pathogenesis of neurodegenerative synucleinopathies. J Biol Chem 275(24):18344–18349
32. Giasson BI, Duda JE, Murray IVJ, Chen Q, Souza JM, Hurtig HI et al (2000) Oxidative damage linked to

neurodegeneration by selective alphasynuclein nitration in synucleonopathy lesions. Science 290:985–989
33. Uversky VN, Yamin G, Souillac PO, Goers J, Glaser CB, Fink AL (2002) Methionine oxidation inhibits fibrillation of human α-synuclein in vitro. FEBS Lett 517(1–3):239–244
34. Li W, West N, Colla E, Pletnikova O, Troncoso JC, Marsh L et al (2005) Aggregation promoting C-terminal truncation of -synuclein is a normal cellular process and is enhanced by the familial Parkinson's disease-linked mutations. Proc Natl Acad Sci 102(6):2162–2167
35. Vicente Miranda H, Szego ÉM, Oliveira LMA, Breda C, Darendelioglu E, De Oliveira RM et al (2017) Glycation potentiates α-synuclein-associated neurodegeneration in synucleinopathies. Brain 140(5):1399–1419
36. Przedborski S, Ischiropoulos H (2005) Reactive oxygen and nitrogen species: weapons of neuronal destruction in models of Parkinson's disease. Antioxid Redox Signal 7(5–6):685–693
37. Schapira a H, Cooper JM, Dexter D, Jenner P, Clark JB, Marsden CD (1989) Mitochondrial complex I deficiency in Parkinson's disease. Lancet 1(8649):1269
38. Devi L, Raghavendran V, Prabhu BM, Avadhani NG, Anandatheerthavarada HK (2008) Mitochondrial import and accumulation of α-synuclein impair complex I in human dopaminergic neuronal cultures and Parkinson disease brain. J Biol Chem 283(14):9089–9100
39. George JM, Jin H, Woods WS, Clayton DF (1995) Characterization of a novel protein regulated during the critical period for song learning in the Zebra Finch. Neuron 15:361–372
40. Murphy DD, Rueter SM, Trojanowski JQ, Lee VM (2000) Synucleins are developmentally expressed, and alpha-synuclein regulates the size of the presynaptic vesicular pool in primary hippocampal neurons. J Neurosci 20(9):3214–3220
41. Abeliovich A, Schmitz Y, Farin I, Choi-lundberg D, Ho W, Castillo PE et al (2000) Mice lacking alpha-synuclein display functional deficits in the nigrostriatal dopamine system. Neuron 25:239–252
42. Davidson WS, Jonas A, Clayton DF, George JM (1998) Stabilization of alpha-synuclein secondary structure upon binding to synthetic membranes. J Biol Chem 273(16):9443–9449
43. Perrin RJ, Woods WS, Clayton DF, George JM (2000) Interaction of human α-synuclein and Parkinson's disease variants with phospholipids: structural analysis using site-directed mutagenesis. J Biol Chem 275(44):34393–34398
44. Eliezer D, Kutluay E, Bussell R, Browne G (2001) Conformational properties of α-synuclein in its free and lipid-associated states. J Mol Biol 307(4):1061–1073
45. Jao CC, Hegde BG, Chen J, Haworth IS, Langen R (2008) Structure of membrane-bound alpha-synuclein from site-directed spin labeling and computational refinement. Proc Natl Acad Sci U S A 105(50):19666–19671
46. Zakharov SD, Hulleman JD, Dutseva EA, Antonenko YN, Rochet J-CC, Cramer WA (2007) Helical alpha-synuclein forms highly conductive ion channels. Biochem Int 46(50):14369–14379
47. Scott D, Roy S (2012) Alpha synuclein inhibits intersynaptic vesicle mobility and maintains recycling-pool homeostasis. J Neurosci 32(30):10129–10135
48. Garcia-Reitböck P, Anichtchik O, Bellucci A, Iovino M, Ballini C, Fineberg E et al (2010) SNARE protein redistribution and synaptic failure in a transgenic mouse model of Parkinson's disease. Brain 133(7):2032–2044
49. Lücke C, Gantz DL, Klimtchuk E, Hamilton JA (2006) Interactions between fatty acids and α-synuclein. J Lipid Res 47(8):1714–1724
50. Sharon R, Goldberg MS, Bar-josef I, Betensky RA, Shen J, Selkoe DJ (2001) Alpha-Synuclein occurs in lipid-rich high molecular weight complexes, binds fatty acids, and shows homology to the fatty acid-binding proteins. Proc Natl Acad Sci U S A 98(16):9110–9115
51. Golovko MY, Faergeman NJ, Cole NB, Castagnet PI, Nussbaum RL, Murphy EJ (2005) Alpha-Synuclein gene deletion decreases brain palmitate uptake and alters the palmitate metabolism in the absence of alpha-Synuclein palmitate binding. Biochemistry 44(23):8251–8259
52. Broersen K, Van Den Brink D, Fraser G, Goedert M, Davletov B (2006) Alpha-Synuclein adopts an alpha-helical conformation in the presence of polyunsaturated fatty acids to hinder micelle formation. Biochemistry 45(51):15610–15616
53. Zhu M, Li J, Fink AL (2003) The association of α-synuclein with membranes affects bilayer structure, stability, and fibril formation. J Biol Chem 278(41):40186–40197
54. Trostchansky A, Lind S, Hodara R, Oe T, Blair IA, Ischiropoulos H et al (2006) Interaction with phospholipids modulates α-synuclein nitration and lipid–protein adduct formation. Biochem J 393(1):343–349
55. Batthyany C, Schopfer FJ, Baker PRS, Durán R, Baker LMS, Huang Y et al (2006) Reversible post-translational modification of proteins by nitrated fatty acids in vivo. J Biol Chem 281(29):20450–20463
56. Schopfer FJ, Batthyany C, Baker PRS, Bonacci G, Cole MP, Rudolph V et al (2009) Detection and quantification of protein adduction by electrophilic fatty acids: mitochondrial generation of fatty acid nitroalkene derivatives. Free Radic Biol Med 46(9):1250–1259
57. Rubbo H, Radi R, Trujillo M, Telleri R, Kalyanaraman B, Barnes S et al (1994) Nitric oxide regulation of superoxide and peroxynitrite-dependent lipid peroxidation: formation of novel nitrogen-containing oxidized lipid derivatives. J Biol Chem 269(42):26066–26075
58. Baker PRS, Lin Y, Schopfer FJ, Woodcock SR, Groeger AL, Batthyany C et al (2005) Fatty acid

transduction of nitric oxide signaling: multiple nitrated unsaturated fatty acid derivatives exist in human blood and urine and serve as endogenous peroxisome proliferator-activated receptor ligands. J Biol Chem 280(51):42464–42475
59. Trostchansky A, Mastrogiovanni M, Miquel E, Rodríguez-Bottero S, Martínez-Palma L, Cassina P et al (2018) Profile of arachidonic acid-derived inflammatory markers and its modulation by nitro-oleic acid in an inherited model of amyotrophic lateral sclerosis. Front Mol Neurosci 11(2):1–11
60. Freeman BA, Pekarova M, Rubbo H, Trostchansky A (2017) Electrophilic nitro-fatty acids: nitric oxide and nitrite-derived metabolic and inflammatory signaling mediators. In: Nitric oxide biology and pathobiology, 3rd edn. Academic Press, London, pp 213–229

Bioactive Lipids in Inflammation After Central Nervous System Injury

12

Rubèn López-Vales and Samuel David

Abstract

Despite the progress made over the last decades to understand the mechanisms underlying tissue damage and neurological deficits after neurotrauma, there are currently no effective treatments in the clinic. It is well accepted that the inflammatory response in the CNS after injury exacerbates tissue loss and functional impairments. Unfortunately, the use of potent anti-inflammatory drugs, such as methylprednisolone, fails to promote therapeutic recovery and also gives rise to several undesirable side effects related to immunosuppression. The injury-induced inflammatory response is complex, and understanding the mechanisms that regulate this inflammation is therefore crucial in the quest to develop effective treatments. Bioactive lipids have emerged as potent molecules in controlling the initiation, coordination, and resolution of inflammation and in promoting tissue repair and recovery of homeostasis. These bioactive lipids are produced by cells involved in the inflammatory response, and their defective synthesis leads to persistent chronic inflammation, tissue damage, and fibrosis. The present chapter discusses recent evidence for the role of some of these bioactive lipids, in particular, eicosanoid and pro-resolving lipid mediators, in the regulation of inflammation after neurotrauma and highlights the therapeutic potential of some of these lipids in enhancing neurological outcomes after CNS injuries.

Keywords

Eicosanoids · Inflammation · Resolution · Specialized pro-resolving mediators · Neurotrauma

12.1 Introduction

Spinal cord and brain injuries lead to devastating neurological deficits and disabilities and represent a significant health and social problem. The neurological deficits comprise not only of loss of sensorimotor function but also bladder, bowel, and sexual dysfunction, as well as kidney infections and cardiac and respiratory problems that have detrimental consequences on psychological and social behavior. They also have a significant

R. López-Vales (✉)
Departament de Biologia Cellular, Fisiologia i Immunologia, Institut de Neurociències, Centro de Investigación Biomédica en Red sobre Enfermedades Neurodegenerativas (CIBERNED), Universitat Autònoma de Barcelona, Barcelona, Catalonia, Spain
e-mail: ruben.lopez@uab.cat

S. David
Centre for Research in Neuroscience, The Research Institute of the McGill University Health Center, Montreal, Quebec, Canada

A. Trostchansky, H. Rubbo (eds.), *Bioactive Lipids in Health and Disease*, Advances in Experimental Medicine and Biology 1127, https://doi.org/10.1007/978-3-030-11488-6_12

impact on their families and are a considerable economic burden.

Injury to the CNS results in an initial zone of damage to neural tissue, which includes glia and neurons, at the site of injury. This primary damage spreads to regions rostral and caudal to the injury epicenter during the days and weeks after the trauma, in a process known as secondary damage. A variety of factors contribute to this secondary damage, such as vascular changes (hemorrhage and ischemia), excitotoxicity, lack of ionic homeostasis, and multiple factors arising from the inflammatory response [5, 22, 39]. Secondary damage in the spinal cord enlarges the area of the lesion into rostral and caudal segments from the injury epicenter, thus increasing neuronal and glia cell death, damage to axons and myelin, and consequently, to greater functional loss. These consequences are especially detrimental in the CNS due to its limited ability for spontaneous self-repair, leading to irreversible disabilities in patients with spinal cord injury (SCI) [29, 36].

Currently, there is no effective clinical therapy to treat CNS injuries. Although regeneration of long and short fiber tracts and their functional reconnection to neural networks and the replacement of dead neuronal and glial cells occurring after neurotrauma are important goals to restore function, prevention of secondary damage to axons, neuronal cell bodies, myelin, and glial cells that follows the initial trauma is important and likely to be more easily amenable to treatment.

The inflammatory response that occurs after CNS injury is one of the main contributors to secondary damage. This pathophysiological response is mediated, to a large extent, by resident microglial cells and by peripheral monocyte-derived macrophages (MDMs) and granulocytes that infiltrate into the lesioned area from the circulation [24, 28, 100]. Although inflammation is a naturally occurring response of the body to fight infections, clear cell and tissue debris, restore tissue homeostasis, and promote repair, when this process is not controlled and/or resolved, it can result in secondary tissue damage and be detrimental [20, 86]. Unlike many other tissues, this is the case of the inflammatory response after neurotrauma, as immune cells remain in the injured CNS for several weeks and months after injury combined with the limited capacity of the CNS to replace and regenerate damaged nerve cells [23, 24, 28]. There is therefore a need to understand the factors that underlie this dysregulated inflammatory response after CNS injuries in order to develop effective therapies to counteract it.

Studies over the last few years have revealed an important multifaceted role for phospholipase A_2 (PLA_2) in various aspects of inflammation in the peripheral and central nervous system. [22, 47, 65, 67]. PLA_2 encompasses a family of enzymes that catalyze the cleavage of the acyl bond at the sn-2 position of membrane phospholipids to release a free fatty acid (such as arachidonic acid (AA), docosahexaenoic acid (DHA), and eicosapentaenoic acid (EPA)) and a lysophospholipid [22, 95]. These fatty acids and lysophospholipids are the precursor of different bioactive lipid mediators that have a broad range of biological functions, including the active control of the inflammatory response [19, 22, 25, 86].

There are currently six major types of PLA_2 enzymes which include $sPLA_2$ (secreted), $cPLA_2$ (calcium-dependent), $iPLA_2$ (calcium-independent), Lp-PLA_2 (lipoprotein-associated, also known as platelet-activating factor acetylhydrolase (PAF-AH)), LPLA2 (lysosomal), and $AdPLA_2$ (adipose) [95]. Lp-PLA_2 and $sPLA_2$ are both secreted enzymes that act on membrane lipids extracellularly, while $cPLA_2$ and $iPLA_2$ hydrolyze intracellular lipids in membrane bilayers including various subcellular organelles [22, 95]. Less is understood about the localization of $LPLA_2$ and $AdPLA_2$.

Multiple forms of PLA_2s are constitutively detected in the spinal [64, 67, 93], suggesting that they play a role in normal physiological function in the CNS. We analyzed the changes in the expression of 14 mammalian PLA_2s in the spinal cord after contusion injury in mice ($cPLA_2$ GIVA, $cPLA_2$ GIVB, $iPLA_2$ GVIA, $iPLA_2$ GVIB, and $sPLA_2$ GIIA, GIIC, IID, GIIE, GIIF, GV, GVII, GX, GXIIA, GIIB) and found that only the transcripts of cPLA2 GIVA, $iPLA_2$ VIA, and

$sPLA_2$ IIA were increased [67]. By using knockout mice or selective inhibitors for these PLA_2, we found that the different PLA_2s play divergent actions after spinal cord injury (SCI). Specifically, we observed that $sPLA_2$ mediated detrimental actions and $iPLA_2$ had neutral effects, while $cPLA_2$ exerted a beneficial role. These varied functions of the PLA_2 superfamily suggest that lipid mediators generated by the different PLA_2s may exert both harmful and beneficial actions after SCI [67]. In line with this, we found administration of a potent pan PLA_2 inhibitor (FKGK2), that blocks all three PLA_2s to about the 90% level, worsened functional outcomes and tissue damage after SCI further indicating that some of the lipids generated by the actions of PLA_2s (possibly $cPLA_2$) are needed for tissue repair [67]. In other studies, we deciphered that $iPLA_2$s also contribute to the onset of inflammation after peripheral neve injury [65] and in a model of multiple sclerosis (experimental autoimmune encephalomyelitis (EAE)) [46, 47]. Interestingly, $iPLA_2$ did not have much effect in SCI but was most effective in EAE [47]. In addition, $cPLA_2$ which was detrimental in SCI inflammation was required in injured peripheral nerve for clearance of myelin and immune cells and promote subsequent axon regeneration [65]. Therefore, the action of the different PLA_2s generates a wide range of bioactive lipid mediators, some of which mediate pro-inflammatory actions, whereas others have anti-inflammatory and pro-resolution, pro-repair features. In the following sections, we will discuss the involvement of the main metabolites derived from the free fatty acids generated by PLA_2s to the initiation and resolution of inflammation after neurotrauma, with special emphasis on SCI.

12.1.1 Role of Eicosanoids in CNS Trauma

As mentioned above, PLA_2s release a free fatty acid (such as arachidonic acid) by hydrolysis of the acyl group at the sn-2 position of membrane phospholipids. Eicosanoids arising from metabolites of arachidonic acid (AA) include prostaglandins (PGs) and thromboxane, which are generated from AA by the enzymatic actions of cyclooxygenase 1 and 2 (COX-1 and 2); and leukotriene A_4 (LTA_4) produced via the actions of 5-lipoxygenase (5-LOX), which is then converted by LTA_4 hydrolase to LTB_4, and LTC_4 synthase to LTC_4. These prostaglandins and thromboxane mediate their diverse effects that include either promoting or inhibiting inflammation via binding to nine G-protein-coupled receptors (DP1, DP2, EP1–4, FP, IP, and TP) that differ, for example, in their ability to change levels of intracellular cyclic AMP, mobilize intracellular Ca^{2+}, and induce phosphoinositol turnover [37]. The role of several of these has been studied in SCI. Leukotrienes also mediate their effects via specific receptors, i.e., LTB_4 acts via LTB_4R2 receptor to promote leukocyte adhesion and extravasation from endothelial cells into the tissue parenchyma [81] and promotes recruitment of cytotoxic T cells [32, 89, 90], as well as expression of IL-1, IL-12, and IFN-γ by monocytes and macrophages [14, 81]. The C_4, D_4, and E_4 leukotrienes are cysteinyl LT (cysLTs) that bind to cysLT1 and cysLT2 receptors that mediate allergic and asthma responses such as vascular and bronchial constriction [30]. Little is known about their role in spinal cord injury [15]. The three classes of eicosanoids therefore mediate inflammatory responses by promoting chemotaxis and activation of immune cells, increasing vascular permeability, and having effects on blood flow.

Role of Prostaglandins in SCI Prostaglandins have diverse inflammatory effects, depending on the type of prostaglandin and the receptors they bind, e.g., PGD2 acts on a variety of immune cells including Th2 cells, eosinophils, and basophils and macrophages [40] to mediate allergic responses, while PGE2 acts on T cells in EAE and mediates pain after nerve injury and contributes to pro-inflammatory responses in other CNS conditions such as cerebral ischemia ; [6; 43; 68; 75]. We will focus here mainly on the role of 15d-PGJ2, PGD2, and PGE2 and the influence of the EP1 and EP2 receptors in SCI because of the work we have done.

15-Deoxy-D-12,14-Prostaglandin J2 (15d-PGJ2) 15d-PGJ2, a cyclopentenone-type prostaglandin of the J2 variety, is a metabolite of PGD2 that is generated by a series of nonenzymatic dehydration steps [83]. It does not require a synthase for its production. Unlike other prostaglandins, 15d-PGJ2 has anti-inflammatory effects. A specific receptor is not identified, but it binds to the DP1 and DP2 receptor as well as to intracellular ligands, peroxisome proliferator-activated receptor gamma (PPARγ), which leads to the inhibition of NF-κB and increase in expression of suppressor of cytokine signaling-1 (SOCS1) and SOCS3 [77, 83]. 15d-PGJ2 plays a role in the resolution of PGE2-mediated inflammation [31] and thus could act like resolvins and neuroprotectins (see later section). Our work on spinal cord contusion injury in adult mice showed that daily intraperitoneal injections of 15d-PGJ2 (200 μg/kg body weight) starting immediately after injury resulted in significant improvement in locomotor recovery [53]. Interestingly, a higher dose (1 mg/kg BW) was detrimental, as has also been seen in EAE [26], a mouse model used widely to study multiple sclerosis. 15d-PGJ_2 was also shown to have anti-inflammatory, beneficial effects in an animal model of cerebral ischemia [78].

Treatment with 15d-PGJ2 also resulted in improved sensory recovery based on the von Frey hair test, using monofilaments with a range of bending forces of 0.12–1.4 g. The sensory responsiveness of 15d-PGJ2-treated mice for the entire range of monofilaments returned to normal by 7 days post-injury, while the vehicle-treated mice remained unresponsive [53]. Histological analysis showed the treatment with 15d-PGJ2 improved neuronal survival, as well as increased serotonergic innervation caudal to the lesion, and reduced myelin loss in the injured spinal cord [53]. These protective effects on secondary damage could be mediated via reduction in expression of MCP-1, a pro-inflammatory chemokine, as well as rapid reduction in NF-κB activation and increased expression of SOCS1 that result in reduction of JAK2 activation [53]. This was also accompanied by reduction in microglial activation [53]. There is also other evidence that 15d-PGJ_2 inhibits macrophage activation and reduces expression of the pro-inflammatory cytokines IL-1β and TNF-α [44]. 15d-PGJ2 is therefore likely to act via multiple signaling pathways to reduce inflammation and mediate protective responses after SCI. With additional testing and development, 15d-PGJ_2 could be a promising therapeutic candidate for clinical testing in SCI.

Prostaglandin D2 PGD2 is generated from PGH2 via the actions of two synthases: hematopoietic PGD synthase (HPGDS) and lipocalin-type PGDS (L-PGDS). Expression of HPGDS is found in mast cells [48] and in activated microglia and astrocytes around plaques in Alzheimer's brain [73] and in macrophages and microglia in the injured spinal cord [79]. L-PGDS on the other hand is expressed by leptomeningeal cells and the choroid plexus [57] as well as by oligodendrocytes [94]. L-PGDS is thought to increase PGD2 levels in the cerebrospinal fluid (CSF) [57] that may be responsible for its effect on sleep and sleep-related disorders [38]. PGD2 generated by HPGDS can mediate a variety of responses, such as allergic responses involving Th2, eosinophil, and basophils [40, 55] and inflammatory responses involving activated microglia and astrocytes [73, 74, 79]. There are also contrary reports showing its detrimental effects in cerebral ischemia [63]. These differences may depend on which of the two receptors (DP1 and DP2) are engaged, the mode of induction of neurotoxicity or other as yet unknown factors. The DP1 receptor has been shown to mediate protective responses in various models of cerebral ischemia [3, 82, 92]. Its role in SCI has yet to be studied; however, DP1 receptor is expressed in astrocytes in the spinal cord, and its expression is markedly increased in these glia after SCI (unpublished data).

HPGDS but not L-PGDS is responsible for the increase in PGD2 after spinal cord injury (about a threefold increase measured using a competitive enzyme-linked immunoassay); in contrast, PGD2 levels remain unchanged in the injured spinal cord in HPGDS null mice [79]. We also reported that after SCI, HPGDS expression is increased at the mRNA and protein level and is highly expressed in CD11b + activated macrophages and microglia, while no differences were noted in the expression levels of L-PGDS [79]. HPGDS null mice also showed significant reduction in secondary damage (myelin and neuronal loss), improvement in serotonergic innervation of the ventral horn below the lesion, as well as significant improvement of locomotor recovery after spinal cord contusion injury, as compared to injured wild-type mice [79]. These results were additionally confirmed using a small molecule inhibitor of HPGDS (HQL-79) given daily by subcutaneous injection for 28 days. Small molecule inhibitors of HPGDS are therefore promising candidates for development for clinical testing in SCI.

Prostaglandin E2 PGE2 is a classic inflammatory prostaglandin that accumulates at sites of inflammation, including CNS ischemia [42; 52] and in neurodegenerative disease [6]. The expression of PGE_2 is also significantly increased over the first 72 h after SCI [80]. Like PGD_2, PGE2 has also been shown to mediate both detrimental and protective effects [1; 2; 4; 10; 51; 62; 70; 91]. These contradictory effects are likely to depend on the type of receptors it binds to and the cell types transducing these responses. PGE2 binds to four G-protein-coupled EP receptors (EP1–4) [12]. We will focus here only on the role of EP1 and EP2 receptors in SCI.

EP1 Receptor Evidence from studies on *EP1* receptor gene knockout mice and pharmacological inhibition of the EP1 receptor show that this receptor mediates neurotoxic effects after cerebral ischemia, NMDA-mediated excitotoxicity, and oxygen-glucose deprivation [2, 51]. The neurotoxic effects are thought to be primarily mediated by increased intracellular Ca^{2+}.

Our work showing a beneficial role of the EP1 receptor in SCI came from a rather long and circuitous route on a broader analysis of the role of various members of the PLA2 family in SCI. As mentioned above, the only PLA_2s that showed increased mRNA expression after spinal cord contusion injury and therefore considered important in injury responses were $cPLA_2$ GIVA, $iPLA_2$ GVIA, and $sPLA_2$ GIIA [67]. Using selective and highly potent small molecule inhibitors to block each of these three groups of PLA_2s, we found that $cPLA_2$ GIVA is protective, while $sPLA_2$ GIIA and to a much lesser extent $iPLA_2$ GVIA are detrimental to locomotor recovery and induce secondary tissue damage [67]. The protective effects of the $cPLA_2$ were very surprising given its pro-inflammatory role and detrimental effects in other CNS conditions such as EAE and cerebral ischemia [11; 47; 88]. We therefore further confirmed this by carrying out SCI in $cPLA_2$ null mice. These mice also showed evidence of increased secondary tissue damage and lack of locomotor recovery [67]. Interestingly, our studies also showed that a small molecule pan-PLA_2 inhibitor (AX115) that partially blocks all three forms of PLA_2 between 45 and 62% was the most effective in promoting locomotor recovery [67]. Surprisingly, treatment with AX115 resulted in upregulation of expression of cPLA2 GIVA but not the other two PLA_2s, as well as increased expression of COX-2 (not 5-LOX), mPGES-1 (not HPGDS) after SCI, suggesting increased expression of PGE2. This was accompanied by increase in expression of the EP1 receptor (not EP2 or EP4). Furthermore, the improvement in locomotor recovery induced by AX115 treatment after SCI was blocked completely when the EP1 receptor was pharmacologically inhibited (SC51089) [67]. Our work therefore suggests that under these conditions, signaling via the EP1 receptor promotes recovery after SCI. Why this bypasses the pro-inflammatory effects via EP1 is still not known. In addition, what underlies the upregulation of only $cPLA_2$ by the pan-PLA_2 inhibitor AX115 is also not known. Compounds like AX115 can therefore be very good therapeutic candidates for treating SCI but need further work.

EP2 Receptor EP2 receptor activation has neurotoxic effects in animal models of Parkinson's disease, Alzheimer's disease, and ALS [6, 45, 58,

59]. On the other hand, EP2 receptor activation promotes neuroprotection in in vitro models of NMDA toxicity and oxygen-glucose deprivation [70, 91]. The protective effects of EP2 receptor signaling were also shown in vivo in cerebral ischemia using *EP2*$^{-/-}$ mice, in which knockout mice showed larger infarct size [62, 70].

Recent work in our group revealed a novel communication between peripheral macrophages and central microglia that reduce microglial activation in vitro and in vivo. A fundamental question in many CNS pathologies is why monocyte-derived macrophages (MDMs) from the circulation enter the CNS after injury or disease, when there are already resident tissue macrophages (microglia) present that can carry out similar functions. In fact, after CNS injury, microglia are the first responders as MDMs only enter the CNS after a delay of 2–3 days [33]. In the first 2 days after spinal cord injury, microglia appear aligned along degenerating axons and show evidence of phagocytosis before the entry of MDMs. However, once MDMs enter in the injury site, they are associated with the degenerating axons, and contacts between microglia and damaged axons decrease [33]. It appears as if the infiltrating MDMs signal microglia to stop phagocytosing. Morphological evidence suggest that these cells are in close contact with each other and in a position to communicate with each other. To address this, we moved to a simpler cell culture system using purified adult mouse microglia and bone marrow-derived macrophages (BMDMs). Adult microglia were plated into culture wells and the BMDMs on to round glass coverslips. After the cultures were established, the two cell types were co-incubated by placing the BMDM containing coverslips into the wells containing microglia. The cells were prevented from touching each other by means of small paraffin spacers. This allowed the cell layers to be in proximity without touching each other. After an overnight incubation, the coverslips were removed, and the ability of the two cell types to phagocytose myelin was assessed using pH-rhodo-tagged myelin and flow cytometry. Surprisingly, co-incubation with macrophages (BMDMs) resulted in suppression of phagocytosis by microglia; and interestingly, under the same conditions, microglia induced an increase in phagocytosis by macrophages (BMDMs) [34]. Under conditions of inflammatory stimulation with LPS, Affymetrix analysis showed that BMDMs induced profound changes in microglial gene expression. 1076 genes were significantly differentially regulated in activated microglia when co-cultured with macrophages (BMDMs). These included reduction in expression of genes in the NF-κB signaling pathway and dysregulation of pathways involved in apoptosis and cell death. Network analysis revealed a cluster of 185 genes that were downregulated in microglia in the presence of BMDMs, which included those that regulate TNF, MyD88, and IL-1β [34]. This suggests that peripheral macrophages may play a role in reducing microglial activation and phagocytic activity. These in vitro results were confirmed in vivo by doing SCI in CCR2 null mice. In these mice, circulating macrophages fail to enter the lesioned spinal cord, and locomotor recovery is impaired, and secondary damage and microglial activation are greater than in SCI in wild-type mice. These in vitro and in vivo experiments suggest that infiltrating macrophages communicate with microglia to suppress microglial activation. Further, in vitro and in vivo work indicates that the macrophage-mediated suppression of microglial activation is mediated by PGE2 acting via EP2 receptors on microglia. In co-cultures in which microglia and BMDMs are incubated together as described above, treating these cultures with an EP2 antagonist prevents macrophage-mediated suppression of myelin phagocytosis by microglia. This effect was also seen when macrophages (BMDMs) from mPGES null mice were used. BMDMs from mPGES null mice are unable to express PGE2. Furthermore, simply treating microglia cultures with an EP2 agonist (Butaprost) suppresses phagocytosis of myelin by microglia, similar to the effects of co-culture with BMDMs. We further confirmed this effect in vivo by injecting pH-rhodo-tagged myelin into the corpus callosum of wild-type mice together with vehicle or EP2 antagonist (PF-04418948). In mice injected with the EP2 antagonist, the area injected with myelin was abundant in activated microglia that were phagocytosing myelin. In contrast, in control mice injected with myelin and vehicle, the area was full of peripheral macrophages from the circulation that were phagocytosing myelin [34]. These findings provide strong evidence that macrophages from the circulation entering sites of CNS damage suppress resident microglia and take over the role of phagocytosing tissue debris and clearing the area of damaged cells to prepare the way for repair. This would appear to make sense, since macrophages from the circulation are the "professional" phagocytes, while microglia are the tissue resident cells that perform key surveillance functions in the CNS that should not be compromised. Their ability to perform surveillance functions would be compromised when microglia retract their cellular processes when activated. This form of cell-cell interactions could also be a way to switch off microglial activation and thus prevent chronic microglia-mediated CNS inflammation. Our work indicates a novel role for PGE2 and the EP2 receptor in such macrophage-microglia interactions. Further work is needed to develop ways in which these interactions can be stimulated locally in neurological conditions to prevent microglia-mediated chronic CNS inflammation.

12.1.2 Role of Specialized Pre-resolving Mediators in the Resolution of Inflammation

Inflammation is a normal physiological response to injury or disease, the purpose of which is to eliminate pathogens and clear the tissue of cell debris. Inflammation is also critical in preparing the tissue for repair and recovery of homeostasis. Acute inflammation, however, must be resolved properly and in a timely fashion; otherwise, it can develop into chronic inflammatory, which in the CNS can be very detrimental [85, 86].

"Resolution of inflammation" is defined as the phase in which immune cells are cleared from the tissue [85, 86]. This is a very tightly coordinated process that involves several steps:

1. Limitation or cessation of inflammatory cell recruitment
2. The counter-regulation of pro-inflammatory factors, such as TNF-α, IL-1β, or iNOS, and by the suppression of the inflammatory pathways, such as NF-κB, JAK/STAT, and MAPK, among others
3. Induction of apoptosis of inflammatory cells and their phagocytosis by macrophages (efferocytosis)
4. Stimulation of macrophage efflux from the tissue, mainly, through lymphatics
5. Promotion of tissue repair without scarring or fibrosis
6. Return to tissue homeostasis, which marks the final step of resolution

Historically, the resolution of inflammation was believed to be a passive process triggered by the dilution of pro-inflammatory mediators at the injury site that would impede the recruitment of leukocytes from the circulation. However, it is now known that resolution is an active and coordinated event controlled by several local mediators known as "specialized pro-resolving mediators" (SPMs) [13, 84, 86].

SPMs are bioactive lipids synthesized during the acute phase of the inflammatory response [13, 84, 86]. These bioactive lipids have been identified in self-limited inflammatory exudates in animal models by using liquid chromatography tandem mass spectrometry (LC-MS/MS)-based analysis. SPMs are produced from omega-6 (AA) and omega-3 fatty acids (DHA and EPA), which are released from phospholipids by the actions of PLA_2 enzymes [19, 22, 50, 85]. These fatty acids are then further catalyzed by the action of LOX enzymes (5-LOX, 12-LOX, and 15-LOX) to produce at least four different SPM families: lipoxins, resolvins, neuroprotectins, and maresins.

Lipoxins (LX) are produced from AA, like the pro-inflammatory leukotriene (LT) LTB4. At present, two lipoxins have been identified: lipoxin A4 (LXA4) and lipoxin B4 (LXB4) [20, 85]. LOX can therefore produce LTB4 and LX from

AA. The cellular localization of LOX, especially 5-LOX, seems to be critical to determine which product, LTB or LX, is generated. 5-LOX action generates LTA4 from AA. Subsequently, LTA4 can then be further catalyzed by LTA4 hydrolase to produce LTB4 or by 12-LOX to produce LX. LTA4 hydrolase is found in the nuclear membrane, and therefore, nuclear localization of LOX-5 is likely to favor the production of LTB4, while the presence of LOX-5 in the cytoplasm is likely to promote the formation of LX, since LOX-12 and LOX-15 are also found in the cytoplasm [50, 97, 98].

The other three families of SPMs are produced from omega-3 fatty acids. E-series resolvins (RvE), which encompass RvE1, RvE2, and RvE3, are generated from EPA, while DHA is the metabolic precursor for the synthesis of D-series resolvins (RvD1-RvD6), protectin D1 (PD1), also known as neuroprotectin D1 (NPD1) when produced in neuronal tissues, and maresins (MaR1 and MaR2) [20, 86]. Apart from LOX enzymes, SPMs can also be produced from the action of COX-2 in the presence of aspirin [20, 86]. This is due to the ability of aspirin to modify the activity profile of COX-2, switching from cyclooxygenase to lipoxygenase activity, leading to the formation of aspirin-triggered SPMs (AT-SPMs, i.e., ET-RvD1). This effect appears to be aspirin-specific, since it is not shared by other nonsteroidal anti-inflammatory drugs (NSAIDs).

SPMs mediate resolution functions by acting on different G-protein-coupled receptors (GPCRs). Understanding the specificity of the receptor-SPM ligand interactions and the cellular distribution of these receptors is key to understanding how inflammation is resolved and key to developing therapeutic strategies. LXA4 signals via ALX/FPR2 [16, 85], although the receptor for LXB4 has not been identified yet. RvE1 and RvE2 mediate their actions mainly by binding to ChemR23, also known as CMKLR1 [7, 16]. However, RvE1 can also act as partial agonist of LTB4 receptor BLT-1 to attenuate LTB4 signaling [7, 16] which most likely explains their ability to potently regulate PMN trafficking to sites of inflammation [86]. RvE1 can also block the TRP family of nociceptors on peripheral sensory neurons to mute inflammatory pain [99]. RvE2 also shares, at least in part, receptors with RvE1 [76]. RvD1, like LXA4, acts through the ALX/FPR2. RvD1 is an agonist for GPR32 receptor [56], which is present in humans but not in mice. RvD2 acts through the GPR18 receptor [56], while RvD3, AT-RvD3, and RvD5 [17, 56] [18] that appear later in the resolution phase also bind to human GPR32 receptor [21]. PD1/NPD1 has been recently reported to signal via GPR37, which is highly expressed in the brain and spinal cord mainly in oligodendrocytes. Interestingly, GPR37 is found in peripheral monocyte-derived macrophages but not in microglia, at least in physiological conditions. Finally, the receptor for maresins has not been identified yet, although a recent work suggests that ALX/FPR2 can be a potential receptor candidate for MaR1 [35].

There are currently very few studies addressing the role of SPMs in CNS injuries. We have recently reported that there is a defective lipid mediator class switch, from pro-inflammatory to pro-resolution lipids, after SCI in mice [28]. In particular, we reported that the levels of the pathway marker for the formation of LXA4, NP1, RvD series, and MaR1 did not increased in the spinal cord until day 14 after injury, while the pathway marker for the formation of RvE series was undetectable. In contrast, the levels of PGE_2 were elevated in the contused spinal cord up to day 7 post-injury [28]. Failure to produce adequate amounts of SPMs has been associated to persistent inflammation in many inflammatory disorders such as asthma, atherosclerosis, and ulcerative colitis, and the administration of SPMs in such conditions reduced inflammation and ameliorated disease symptomatology [85]. This suggests that approaches aimed at increasing SPM levels in the injured spinal cord could mitigate inflammation and be of therapeutic potential.

The first evidence suggesting the helpful action of SPMs in neurotrauma comes from studies demonstrating the beneficial effects of omega-3 PUFA. In a model of spinal cord hemisection, a single intravenous administration of a very low dose of DHA (250 nmol/kg) conferred protection against neuronal and oligodendrocyte loss and promoted functional recovery [54]. Later studies also demonstrated the protective actions of DHA and EPA in more clinically relevant models of SCI, especially, when single injection of DHA was combined with DHA-enriched diet (400 mg/kg/day) for 6 weeks [41]. Similar findings have been also observed with the administration of EPA in SCI [60]. The therapeutic efficacy of omega-3 PUFA was further supported by experiments using mice that express the *fat-1* gene encoding for omega 3 fatty acid desaturase from *Caenorhabditis elegans*, which leads to an increase in endogenous omega-3 PUFAs. These mice showed improved neurological recovery and tissue sparing after SCI [61]. The beneficial actions of omega-3 PUFA have been also reported after traumatic brain injury [8, 72]. Contrary to the therapeutic potential of omega-3 fatty acids on SCI, administration of AA (omega-6 PUFA) worsened functional and histopathological outcomes [54]. This is important, as the levels of AA are increased in the CNS and plasma after SCI, while the levels of DHA are reduced, suggesting that this switch in PUFA availability may contribute to the physiopathology of CNS injury [66]. Indeed, the administration of fenretinide, a semi-synthetic analog of retinoic acid, normalized the ratio of DHA/AA in both, plasma and spinal cord, after SCI and led to improved locomotor recovery [66].

Several reports suggest that the neuroprotective actions of omega-3 PUFA in neurotrauma are due to their ability to reduce oxidative stress, apoptosis, and inflammation, modulate ion channel functions, and act as a ligand for the retinoid X receptor [66, 71]. However, the beneficial effects of omega-3 PUFA in SCI are also likely to be mediated by the generation of SPMs.

We recently obtained clear evidence, for the first time, that support the concept that the inappropriate biosynthesis of SPMs after SCI hampers resolution of inflammation and contributes to the pathology and poor functional outcome after SCI [28]. We demonstrated that daily intravenous administration of MaR1 (1 μg) starting at 1 h after SCI for 7 days resulted in significant locomotor recovery and reduced tissue damage in mice [28]. Importantly, we also uncovered that administration of MaR1 fosters different steps involved in this process. First, we found that MaR1 accelerated and enhanced neutrophil clearance in the lesioned spinal cord and also reduced the accumulation of macrophages [28]. Second, MaR1 reduced the protein levels of pro-inflammatory cytokines in the contused cord and turned off several JAK/STAT and MAPK intracellular signaling, but did not abrogate NF-κB activation [28]. Third, MaR1 also drove macrophages toward a more anti-inflammatory phenotype and stimulated the phagocytosis of neutrophils [28] .

Aberrant production of SPMs has also been observed in the cerebrospinal fluid (CSF) of individuals suffering from Alzheimer's disease (AD) [96]. Moreover, LXA4 levels in postmortem samples of hippocampal tissue from AD patients were found to be significantly lower than in age-matched non-AD subjects. Interestingly, the decrease in LXA4 levels correlated with the degree of cognitive deficits and accumulation of tau protein. In line with this, administration of LXA_4 reduced inflammation, amyloid plaques formation, and tau phosphorylation and improved the cognitive performance in the 3xTg-AD mouse model [27]. Similar beneficial results have also been reported after the administration of LXA4 and RvE_1 alone in another model of AD (5xFAD mice), and that the combination of both SPMs had a more potent effect [49]. Several studies have also shown that LXA4 and NP1 reduce inflammation and tissue damage in animal models of stroke [9, 69, 87]. Altered levels of SPMs have been detected in the CSF of patients with multiple sclerosis. However, to the best of our knowledge, the role of SPM in this neuroinflammatory condition has not been studied.

12.2 Concluding Remarks

For decades, lipids were considered mere constituents of cellular membranes and efficient energy sources. We know now that lipids are also key molecules that act as intracellular and extracellular messengers that are implicated in the control of a broad spectrum of physiological processes. Our understanding of the mechanisms underlying secondary damage after neurotrauma has made major strides recently and has led to the identification of bioactive lipids generated by PLA_2s as crucial regulators of the inflammatory response. During inflammation, various families of lipids are temporally and spatially synthesized so that inflammation is efficiently initiated, coordinated, and terminated. Classically, eicosanoids have been considered as potent initiators of inflammation since they mediate events that promote the recruitment of immune cells to the injury site and the efficient removal of cell debris, pathogens, and dead cells. When leukocytes successfully execute their functions, SPMs are synthesized in order to resolve inflammation and restore tissue homeostasis. Our recent studies have also revealed some unexpected anti-inflammatory roles of the EP1 and EP2 receptor in SCI. The latter in particular were found to play a role in signaling between infiltrating macrophages and resident microglia that leads to suppression of microglial activation and thus microglia-mediated CNS inflammation. More work is needed to see how such interactions can be fostered therapeutically to reduce inflammation after CNS injury.

Recent evidence suggests that injury to the CNS results in an imbalance in the production of eicosanoids versus SPMs. Although we do not fully understand yet why injury-induced inflammation persists over time in CNS tissue, such aberrant switch in the synthesis of these bioactive lipids might be one of the main factors. One can consider the direct approach of administering drugs to inhibit eicosanoid production, such as COX-2 or LOX inhibitors. However, eicosanoids also mediate potent anti-inflammatory actions depending on the receptors they signal through or the cell type involved. Moreover, COX-2 and LOX are also the enzymes involved in the synthesis of SPMs. Therefore, the use of inhibitors of drugs that target the production of eicosanoid may initially dampen inflammation; however, they may also block their anti-inflammatory actions and exacerbate the deficit in the production of SPMs at later stages of neurotrauma. Another alternative is to treat with omega-3 PUFA or drugs that normalize the ratio of AA/SPMs as discussed above to restore this lipid imbalance. These approaches have demonstrated to contain inflammation and tissue damage after CNS injuries. However, the limited production of SPMs in the injured CNS is unlikely to be mediated by the restricted bioavailability of omega-3 PUFA but to improper induction of LOX enzymes, since DHA is one of the most abundant lipids in the CNS.

We therefore propose that the resolution of inflammation after CNS injuries can be promoted by the exogenous administration of drugs that potentiate the anti-inflammatory action of some eicosanoid (eicosanoid receptor agonist or antagonist) and also the delivery of SPMs, rather than simply inhibiting eicosanoid production or increasing omega-3 PUFA levels. Although very few studies have addressed the efficacy of some of these agonist/antagonist of eicosanoid receptors and SPMs after neurotrauma, recent studies highlight their therapeutic potential. Further work is needed to understand of how imbalance between harmful and beneficial lipids is produced in CNS injuries and how anti-inflammatory eicosanoid and SPM mediate their actions on immune cells to contain and resolve inflammation.

Acknowledgments This work has been supported by grants from the Spanish Ministry of Economy and Competitiveness (SAF2016-79774-R), International Foundation for Research in Paraplegia, and Wings for Life Foundation and by funds from the Fondo de Investigación Sanitaria of Spain (TERCEL and CIBERNED) to RLV and from the Canadian Institutes of Health Research (MOP 14828) and the Multiple Sclerosis Society of Canada to SD.

References

1. Ahmad AS, Ahmad M, de Brum-Fernandes AJ, Dore S (2005) Prostaglandin EP4 receptor agonist protects against acute neurotoxicity. Brain Res 1066:71–77
2. Ahmad AS, Saleem S, Ahmad M, Dore S (2006a) Prostaglandin EP1 receptor contributes to excitotoxicity and focal ischemic brain damage. Toxicol Sci 89:265–270
3. Ahmad AS, Ahmad M, Maruyama T, Narumiya S, Dore S (2010) Prostaglandin D2 DP1 receptor is

beneficial in ischemic stroke and in acute excitotoxicity in young and old mice. Age (Dordr) 32:271–282

4. Ahmad M, Saleem S, Zhuang H, Ahmad AS, Echeverria V, Sapirstein A, Dore S (2006b) 1-hydroxyPGE reduces infarction volume in mouse transient cerebral ischemia. Eur J Neurosci 23:35–42
5. Ahuja CS, Wilson JR, Nori S, Kotter MRN, Druschel C, Curt A, Fehlings MG (2017) Traumatic spinal cord injury. Nat Rev Dis Primers 3:17018
6. Andreasson K (2010) Emerging roles of PGE2 receptors in models of neurological disease. Prostaglandins Other Lipid Mediat 91:104–112
7. Arita M, Ohira T, Sun YP, Elangovan S, Chiang N, Serhan CN (2007) Resolvin E1 selectively interacts with leukotriene B4 receptor BLT1 and ChemR23 to regulate inflammation. J Immunol 178:3912–3917
8. Bailes JE, Mills JD (2010) Docosahexaenoic acid reduces traumatic axonal injury in a rodent head injury model. J Neurotrauma 27:1617–1624
9. Bazan NG, Eady TN, Khoutorova L, Atkins KD, Hong S, Lu Y, Zhang C, Jun B, Obenaus A, Fredman G, Zhu M, Winkler JW, Petasis NA, Serhan CN, Belayev L (2012) Novel aspirin-triggered neuroprotectin D1 attenuates cerebral ischemic injury after experimental stroke. Exp Neurol 236:122–130
10. Bilak M, Wu L, Wang Q, Haughey N, Conant K, St Hillaire C, Andreasson K (2004) PGE2 receptors rescue motor neurons in a model of amyotrophic lateral sclerosis. Ann Neurol 56:240–248
11. Bonventre JV, Huang Z, Taheri MR, O'Leary E, Li E, Moskowitz MA, Sapirstein A (1997) Reduced fertility and postischaemic brain injury in mice deficient in cytosolic phospholipase A2. Nature 390:622–625
12. Breyer RM, Bagdassarian CK, Myers SA, Breyer MD (2001) Prostanoid receptors: subtypes and signaling. Annu Rev Pharmacol Toxicol 41:661–690
13. Buckley CD, Gilroy DW, Serhan CN (2014) Proresolving lipid mediators and mechanisms in the resolution of acute inflammation. Immunity 40:315–327
14. Caughey GE, Pouliot M, Cleland LG, James MJ (1997) Regulation of tumor necrosis factor-alpha and IL-1 beta synthesis by thromboxane A2 in nonadherent human monocytes. J Immunol 158:351–358
15. Chen XG, Hua F, Wang SG, Xu YY, Yue HT, Sun J (2018) Zafirlukast in combination with pseudohypericin attenuates spinal cord injury and motor function in experimental mice. Drug Des Devel Ther 12:2389–2402
16. Chiang N, Serhan CN (2017) Structural elucidation and physiologic functions of specialized proresolving mediators and their receptors. Mol Asp Med 58:114–129
17. Chiang N, Dalli J, Colas RA, Serhan CN (2015) Identification of resolvin D2 receptor mediating resolution of infections and organ protection. J Exp Med 212:1203–1217
18. Chiang N, Fredman G, Backhed F, Oh SF, Vickery T, Schmidt BA, Serhan CN (2012) Infection regulates pro-resolving mediators that lower antibiotic requirements. Nature 484:524–528
19. Chiurchiu V, Leuti A, Maccarrone M (2018) Bioactive lipids and chronic inflammation: managing the fire within. Front Immunol 9:38
20. Dalli J, Serhan CN (2018) Immunoresolvents signaling molecules at intersection between the brain and immune system. Curr Opin Immunol 50:48–54
21. Dalli J, Winkler JW, Colas RA, Arnardottir H, Cheng CY, Chiang N, Petasis NA, Serhan CN (2013) Resolvin D3 and aspirin-triggered resolvin D3 are potent immunoresolvents. Chem Biol 20:188–201
22. David S, Greenhalgh AD, Lopez-Vales R (2012a) Role of phospholipase A2s and lipid mediators in secondary damage after spinal cord injury. Cell Tissue Res 349:249–267
23. David S, Lopez-Vales R, Wee Yong V (2012b) Harmful and beneficial effects of inflammation after spinal cord injury: potential therapeutic implications. Handb Clin Neurol 109:485–502
24. David S, Kroner A, Greenhalgh AD, Zarruk JG, Lopez-Vales R (2018) Myeloid cell responses after spinal cord injury. J Neuroimmunol 321:97–108
25. Dennis EA, Norris PC (2015) Eicosanoid storm in infection and inflammation. Nat Rev Immunol 15:511–523
26. Diab A, Deng C, Smith JD, Hussain RZ, Phanavanh B, Lovett-Racke AE, Drew PD, Racke MK (2002) Peroxisome proliferator-activated receptor-gamma agonist 15-deoxy-Delta(12,14)-prostaglandin J(2) ameliorates experimental autoimmune encephalomyelitis. J Immunol 168:2508–2515
27. Dunn HC, Ager RR, Baglietto-Vargas D, Cheng D, Kitazawa M, Cribbs DH, Medeiros R (2015) Restoration of lipoxin A4 signaling reduces Alzheimer's disease-like pathology in the 3xTg-AD mouse model. J Alzheimers Dis 43:893–903
28. Francos-Quijorna I, Santos-Nogueira E, Gronert K, Sullivan AB, Kopp MA, Brommer B, David S, Schwab JM, Lopez-Vales R (2017) Maresin 1 promotes inflammatory resolution, neuroprotection, and functional neurological recovery after spinal cord injury. J Neurosci 37:11731–11743
29. Frisen J (2016) Neurogenesis and gliogenesis in nervous system plasticity and repair. Annu Rev Cell Dev Biol 32:127–141
30. Funk CD, Chen XS, Johnson EN, Zhao L (2002) Lipoxygenase genes and their targeted disruption. Prostaglandins Other Lipid Mediat 68-69:303–312
31. Gilroy DW, Colville-Nash PR, Willis D, Chivers J, Paul-Clark MJ, Willoughby DA (1999) Inducible cyclooxygenase may have anti-inflammatory properties. Nat Med 5:698–701
32. Goodarzi K, Goodarzi M, Tager AM, Luster AD, von Andrian UH (2003) Leukotriene B4 and BLT1 control cytotoxic effector T cell recruitment to inflamed tissues. Nat Immunol 4:965–973
33. Greenhalgh AD, David S (2014) Differences in the phagocytic response of microglia and peripheral

macrophages after spinal cord injury and its effects on cell death. J Neurosci 34:6316–6322
34. Greenhalgh AD, Zarruk JG, Healy LM, Baskar Jesudasan SJ, Jhelum P, Salmon CK, Formanek A, Russo MV, Antel JP, McGavern DB, McColl BW, David S (2018) Peripherally derived macrophages modulate microglial function to reduce inflammation after CNS injury. PLoS Biol 16:e2005264
35. Gu J, Luo L, Wang Q, Yan S, Lin J, Li D, Cao B, Mei H, Ying B, Bin L, Smith FG, Jin SW (2018) Maresin 1 attenuates mitochondrial dysfunction through the ALX/cAMP/ROS pathway in the cecal ligation and puncture mouse model and sepsis patients. Lab Investig 98:715–733
36. Hapak SM, Ghosh S, Rothlin CV (2018) Axon regeneration: antagonistic signaling pairs in neuronal polarization. Trends Mol Med 24:615–629
37. Hata AN, Breyer RM (2004) Pharmacology and signaling of prostaglandin receptors: multiple roles in inflammation and immune modulation. Pharmacol Ther 103:147–166
38. Hayaishi O, Urade Y (2002) Prostaglandin D2 in sleep-wake regulation: recent progress and perspectives. Neuroscientist 8:12–15
39. Hilton BJ, Moulson AJ, Tetzlaff W (2017) Neuroprotection and secondary damage following spinal cord injury: concepts and methods. Neurosci Lett 652:3–10
40. Hirai H, Tanaka K, Yoshie O, Ogawa K, Kenmotsu K, Takamori Y, Ichimasa M, Sugamura K, Nakamura M, Takano S, Nagata K (2001) Prostaglandin D2 selectively induces chemotaxis in T helper type 2 cells, eosinophils, and basophils via seven-transmembrane receptor CRTH2. J Exp Med 193:255–261
41. Huang WL, King VR, Curran OE, Dyall SC, Ward RE, Lal N, Priestley JV, Michael-Titus AT (2007) A combination of intravenous and dietary docosahexaenoic acid significantly improves outcome after spinal cord injury. Brain J Neurol 130:3004–3019
42. Iadecola C, Niwa K, Nogawa S, Zhao X, Nagayama M, Araki E, Morham S, Ross ME (2001) Reduced susceptibility to ischemic brain injury and N-methyl-D-aspartate-mediated neurotoxicity in cyclooxygenase-2-deficient mice. Proc Natl Acad Sci U S A 98:1294–1299
43. Ikeda-Matsuo Y, Ota A, Fukada T, Uematsu S, Akira S, Sasaki Y (2006) Microsomal prostaglandin E synthase-1 is a critical factor of stroke-reperfusion injury. Proc Natl Acad Sci U S A 103:11790–11795
44. Jiang C, Ting AT, Seed B (1998) PPAR-gamma agonists inhibit production of monocyte inflammatory cytokines. Nature 391:82–86
45. Jin J, Shie FS, Liu J, Wang Y, Davis J, Schantz AM, Montine KS, Montine TJ, Zhang J (2007) Prostaglandin E2 receptor subtype 2 (EP2) regulates microglial activation and associated neurotoxicity induced by aggregated alpha-synuclein. J Neuroinflammation 4:2
46. Kalyvas A, David S (2004) Cytosolic phospholipase A2 plays a key role in the pathogenesis of multiple sclerosis-like disease. Neuron 41:323–335
47. Kalyvas A, Baskakis C, Magrioti V, Constantinou-Kokotou V, Stephens D, Lopez-Vales R, Lu JQ, Yong VW, Dennis EA, Kokotos G, David S (2009) Differing roles for members of the phospholipase A2 superfamily in experimental autoimmune encephalomyelitis. Brain 132:1221–1235
48. Kanaoka Y, Fujimori K, Kikuno R, Sakaguchi Y, Urade Y, Hayaishi O (2000) Structure and chromosomal localization of human and mouse genes for hematopoietic prostaglandin D synthase. Conservation of the ancestral genomic structure of sigma-class glutathione S-transferase. Eur J Biochem 267:3315–3322
49. Kantarci A, Aytan N, Palaska I, Stephens D, Crabtree L, Benincasa C, Jenkins BG, Carreras I, Dedeoglu A (2018) Combined administration of resolvin E1 and lipoxin A4 resolves inflammation in a murine model of Alzheimer's disease. Exp Neurol 300:111–120
50. Kasikara C, Doran AC, Cai B, Tabas I (2018) The role of non-resolving inflammation in atherosclerosis. J Clin Invest 128:2713–2723
51. Kawano T, Anrather J, Zhou P, Park L, Wang G, Frys KA, Kunz A, Cho S, Orio M, Iadecola C (2006) Prostaglandin E2 EP1 receptors: downstream effectors of COX-2 neurotoxicity. Nat Med 12:225–229
52. Kempski O, Shohami E, von Lubitz D, Hallenbeck JM, Feuerstein G (1987) Postischemic production of eicosanoids in gerbil brain. Stroke 18:111–119
53. Kerr BJ, Girolami EI, Ghasemlou N, Jeong SY, David S (2008) The protective effects of 15-deoxy-delta-(12,14)-prostaglandin J2 in spinal cord injury. Glia 56:436–448
54. King VR, Huang WL, Dyall SC, Curran OE, Priestley JV, Michael-Titus AT (2006) Omega-3 fatty acids improve recovery, whereas omega-6 fatty acids worsen outcome, after spinal cord injury in the adult rat. J Neurosci 26:4672–4680
55. Kostenis E, Ulven T (2006) Emerging roles of DP and CRTH2 in allergic inflammation. Trends Mol Med 12:148–158
56. Krishnamoorthy S, Recchiuti A, Chiang N, Yacoubian S, Lee CH, Yang R, Petasis NA, Serhan CN (2010) Resolvin D1 binds human phagocytes with evidence for proresolving receptors. Proc Natl Acad Sci U S A 107:1660–1665
57. Kunz T, Marklund N, Hillered L, Oliw EH (2002) Cyclooxygenase-2, prostaglandin synthases, and prostaglandin H2 metabolism in traumatic brain injury in the rat. J Neurotrauma 19:1051–1064
58. Liang X, Wang Q, Hand T, Wu L, Breyer RM, Montine TJ, Andreasson K (2005) Deletion of the prostaglandin E2 EP2 receptor reduces oxidative damage and amyloid burden in a model of Alzheimer's disease. J Neurosci 25:10180–10187
59. Liang X, Wang Q, Shi J, Lokteva L, Breyer RM, Montine TJ, Andreasson K (2008) The prostaglandin E2 EP2 receptor accelerates disease progression

and inflammation in a model of amyotrophic lateral sclerosis. Ann Neurol 64:304–314
60. Lim SN, Huang W, Hall JC, Ward RE, Priestley JV, Michael-Titus AT (2010) The acute administration of eicosapentaenoic acid is neuroprotective after spinal cord compression injury in rats. Prostaglandins Leukot Essent Fat Acids 83:193–201
61. Lim SN, Gladman SJ, Dyall SC, Patel U, Virani N, Kang JX, Priestley JV, Michael-Titus AT (2013) Transgenic mice with high endogenous omega-3 fatty acids are protected from spinal cord injury. Neurobiol Dis 51:104–112
62. Liu D, Wu L, Breyer R, Mattson MP, Andreasson K (2005) Neuroprotection by the PGE2 EP2 receptor in permanent focal cerebral ischemia. Ann Neurol 57:758–761
63. Liu M, Eguchi N, Yamasaki Y, Urade Y, Hattori N, Urabe T (2009) Protective role of hematopoietic prostaglandin D synthase in transient focal cerebral ischemia in mice. Neuroscience 163:296–307
64. Liu NK, Zhang YP, Titsworth WL, Jiang X, Han S, Lu PH, Shields CB, Xu XM (2006) A novel role of phospholipase A2 in mediating spinal cord secondary injury. Ann Neurol 59:606–619
65. Lopez-Vales R, Navarro X, Shimizu T, Baskakis C, Kokotos G, Constantinou-Kokotou V, Stephens D, Dennis EA, David S (2008) Intracellular phospholipase A(2) group IVA and group VIA play important roles in Wallerian degeneration and axon regeneration after peripheral nerve injury. Brain 131:2620–2631
66. Lopez-Vales R, Redensek A, Skinner TA, Rathore KI, Ghasemlou N, Wojewodka G, DeSanctis J, Radzioch D, David S (2011a) Fenretinide promotes functional recovery and tissue protection after spinal cord contusion injury in mice. J Neurosci 30:3220–3226
67. Lopez-Vales R, Ghasemlou N, Redensek A, Kerr BJ, Barbayianni E, Antonopoulou G, Baskakis C, Rathore KI, Constantinou-Kokotou V, Stephens D, Shimizu T, Dennis EA, Kokotos G, David S (2011b) Phospholipase A2 superfamily members play divergent roles after spinal cord injury. FASEB J 25:4240–4252
68. Ma W, Quirion R (2008) Does COX2-dependent PGE2 play a role in neuropathic pain? Neurosci Lett 437:165–169
69. Marcheselli VL, Hong S, Lukiw WJ, Tian XH, Gronert K, Musto A, Hardy M, Gimenez JM, Chiang N, Serhan CN, Bazan NG (2003) Novel docosanoids inhibit brain ischemia-reperfusion-mediated leukocyte infiltration and pro-inflammatory gene expression. J Biol Chem 278:43807–43817
70. McCullough L, Wu L, Haughey N, Liang X, Hand T, Wang Q, Breyer RM, Andreasson K (2004) Neuroprotective function of the PGE2 EP2 receptor in cerebral ischemia. J Neurosci 24:257–268
71. Michael-Titus AT, Priestley JV (2014) Omega-3 fatty acids and traumatic neurological injury: from neuroprotection to neuroplasticity? Trends Neurosci 37:30–38
72. Mills JD, Bailes JE, Sedney CL, Hutchins H, Sears B (2011) Omega-3 fatty acid supplementation and reduction of traumatic axonal injury in a rodent head injury model. J Neurosurg 114:77–84
73. Mohri I, Kadoyama K, Kanekiyo T, Sato Y, Kagitani-Shimono K, Saito Y, Suzuki K, Kudo T, Takeda M, Urade Y, Murayama S, Taniike M (2007) Hematopoietic prostaglandin D synthase and DP1 receptor are selectively upregulated in microglia and astrocytes within senile plaques from human patients and in a mouse model of Alzheimer disease. J Neuropathol Exp Neurol 66:469–480
74. Mohri I, Taniike M, Taniguchi H, Kanekiyo T, Aritake K, Inui T, Fukumoto N, Eguchi N, Kushi A, Sasai H, Kanaoka Y, Ozono K, Narumiya S, Suzuki K, Urade Y (2006) Prostaglandin D2-mediated microglia/astrocyte interaction enhances astrogliosis and demyelination in twitcher. J Neurosci 26:4383–4393
75. Murakami M, Nakatani Y, Tanioka T, Kudo I (2002) Prostaglandin E synthase. Prostaglandins Other Lipid Mediat 68-69:383–399
76. Oh SF, Dona M, Fredman G, Krishnamoorthy S, Irimia D, Serhan CN (2012) Resolvin E2 formation and impact in inflammation resolution. J Immunol 188:4527–4534
77. Park EJ, Park SY, Joe EH, Jou I (2003) 15d-PGJ2 and rosiglitazone suppress Janus kinase-STAT inflammatory signaling through induction of suppressor of cytokine signaling 1 (SOCS1) and SOCS3 in glia. J Biol Chem 278:14747–14752
78. Pereira MP, Hurtado O, Cardenas A, Bosca L, Castillo J, Davalos A, Vivancos J, Serena J, Lorenzo P, Lizasoain I, Moro MA (2006) Rosiglitazone and 15-deoxy-Delta12,14-prostaglandin J2 cause potent neuroprotection after experimental stroke through noncompletely overlapping mechanisms. J Cereb Blood Flow Metab 26:218–229
79. Redensek A, Rathore KI, Berard JL, Lopez-Vales R, Swayne LA, Bennett SA, Mohri I, Taniike M, Urade Y, David S (2011) Expression and detrimental role of hematopoietic prostaglandin D synthase in spinal cord contusion injury. Glia 59:603–614
80. Resnick DK, Nguyen P, Cechvala CF (2001) Regional and temporal changes in prostaglandin E2 and thromboxane B2 concentrations after spinal cord injury. Spine J 1:432–436
81. Rocha PN, Plumb TJ, Coffman TM (2003) Eicosanoids: lipid mediators of inflammation in transplantation. Springer Semin Immunopathol 25:215–227
82. Saleem S, Zhuang H, de Brum-Fernandes AJ, Maruyama T, Narumiya S, Dore S (2007) PGD(2) DP1 receptor protects brain from ischemia-reperfusion injury. Eur J Neurosci 26:73–78
83. Scher JU, Pillinger MH (2005) 15d-PGJ2: the anti-inflammatory prostaglandin? Clin Immunol 114:100–109

84. Schwab JM, Chiang N, Arita M, Serhan CN (2007) Resolvin E1 and protectin D1 activate inflammation-resolution programmes. Nature 447:869–874
85. Serhan CN (2014) Pro-resolving lipid mediators are leads for resolution physiology. Nature 510:92–101
86. Serhan CN, Levy BD (2018) Resolvins in inflammation: emergence of the pro-resolving superfamily of mediators. J Clin Invest 128:2657–2669
87. Sobrado M, Pereira MP, Ballesteros I, Hurtado O, Fernandez-Lopez D, Pradillo JM, Caso JR, Vivancos J, Nombela F, Serena J, Lizasoain I, Moro MA (2009) Synthesis of lipoxin A4 by 5-lipoxygenase mediates PPARgamma-dependent, neuroprotective effects of rosiglitazone in experimental stroke. J Neurosci 29:3875–3884
88. Tabuchi S, Uozumi N, Ishii S, Shimizu Y, Watanabe T, Shimizu T (2003) Mice deficient in cytosolic phospholipase A2 are less susceptible to cerebral ischemia/reperfusion injury. Acta Neurochir Suppl 86:169–172
89. Tager AM, Luster AD (2003) BLT1 and BLT2: the leukotriene B(4) receptors. Prostaglandins Leukot Essent Fat Acids 69:123–134
90. Tager AM, Bromley SK, Medoff BD, Islam SA, Bercury SD, Friedrich EB, Carafone AD, Gerszten RE, Luster AD (2003) Leukotriene B4 receptor BLT1 mediates early effector T cell recruitment. Nat Immunol 4:982–990
91. Takadera T, Yumoto H, Tozuka Y, Ohyashiki T (2002) Prostaglandin E(2) induces caspase-dependent apoptosis in rat cortical cells. Neurosci Lett 317:61–64
92. Thura M, Hokamura K, Yamamoto S, Maeda M, Furuta K, Suzuki M, Ibaraki K, Umemura K (2009) GIF-0173 protects against cerebral infarction through DP1 receptor activation. Exp Neurol 219:481–491
93. Titsworth WL, Cheng X, Ke Y, Deng L, Burckardt KA, Pendleton C, Liu NK, Shao H, Cao QL, Xu XM (2009) Differential expression of sPLA2 following spinal cord injury and a functional role for sPLA2-IIA in mediating oligodendrocyte death. Glia 57:1521–1537
94. Urade Y, Kaneko T, Fujimoto N, Watanabe Y, Mizuno N, Hayaishi O (1985) Purification, characterization, and immunohistochemistry of rat brain prostaglandin D synthetase. Adv Prostaglandin Thromboxane Leukot Res 15:549–551
95. Vasquez AM, Mouchlis VD, Dennis EA (2018) Review of four major distinct types of human phospholipase A2. Adv Biol Regul 67:212–218
96. Wang X, Zhu M, Hjorth E, Cortes-Toro V, Eyjolfsdottir H, Graff C, Nennesmo I, Palmblad J, Eriksdotter M, Sambamurti K, Fitzgerald JM, Serhan CN, Granholm AC, Schultzberg M (2015) Resolution of inflammation is altered in Alzheimer's disease. Alzheimers Dement 11(40-50):e41–e42
97. Wong A, Cook MN, Foley JJ, Sarau HM, Marshall P, Hwang SM (1991) Influx of extracellular calcium is required for the membrane translocation of 5-lipoxygenase and leukotriene synthesis. Biochemistry 30:9346–9354
98. Woods JW, Evans JF, Ethier D, Scott S, Vickers PJ, Hearn L, Heibein JA, Charleson S, Singer II (1993) 5-lipoxygenase and 5-lipoxygenase-activating protein are localized in the nuclear envelope of activated human leukocytes. J Exp Med 178:1935–1946
99. Xu ZZ, Zhang L, Liu T, Park JY, Berta T, Yang R, Serhan CN, Ji RR (2010) Resolvins RvE1 and RvD1 attenuate inflammatory pain via central and peripheral actions. Nat Med 16:592–597, 591p following 597
100. Zarruk JG, Greenhalgh AD, David S (2018) Microglia and macrophages differ in their inflammatory profile after permanent brain ischemia. Exp Neurol 301:120–132

Perspectives

This book analyzes key topics involving bioactive lipids and their role in normal signaling and the mechanisms of disease. The first part is focused on the structure, characterization and physicochemical properties of bioactive lipids. A special emphasis is given on the diffusion and transport of reactive species across cell membranes and analysis of lipid oxidation products. With respect to the analytical technologies applied in lipid analysis, mass spectrometry stands central with a shift from classic structural studies toward large panel targeted lipidomics approaches. The latter allowing for a more quantitative view of clinical data sets in correlation with disease mechanisms. The second part analyze signaling cascades mediated by bioactive lipids, focusing on key biologically-relevant mechanisms that involve peroxisome proliferator-activated receptors, xanthine oxidase and unfolded protein response. Related with this topic, several chapters put these observations in the context of inflammatory diseases. The role of arachidonic acid nitration in macrophages and platelets is analyzed in detail since these modifications could divert this fatty acid from normal metabolism to unraveled pathways. Then, the role of dyslipidemia in ischemia-reperfusion injury is updated as well as the significance of epicardial adipose tissue in cardiovascular disease. Finally, efforts are focused on understand the link between bioactive lipids and neurodegenerative diseases. While activation of signaling pathways by bioactive lipids in hippocampus is analyzed, nitroalkylation of proteins in Central Nervous System and implications in Parkinson Disease as well as the metabolism and biological effects of ω-3 and ω-6 fatty acids in spinal cord injury. We hope these mechanistic observations on the role of bioactive lipids in health and disease serve a perspective to improve the existing treatments or propose new lipid-based pharmacology.

Andrés Trostchansky and Homero Rubbo

A. Trostchansky, H. Rubbo (eds.), *Bioactive Lipids in Health and Disease*, Advances in Experimental Medicine and Biology 1127, https://doi.org/10.1007/978-3-030-11488-5

Index

A. Trostchansky, H. Rubbo (eds.), *Bioactive Lipids in Health and Disease*, Advances in Experimental Medicine and Biology 1127, https://doi.org/10.1007/978-3-030-11488-6

The manufacturer's authorised representative in the EU is Springer Nature Customer Service Centre GmbH, Europaplatz 3, 69115 Heidelberg, Germany. If you have any concerns regarding our products, please contact ProductSafety@springernature.com

Printed and bound by CPI Group (UK) Ltd, Croydon, CR0 4YY
28/09/2026
02224077-0020